A Colour Handbook on Practicals in Veterinary Medicine

NIPA® GENX ELECTRONIC RESOURCES & SOLUTIONS P. LTD.
New Delhi-110 034

About the Authors

Prof. (Dr.) J. P. Varshney - the author of the book, is a renowned veterinarian with professional experience of more than 58 years. He had been a teacher, researcher and clinician honored with so many awards including Canine Excellence, Best Teacher and Lifetime Achievement Awards. Besides scientific publications he is the author of books titled "Equine Parasitic Diseases and Their Management (ICAR-IVRI Izatnagar Publication, 2003), "Research Findings-Homeopathic Bio efficacy and Management of Animal Health (Sintex International limited, Kalol, 2007), "Electrocardiography in Veterinary Medicine" (Springer Nature, Singapore, 2020), "Handbook of Exotic Pet Medicine" (Brillion Publishing, New Delhi, 2021), "Ultrasound in Veterinary Medicine: Fundamentals and Applications "(NIPA, New Delhi, 2022), "Heart Failure in Animals: Current Concepts "(NIPA, New Delhi, 2023), "Electrocardiogram Made Easy for Veterinary Students and Practitioners" (NIPA, New Delhi, 2023), "Clinical Diagnostic Processes in Canine Medicine" (NIPA, New Delhi, 2025), and "Homeopathy in Veterinary Medicine" (NIPA, New Delhi, 2025)

Dr. Sujata Turkar, Professor (Veterinary Medicine), College of Veterinary Sciences, Guru Angad Dev Veterinary and Animal Sciences University, Ludhiana, Punjab, has a professional experience of over 18 years. She is very popular among UG and PG students for her excellent teaching, guiding and clinical skills. Her commitment to Veterinary Medicine extends beyond the classroom, as she is actively involved in research, collaboration with experts, and publishing research findings in scientific journals of repute. This unique blend of teaching and clinical skill and research accomplishments makes her a distinguished reputed personality in the field of Veterinary Medicine.

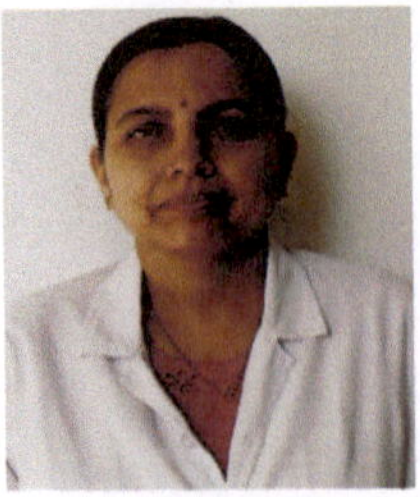

Dr. Sarita Devi, Associate Professor (Veterinary Medicine), College of Veterinary Science & Animal Husbandry, Kamdhenu University (KU), Junagadh, Gujarat has a clinical, teaching and research experience of over 12 years in the field of Veterinary Medicine. She is the recipient of ISVM Intas Young Scientist Award-2009 and AVA Roy gold medal -2023 for her outstanding contribution in the field of Veterinary Medicine. She has publications in national and international scientific journals of repute besides book chapters to her credit.

A Colour Handbook on Practicals in Veterinary Medicine

J. P. Varshney
B. V. Sc. and A. H. ; M. V. Sc. (Veterinary Medicine); Ph. D. (Veterinary Medicine)
Ex. Principal Scientist (Veterinary Medicine, IVRI Izatnagar)
Senior Consultant (Veterinary Medicine)
Nandini Veterinary Hospital
Ghod-Dod Road, Surat-395001, Gujarat, India

Sujata Turkar
B. V. Sc. and A. H. ; M. V. Sc. (Veterinary Medicine); Ph. D. (Veterinary Medicine)
Professor (Veterinary Medicine)
Department of Veterinary Medicine
Guru Angad Dev Veterinary and Animal Sciences University
Ludhiana - 141004, Punjab, India

Sarita Devi
B. V. Sc. and A. H. ; M. V. Sc. (Veterinary Medicine); Ph. D. (Veterinary Medicine)
Associate Professor (Veterinary Medicine)
Department of Veterinary Medicine
College of Veterinary Science & Animal Husbandry, Junagadh
Kamdhenu University (KU)
Gandhinagar, Gujarat

NIPA® GENX ELECTRONIC RESOURCES & SOLUTIONS P. LTD.
New Delhi-110 034

NIPA® GENX ELECTRONIC RESOURCES & SOLUTIONS P. LTD.

101,103, Vikas Surya Plaza, CU Block
L. S. C. Market, Pitam Pura, New Delhi-110 034
Ph : +91-11-43860225, Mob. : +91 9717133558, 9540816132
E-mail: newindiapublishingagency@gmail. com
Website: www. nipaersources. com

Print ISBN: 978-93-5887-896-7
ebook ISBN: 978-93-5887-399-3

Composed and Designed by NIPA®.

Dedicated
to
My Late Parents
(Smt. Ramwati Devi and Devi Ram Varshney)
and
My Teachers
for
Shaping Me What I am

J. P. Varshney

Preface

Practicals in veterinary medicine are crucial as they transform the theoretical knowledge,gained in lecture theatres, into real-world competence; build essential hand-on skill; develop clinical reasoning, competence and dexterity for procedures vital for handling clinical problems of different species safely and effectively; infuse confidence in handling animals safely in critical time eventually producing proficient qualified veterinarians prepared for diverse careers and enhancing welfare and quality of life of the animals. Practicals bridge the gap from classroom to clinic, giving opportunities for repeated practice with feedback using simulators and live animals before facing real complex cases in clinics. In fact, practicals make the knowledge, acquired in the classroom, functional for providing quality care to animals.

Clinical examination is the most practical and integral part of the process of making diagnosis in all systems of medicine both in humans and animals. Required information for ascertaining the disease or diseases producing clinical signs and symptoms is made available by the detailed systematic clinical examination. In fact, accurate diagnosis is not possible without a proficient and systematic clinical examination. Failing which, effective treatment and control for optimum animal welfare and production cannot be achieved. Clinical examination requires fundamental knowledge of anatomy, physiology, pathology and animal behavior; skills in the methods and techniques of clinical examination; and clinical sign and pathogenesis of the diseases for making diagnosis. A veterinarian has to be proficient and methodical in physical examination (general inspection, palpation, percussion and auscultation) techniques for capturing clinical signs of abnormalities.

The aim of this book is to provide requisite basic knowledge for training the veterinary scholars in client communication skill; animal handling and restrain techniques ; history and physical examination procedures for identifying signs and symptoms of ailing body organs ; special examination procedures pertaining to gastrointestinal, urinary, cardio-pulmonary, nervous, musculo- skeletal system, eye and ear, skin, endocrines and metabolic diseases; blood transfusion; specialized diagnostic techniques (ultrasonography, electrocardiography, endoscopy) : herd health, udder health, foot health, peri-parturient health; and vaccination procedures. It is essential for the veterinary students to become well versed with the practical aspects of the diagnostic techniques related to different body systems for a complete physical examination. The information regarding advance diagnostic methodology will broaden the horizon of the would be veterinarians and pave the way for making correct referrals for an exact diagnosis for better outcome.

J. P. Varshney
Sujata Turkar
Sarita Devi

Acknowledgements

I am grateful to my wife Jai Prabha Varshney for her exemplary support in all odd and even conditions. I take this opportunity to express thanks to my daughters Pratibha, Ritu and Prabhanshi; sons Atul, Prabhat and Ajit; and grandchildren Agrima, Vedaant, Devansh and Anvi for their love, emotions, care and concern.

Thanks are also due to Shri Nayan N. Bharatia, Managing trustee and Board of Trustees, Nandini Veterinary Hospital, Surat for their encouragement and facilities.

J. P. Varshney

The completion of this assignment would not have been possible without the blessings of my parents and in-laws. I am grateful for the unwavering support of my family; especially my husband Dr. Raja Ram Singh and daughters, Sanika and Riddhisha, who have encouraged me at every step and their patience and understanding provided the strength I needed to see this project through to completion. I would like to express my deepest gratitude to all those who contributed their expertise, time, and encouragement throughout this journey. Additionally, I am immensely thankful to my colleagues and mentors who shared their invaluable insights and experiences in Veterinary Medicine. I am particularly grateful to PG students for their help and sharing their photographs.

Sujata Turkar

I acknowledge my father, late Shri Lal Chand, whose vision and guidance laid the foundation for my academic journey. His strong belief in the transformative power of education nurtured my growth and motivated me to strive for excellence. I am deeply thankful to my husband, Dr. R. D. Singh, and my lovely daughters, Abhya and Krishya, for their constant support, encouragement, and resilience through all situations. Their presence in my life is my ultimate strength and inspiration guiding me through every challenge on this journey.

Sarita Devi

Contents

Part III: Skin, Endocrine, Metabolic and Deficiency Diseases

List of Abbreviations

%	:	Percent
@	:	At the rate of
<	:	Less than
>	:	More than
±	:	Plus minus
µL	:	Microliter
0^0F	:	Degree Fahrenheit
2,3-DPG	:	2,3-disphosphoglycerate
ACD	:	Acid- citrate-dextrose
AFB	:	Acid Fast Bacillus
AI	:	Artificial Insemination
ALT	:	Alanine aminotransferase
AST	:	Aspartate aminotransferase or Aspartate transaminase
ATP	:	Adenosine triphosphate
B. Wt.	:	Body Weight
BAL	:	Bronchoalveolar lavage
BC	:	Bacterial Culture
BCG	:	Bacillus Calmette-Guerin
BCS	:	Body Condition Score
BFT	:	Back Fat Thickness
BHB	:	Beta-hydroxybutyrate
bpm	:	Beats per minute
BQ	:	Black Quarter
BRSV	:	Bovine Respiratory Syncytial Virus
BSE	:	Bovine Spongiform Encephalopathy
BSP	:	Bromsulphalein
BTB	:	Bromothymol Blue
BU	:	Blood Urea
BUN	:	Blood Urea Nitrogen
BVD	:	Bovine Viral Diarrhoea
BVDV	:	Bovine Viral Diarrhea Virus
BWG	:	Birmingham Wire Gauge
$CaCl_2$	:	Calcium Chloride
CAE	:	Caprine arthritis-encephalitis
CAEV	:	Caprine Arthritis Encephalitis Virus
CAV	:	Canine Adeno Virus
CBC	:	Complete Blood Count
CCD	:	Charge Coupled Device
CD	:	Canine Distemper

CK	:	Creatine kinase
CKD	:	Chronic Kidney Disease
cm	:	Centimeter
CMPT	:	Compton Metabolic Profile Test
CMT	:	California Mastitis test
CN	:	Cranial Nerve
CNS	:	Central Nervous System
Co	:	Cobalt
CO_2	:	Carbon di oxide
CPDA-1	:	Citrate-phosphate-dextrose-adenine-1
CPU	:	Central Processing Unit
CPV	:	Canine Parvo Virus
CRF	:	Chronic Renal Failure
CRT	:	Capillary Refill Time
CSF	:	Cerebrospinal Fluid
CT	:	Computed Tomography
CT scan	:	Computed Tomography Scan
cTn-I	:	Cardiac Troponin-I
Cu	:	Copper
CysC	:	Cystatin C
DAHD	:	Department of Animal Husbandry and Dairying
DCAB	:	Dietary cation-anion balance
DEA	:	Dog Erythrocyte Antigen
DJD	:	Degenerative Joint Disease
dl	:	Deciliter
DM	:	Diabetes Mellitus
DMI	:	Dry Matter Intake
DNA	:	Deoxyribonucleic acid
DTM	:	Dermatophyte Test Medium
e. g.	:	Exempli gratia (Latin word meaning for example
EC	:	Electrical Conductivity
ECF	:	Extracellular fluid
ECG/EKG	:	Electrocardiogram
EDTA	:	Ethylenediamine tetraacetic acid
EIPH	:	Exercise-induced pulmonary hemorrhage
ELISA	:	Enzyme-Linked Immunosorbent Assay
Etc.	:	et cetra (Latin word meaning and other things)
FCV	:	Feline Calci Virus
FHV	:	Feline Herpes Virus
FIC	:	Feline Idiopathic Cystitis
Fig.	:	Figure
FLUTD	:	Feline Lower Urinary Tract Disease
FMD	:	Foot and Mouth Disease
FPV	:	Feline Panleucopenia Virus
g	:	Gram
g-GT	:	gamma-glutamyl transferase
GI	:	Gastro-intestinal
GLDH	:	Glutamate Dehydrogenase
H_2O	:	Dihydrogen monoxide (Water)

Hb	:	Hemoglobin
HCO_3	:	Bicarbonate
HE	:	Hepatic encephalopathy
Hg	:	Mercury
hr	:	Hour
hrs	:	Hours
HS	:	Hemorrhagic Septicemia
IBR	:	Infectious Bovine Rhinotracheitis
ICF	:	Intracellular fluid
IFAT	:	Indirect Fluorescent Antibody Tests
Ig G	:	Immunoglobulin G
IM	:	Intramuscular
IMI	:	Intramammary Infection
IO	:	Intraosseous
IOP	:	Intra Ocular Pressure
IP	:	Intraperitoneal
IRIS	:	International Renal Interest Society
IU	:	International Unit
IV	:	Intravenous
KCl	:	Potassium Chloride
KCS	:	Keratoconjunctivitis sicca
Kg	:	Kilogram
KOH	:	Potassium hydroxide
L	:	Liter
LA	:	Left Arm (left fore limb)
LD	:	Lactate dehydrogenase
LDA	:	Left Displacement of Abomasum
LL	:	Left leg (left hind limb)
LMN	:	Lower Motor neuron
mEq	:	Milliequivalent
MER	:	Myeloid to Erythroid Ratio
mg	:	Milligram
MHz	:	Megahertz
min.	:	Minute
ml	:	Mililiter
MLV	:	Modified live vaccine
mm	:	Millimeter
mmol	:	Millimole
MPT	:	Metabolic Profile test
MRI	:	Magnetic Resonance Imaging
MSCs	:	Mesenchymal stem cells
MUN	:	Milk Urea Nitrogen
mV	:	Millivolt
NaCl	:	Sodium chloride
NAGs	:	N-acetyl-b-D-glucosaminidase
NEB	:	Negative Energy Balance
NEFA	:	Non-esterified Fatty Acid
nm	:	Nano meter
NRS	:	Numerical Rating Scores

NT pro –BNP : N-Terminal Pro – B-type Natriuretic Peptide
°C : Degree centigrade
OCD : Osteochondritis dissecans
OCT : Ornithine Carbamoyl-Transferase
°F : Degree Fahrenheit
PCR : Polymerase Chain Reaction
PCV : Packed Cell Volume
PDA : Patent Ductus Arteriosus
PDH : Pituitary dependent hyperadrenocorticism
PEM : Polioencephalomalacia
PFT : Pulmonary Function test
pH : Potential of Hydrogen
PI 3 : Parainfluenza 3
PI : Parainfluenza
PLR : Pupillary Light Reflex
PMI : Point of Maximum Intensity
PNS : Peripheral Nervous System
PPR : Peste des Petits Ruminants
PSS : Porto systemic shunt
PTF : Peritoneal Fluid
RA : Right Arm (right fore limb)
RBC : Red Blood Cell
RDA : Right Displacement of Abomasum
RL : Right Leg (right hind limb)
RLN : Recurrent laryngeal neuropathy
rpm : Revolutions per minute
RSV : Respiratory Syncytial Virus
s : Second
SAAG : Serum-Ascites Albumin Gradient
SAP : Serum Alkaline Phosphatase
SCC : Somatic Cell Count
SD : Standard deviation
SDH : Sorbitol Dehydrogenase
SDMA : Serum Symmetric Dimethyl Arginine
Se : Selenium
SG : Specific gravity
SGOT : Serum Glutamic-oxaloacetic Transaminase
SGPT : Serum Glutamic-pyruvic Transaminase
STT : Schirmer Tear Test
TBW : Total Body Water
TLI : Trypsin Like Immunoreactivity
TPP : Total Plasma Protein
TUC : Transurethral Cystoscopy
UAPs : Ununited Anconeal Processes
UB : Urinary Bladder
UCGOC : Ununited Caudal Glenoid Ossification Center
UK : United Kingdom
UMN : Upper Motor Neuron
USA : United States of America

USG	:	Ultrasonography
UTI	:	Urinary Tract Infection
UV	:	Ultra Violet
VAS	:	Visual Analogue Scales
X-ray	:	Radiography
Zn	:	Zinc
μm	:	Micrometer

List of Tables

List of Figures

Part I
General Systemic States and Diseases of Digestive and Urinary System

1

Client Communication and Handling

Client Communication

Client communication is the key to success in practice as it is an important skill to narrow down the gap between owner of the animal and the veterinarian. In other words, it is a meaningful exchange of information between attending veterinarian and the owner of the ailing animal for improving health care outcome. In client centered world (where client satisfaction is in the core of practice), communication with client has assumed greater significance whether it is related to educating him about preventive health care; explaining about disease diagnosis, prognosis and treatment options; or explaining about chronic disease condition and its outcome. Communication also influences the wellbeing of the patient. Therefore, development of trust in client-veterinarian relationship is crucial as it not only enhances the quality of history taking but also help in better clinical diagnosis and subsequent clinical care.

Importance of Communication

- Client experience and feedback is important for providing good clinical care.
- Treating the animal and comforting the owner is a huge task. This can be eased by clear communication.
- A good communication helps in calming the anxiety of the owner leading to client satisfaction.
- A poor communication can make the simple case worst. Open and effective communication can make the handling of complex case easy.
- Risk sharing with owner becomes easy with effective and open communication. Owners should know the risks involved, complications and side effects of the treatment if any.

Ways of Communication

There are different ways of communication such as face-to-face verbal communication, text messaging, phone calls, e-mails or patient portals. Text messaging is easy, convenient, non-intrusive, time saving and cost-effective means of communication. Phone call are another popular means of traditional communication. In today's world, E-mails are an effective means of documented communication at the ease of both parties (client and veterinarian). Patient portals are means of secure communication where patient information and the information related to investigations, diagnosis and treatment etc. is kept secure. As compare to other means of communication, patient portal is a little difficult.

Means of Improving Communication

- Building rapport with the owner.
- Patient and careful listening with noting the important points.
- Empathetic attitude towards owners and patient.
- Clear instructions without any ambiguity.
- Answering all queries/doubts of the client.
- Making right referral, wherever needed, should not be hesitated.
- All communications (verbal, written or digital) should be in simple language preferably in local language.

Building Rapport with the Owner

- Using open body language (eye contact, uncrossing arm, leaning in slightly when talking).
- Talking to owner as a person rather than a client asking about his day-to-day life, welfare of the owner as well as his family, and his pets/animals.
- Mentioning of common things with the client.
- Remembering small details from previous conversation.

Advantages of Effective Communication

- Enhances client satisfaction.
- Yields better health outcome.
- Builds trust and better relationship.
- Prevents and resolves problem easily.
- Avoids miscommunication and misconceptions.
- Increases compliance rate

Handling

Animals should always be approached in a calm and confident manner. Exaggerated or sudden movements (waving of hands and arms) need to be avoided while approaching the animals. It is better to remain talking with the animal in soft tones and quiet voice during the process of approaching as it creates a non-threatening environment and reinforce a caring attitude in the handler.

Handling Methods

Most crucial aspect of handling an animal is choosing an appropriate method of restrain. Correct position of the handler and animal is necessary to achieve quick and secure restrain. The animal should be completely immobile to avoid struggling as struggling increases anxiety and may inflict injury to animal and/ or handler. During restrain, force should be just enough to hold animal firmly (not too much to cause discomfort).The animal should be observed for its response or behavior during handling so that adjustment could be made in the event of discomfort.

Laboratory Animals

Handling in case of laboratory animals is most important as aversive handling is likely to develop anxiety and exaggerated stress. Stress is detrimental to animal welfare and animals may develop defensive aggression.

Rabbits are restrained by firmly gripping the ears and loose skin at the back of the neck by one hand and placing the other hand under the hind quarter to support the weight and lifting gently. For anesthesia or inoculation rabbits can be wrapped in a roller towel or placed in a special box.

Guinea pigs are restrained by placing one hand across the back of the animal with thumb behind the shoulder and other fingers well forward on the opposite site. The animal is lifted gently supporting its weight with other hand placing the palm under hind quarters.

Mouse is handled by gripping on the mid tail by left hand, raising hind limbs and a fold of skin near the head is grasped with the thumb and fingers of the right hand.

While handling the birds, thick leather gloves should be worn. The legs should always be grabbed as close to the body as is possible (i.e. above the stifle joints) to avoid iatrogenic fractures. When transporting a bird, secure the legs with an underhanded grip, cover the bird's head with a towel and place the bird's back to your chest. Safe handling is achieved by controlling the bird's head, feet, legs and wings.

Farm Animals

Most of the farm animals respond to their routine, are calm. Loud noise, quick movements should be avoided while approaching the animals. Animals need to be handled patiently and should not be prod when it has nowhere to go. Animals should be respected and so not fear it. Before reaching to large animal, move slowly and deliberately around the animal and start touching gently rather than shoving or bumping. An escape route should always be there while working with large animals. Individual cattle/buffalo for detail examination or minor surgical procedures can be restrained in cattle crush (Fig.1 A). Cattle response to handling is variable. Most cattle are not aggressive. But their butting (swinging of head), kicking (forward, side), trampling (individuals blocking the path of the animal) and biting (rare) are injurious to persons requiring proper restrain and handling. Cattle can be restrained using nose tongs and rings (to control head); flank fold grip (to prevent the animal from kicking and jumping around) and figure-eight rope (Fig. 1.B). In flank fold grip method, the skin fold on the flank, in front of the knee joint or above the udder is grasped and pulled upwards. In figure-eight method, the rope is placed above the hock joint and tightened to exert pressure on Achilles tendon (Fig. 1B).

A B

Fig. 1. Restrain of cattle. (A) Restrain in cattle crush (B), and by a rope tying in figure -eight shape.

Sheep and goats are not as dangerous as cattle but can harm. They do not kick, strike or bite. Goats are more active than sheep. An individual sheep can be easily caught. Wool should never be grasped. Goats are placed in lateral recumbency (Fig. 2) for examination.

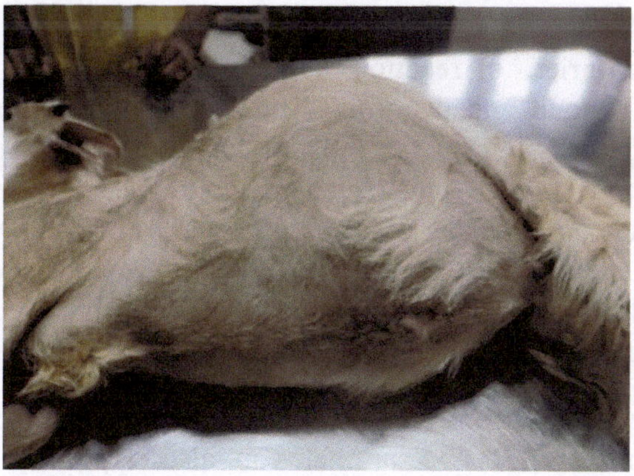

Fig. 2. Restrain of goat in lateral recumbency on the table.

Horses

Horses are accustomed to handling but can inflict injuries by biting, kicking (backward) and striking using front limbs if handlers are not cautious. Loud noise and sudden movement should be avoided when approaching a horse. Approach should be made by gently putting hand on the horse. Avoid approaching the horse from behind. Avoid yourself to be pinned between two horses or wall and horse. Horses should not be placed in cattle chutes. Horses can be restrained using halter, chains or twitches. For putting halter, horse is to be approached from the left side in the area of shoulder. Unbuckled or unsnapped halter is carried in the left hand. A lead shank is used to catch the horse. This is accomplished by placing the lead around the neck and holding both ends as a noose, while the left hand puts the halter in place. Chains increase the amount of restraint through increased pressure. A chain can be placed in the mouth over the tongue just touching the commissure of the lips. It is considered most severe restrain method. Twitching is another method of horse restrain. Twitching consists of applying pressure to certain parts of the horse body such as lips, or ears with either hand or custom-designed devices. Hand twitching should not be applied to horse ears as it may damage ear cartilage. Mechanical twitch is applied on lips. Horse may resist by raising or flipping its head up and down. Lifting of one limb can assist examination of weight bearing limb. Small foals are restrained by an arm in front of their chest and the other arm behind the rump.

Dogs and Cats (companion animals)

Dogs display a variety of reaction, varying from calm and cooperative to aggression, to restrain and handling. The dog should be socialized before handling; approached slowly from the side calling his name and continuously talking with him in soft, low pitched and smooth voice. Owner should always be allowed to stay with the dog while holding and lifting. Collars, leashes with or without collar, muzzles, head snare, chains, chest belt, choke collar, and Elizabethan collar (Fig.3) allows easier control. Gauge muzzle / plastic muzzle (Fig.4) can also be used to restrain dogs.

First allow the dog to approach you. Even fearful dogs need reassurance that you are not a threat. Examiner can also approach the dog. After crouching or kneeling down on the floor, the dog is coaxed slowly with continued verbal reassurance. The dog should be approached carefully and slowly keeping the eye contact. Dog showing signs of aggression or fear require precautions. Hold out hand slowly at a level not higher than the dog's nose (as dog perceive gestures above its nose as threatening, and respond aggressively). As dog accept your handling without fear, leash is slipped over its neck or lift the dog on the examination table.

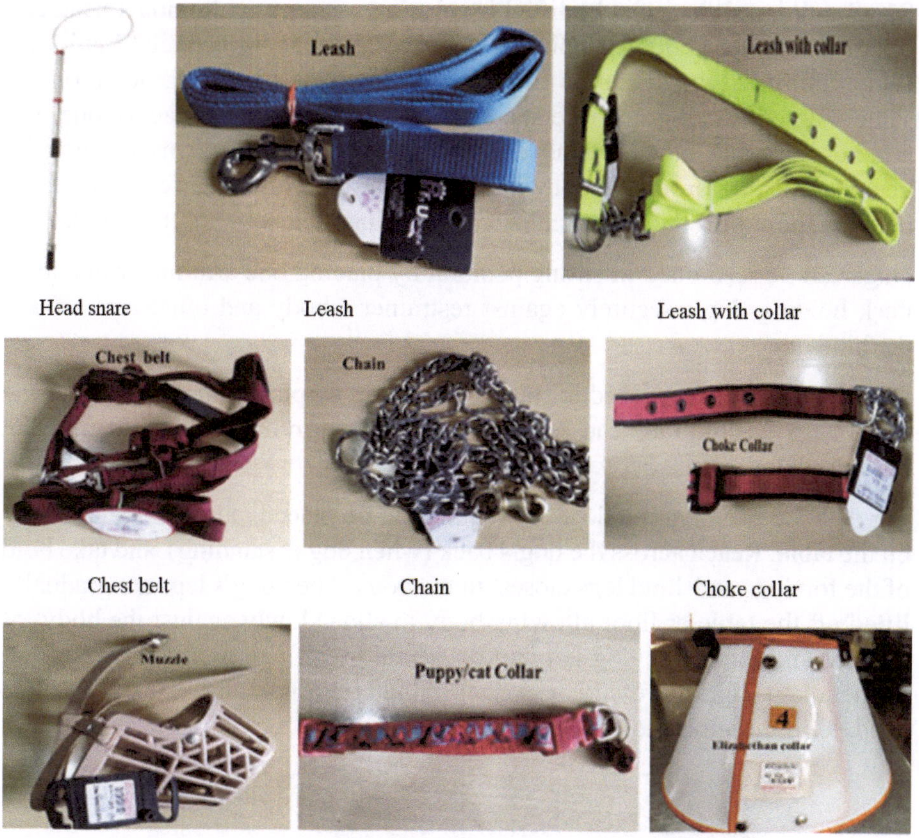

Fig. 3. Handling and restrain accessories for dogs and cats.

Fig. 4. Restraining dogs using plastic muzzle (A) or gauge (B).

Small uninjured dogs can be lifted by placing one arm under dog's head and other arm under abdomen pointing forward and fingers between hind legs. Large uninjured dogs' can be lifted by two persons crouching down on the same side of the dog. One person restrains dog's head by placing one arm around neck and other around chest. Another person restrains the dog by keeping one arm around abdomen and other around hind quarters. Then both persons should lift the dog simultaneously using their legs not their backs.

Dogs can be restrained in sitting position by placing one arm under the dog's neck holding dog's securely against restrainer's body and other arm around the hind quarter.

Dogs can also be restrained in standing position by placing one arm under the dog's neck to hold the head securely and other arm under the abdomen and pulling it close to the body.

For detailed examination dogs are required to be placed in lateral recumbency on the table. Reach across the dog's back (when dog is standing) and take hold of the forelegs and hind legs closest to the body. Then dog's legs are gradually lifted off the table or floor allowing body to slide slowly against the body of the restrainer until the dog is lying on its side with feet pointing away from the handler (Fig.5). Pressure is exerted using fore arm on the side of the dog's head, to immobilize the head of the dog.

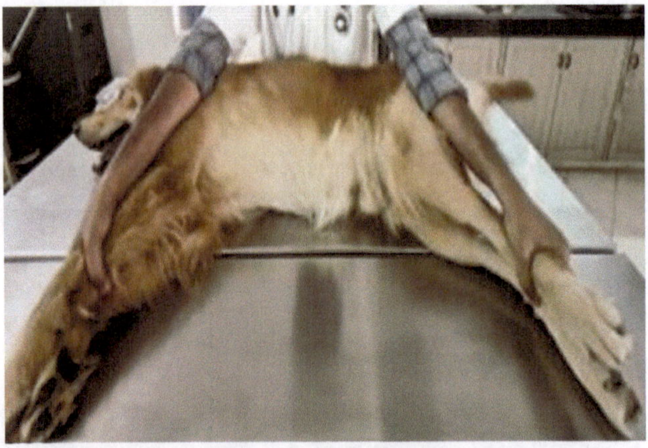

Fig. 5. Restraining a dog in lateral recumbency on the table.

Cat is unpredictable and difficult to handle. They defend themselves by biting and scratching. Before opening cage of the cats, doors and windows should be closed. Cats can be restrained and placed on the table for examination (Fig.6) in the manner similar to dogs.

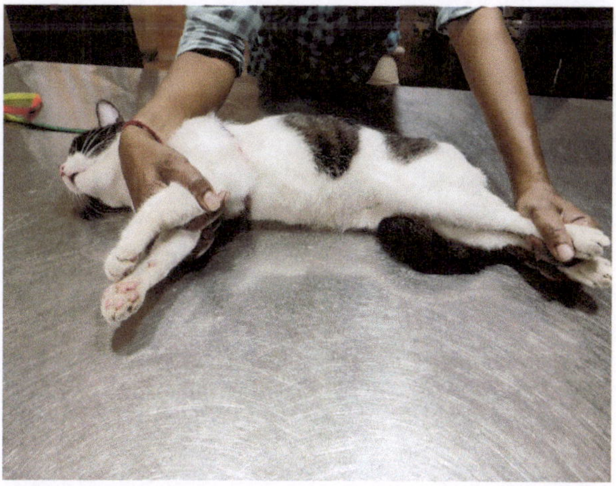

Fig. 6. Restraining a cat in lateral recumbency on the table.

In general, small animals can be restrained by placing them on the table in standing position with owner holding the animal and talking. Small animals should be muzzled to prevent biting.

2

History Taking and Physical Examination: Bovines

History and History Taking

History in medical/ veterinary terminology is a record of information about the health, illness (past and present), diet, feeding, management,preventive health schedules (vaccination,deworming, dipping, sprays) and hygiene etc. History taking and physical examination is the important step in making a correct diagnosis as it provides valuable information to determine the cause of clinical manifestations. The main purpose of taking history is to get detailed information about the animal, its activities, behavior, diet, health care practices, nutritional status, housing, development of abnormal signs and treatment given, if any. The animal may be cared by owner himself or by an attendant. In the event of animal care given by an attendant, information obtained from the attendant may not be complete and fully reliable as sometimes attendant may not provide correct events and complete picture to hide his mistakes. For obtaining right information a good rapport is to be made with the owner or the attendant. A complete history not only suggest diagnostic possibilities but can also suggest diagnostic probabilities. For example a calf or heifer is unlikely to have Johne's disease or an adult cow is more likely to suffer from parturient paresis than a first calf heifer. History can also provide a lead for further examination or which system is to be paid special attention.

Method of Taking History

In history taking there is an involvement of veterinarian and client calling for an healthy pleasant veterinarian-client relationship. It can be strengthened in following manner.

- Start with usual greetings of the day.
- Introduce himself or herself in a friendly manner.
- Asking the owner, how can I help you?
- Handling the owner/attendant diplomatically and politely.

- Language of conversation with the owner/attendant should be simple.
- Technical jargon should be avoided as far as possible. Only utmost necessary technical words should be used.
- Provide enough time to owner to speak out. So that detail information about the animal and its sickness can be obtained.
- Be a patient listener. First allow the owner/ attendant to speak out whatever he/she wants to speak with out interrupting.
- Create a friendly environment so that owner is comfortable in divulging the minute details.
- Try to obtain detailed information about the animal.
- Herdsman/ attendants may attempt to disguise their neglect by changing chronology of event or hiding some event. In such cases own assessment should be exercised.
- Human fallibility should be considered as there may be insufficient time, importance of the particular factors might not be appreciated or there might be some misunderstandings.
- Information obtained should be correlated during physical examination.
- A second round of queries can be made at later stage during systemic examination.
- Direct questions should be avoided as "Did the calf scour?" or "Is there fever?"
- Do not simply ask for a complete history or what has happened.
- Owner do not describe signs in their correct sequence. Veterinarian has to establish chronology of the event.
- Complete history includes patient data, past and present disease history, vaccination and deworming status and management and nutritional history. It is advisable to take disease history first as it reflects concern to animal's illness.
- It is advisable to follow a set routine in history taking for accuracy and completeness.

Components of History

- Patient data or patient information.
- Disease history (present, past illnesses, duration, treatment if given any).
- Management history (nutrition, climate, general management).

i. *Patient Data*

- It is a prerequisite for patient identification, keeping record, for sending material to laboratory investigations.
- It should invariably include name/ number of the animal, color, marking, or any other identification mark if any, sex, age, species, breed and body weight.

ii. *Disease History*

- Whether a single animal or a group of animals is suffering in the herd.
- Disease history should include present disease as well as previous exposure.
- The sequence of development of present illness needs to be enquired. If more than one animal in the herd is affected/sick, a typical case is to be selected and variations in temperature, pulse, respiration, food and water intake, milk production, urination, defecation, gait, activity, posture, voice, and odor are to be noted. In fatal cases pathological or necropsy examination information should also be taken in to consideration. The behavior of the animal before death and time lapse between appearance of symptoms and death should also be enquired and recorded . Information about prior surgical/medical procedures (castration, docking, shearing, dipping/ spraying or vaccination) may provide some lead about the disease. Information about prophylactic and disease control procedures, as well previous treatment given is valuable in the diagnosis of the disease.
- History of adding some animal and previous exposure of the herd is also important. Is the herd a closed herd or animals are introduced frequently in the herd? Information about recent transit, culling rate in the herd and previous disease provide valuable information in making diagnosis.

iii. *Management History*

- It includes information about nutrition, indoor/grazer, breeding policy and practices, housing, climate, transport and general management.
- Nutritional history includes information about whether the animal is at pasture or hand fed/stall fed, quality and quantity of the feed being given. Animals at pasture have a little different problems as their diet is less controlled and are at greater risk of parasitic infestation or toxicity. In pasturing animals information about rotational grazing, pasture fertilizer application, top dressing, mineral supplementation, dung disposal or vegetation changes brought by rain or drought may provide important clues for diagnosis.

- Stall-fed or hand –fed animals are exposed to dietary mistake owing to human error. Horses on diet having more grain are predisposed to *osteodystrophia fibrosa.* Similarly horses fed on high carbohydrate diet during rest have problem of azoturia. Cattle or goats on high grain ration suffer from rumen acidosis. Pigs fed in large number with inadequate trough space may suffer from overheating. Swine suffer from salt poisoning when their water intake is inadequate.
- Breeding history is valuable with regards to inherited diseases, and reproductive problems .
- In single animal breeding and parturition history may help to include or exclude some diagnostic possibilities. Pregnancy toxemia occurs in sheep in late pregnancy. Acetonemia in cows occurs primarily 2- 6 weeks post parturition.
- Climate also plays an important role in causation of disease. Foot rot in cattle and sheep is more common in warm wet summers. Humid conditions favor the diseases spread by insect vectors. In pasturing animals hypomagnesaemia may develop during cold wet season. In hot humid climate horses may suffer from anhidrosis.
- Milking shed hygiene plays an important role in the prevalence of mastitis.
- Enterotoxaemia is more common in finishing lambs and pigs. Parturient paresis is seen in milking cows. Obstructive urolithiasis is commonly seen in lambs and steers in feedlots.

Examination of Environment

- The examination of environment plays a significant role in disease diagnosis.
- Whether the animals are kept totally outside, inside, totally confined inside and outside?
- In animals kept totally outside, topography of grasslands, pastures and wooded area and soil type plays role in the occurrence of diseases. Cattle in flat, treeless plains are predisposed to lactation tetany in inclement weather. Insect borne diseases are common in animals on low marshy areas. Leptospirosis is more prevalent in animals in damp conditions.
- Copper and cobalt deficiencies are more common in animals on littoral sand. Copper deficiency/molybdenum excess is seen on peat soils.
- Overcrowding (high population density) predisposes animals to infections.

- Indoor environment viz. sanitation and hygiene, population density, quality of floor, floor plan and ventilation in sheds is also an important predisposing factors to diseases in animals.

Examination of the Patient

A complete systematic examination of the patient includes history taking, examination of the environment, physical examination and laboratory investigations. History taking and examination of environment has been detailed above. All animals may not require a complete clinical examination because of simplicity of some diseases. Examination of a patient involves general and physical examination. General examination is conducted from a distance while physical examination is conducted on the animal at close quarters.

General Examination

- Behavior, general appearance, voice, appetite, posture, gait, conformation, urination and defecation are observed from a distance. Separation of animal from the group is suggestive of illness. Mania and frenzy, violent licking and kicking at its belly, rolling and bellowing are excitation states. Voice is hoarse in rabies and weak in gut edema. Appetite can be judged by observing animal's response to offering feed. Abnormalities of prehension, mastication and swallowing should also be noted. In ruminants, abnormalities of rumination, regurgitation, eructation and cud dropping need to be given special attention. Straining (tenesmus). is indicative of constipation, rectal paralysis or stenosis. Severe diarrhea or anal sphincter paralysis is associated with involuntary defecation. Cystitis, urethra inflammation, partial obstruction of urinary tract are associated with difficult urination. In horses shifting weight on limbs suggests laminitis; arching of the back with limbs tucked under body is suggestive of abdominal pain; and dog sitting posture with rolling and kicking at belly is seen in acute abdominal pain associated with acute gastric dilatation on grain engorgement. Abducted elbow indicates chest pain. Sternal recumbency with hind limbs extended behind in a frog like attitude is suggestive of hypocalcaemia in sheep and bilateral hip dislocation in cattle. Muscle weakness is associated with inability to rise. Gait abnormalities also reflects abnormalities as cerebellar ataxia affects all qualities of limb movements. High stepping gait in sheep is suggestive of louping ill. Walking in circles may be due to rotation or deviation of head. It becomes a permanent state in listeriosis. Compulsive walking is seen in encephalomyelitis and hepatic insufficiency in horses.

The body condition of the animal can be classified as normal, obese, thin or emaciated . Conformation of the animal is based on symmetry and the shape and size of the different body regions. Skin is also be inspected from a distance for any abnormality such as change in hair, wool, sweating or presence of any lesion. Diffuse alopecia is seen in hyperkeratosis. In ringworm (dermatophyte infection) alopecia is diffuse occurring in discrete areas. Sweating is diminished in anhidrosis in horses.

- Inspection of body regions (head, neck, thorax, abdomen, external genitalia, mammary glands and limbs) is also done without handling the animal before physical examination is under taken.

  The head (Fig.7) is looked for symmetry and configuration of the bony structure. Forehead dooming indicates hydrocephalus. In cases of osteodystrophia fibrosa, swelling of maxillae and mandibles is evident. Tetanus is associated with rigidity of ears, third eye lid prolapse and dilatation of nostrils. Rotation of head is associated with defect of vestibular apparatus on one side. Opisthotonus is associated with tetanus, strychnine poisoning, acute lead poisoning, hypomagnesemic tetany, polyencephalomalacia or encephalitis. Eye are noted for presence of the discharge, protrusion of eye ball (orbital lymphomatosis) and retraction of the bulb (dehydration). Nostrils are visualized for dilatation as well as discharge, Frothing or salivation at mouth is suggestive of painful conditions of the mouth, pharynx or jaw tremors. In actinobacillosis or strangle, there is inflammatory swelling below jaw. Edematous swelling below jaw indicates, severe anemia, protein malnutrition, and congestive heart failure.

  Neck is looked for swelling of salivary glands, lymph nodes, guttural pouches (in horse) or any other soft tissue ; goiter enlargement; jugular pulse, or jugular vein engorgement.

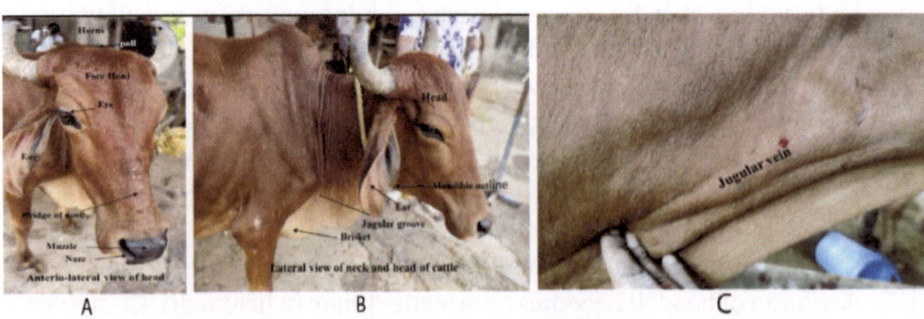

Fig.7. Showing head (A), neck (B) and jugular furrow (C) of a cattle.

Thorax (Fig.8) is inspected for symmetry; rate, rhythm, depth and type of respiration. Respiration rate varies from 8-10/min in horses, 10-13/min.in cattle, 10-20/min. in sheep and pigs and 25-35/min.in goats. Normal respiration is a three cycle of almost equal length viz. inspiration, expiration and pause. In most diseases of lungs there is no pause. Cheyne-Stroke respiration, characterized by a gradual increase and then a gradual decrease in depth, is seen in advanced stage of renal and cardiac diseases. Biot's breathing is characterized by alternating periods of hyperpnea and apnea (unequal length) is seen in meningitis affecting medullary region. In electrolyte and acid-base imbalance apnea is followed by short burst of hyperventilation. Respiration depth is reduced in painful conditions of chest or diaphragm. Normal respiration is associated with movement of both thorax and abdomen. Marked increase in abdominal movement is associated with pleurisy and in paralysis of intercostal muscles. The presence of respiratory noises such as coughing (pharynx irritation), trachea and bronchi, sneezing, wheezing, snoring, roaring or grunting are to be noted as they indicate painful labored breathing.

Abdomen (Fig.9) is inspected for variations in size. An increase in abdomen size may be due to many factors such as ingestion of an excess feed and/or water, presence of fluid, feces, fat or gas, foetus or neoplasm. Visible decrease in size of the abdomen (gaunt) may be seen in starvation, severe diarrhea, chronic diseases associated with decreased appetite. Umbilical hernia may be visible on inspection of ventral abdomen. Grossly enlarged asymmetrical swelling of the flank may indicate hernia. Ruminal movements can be seen on the inspection of left para-lumbar fossa and flank. But their detailed examination needs palpation, percussion and auscultation.

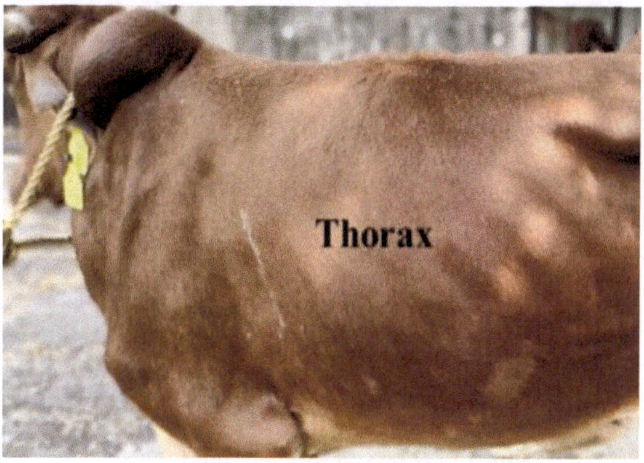

Fig. 8. Showing thorax of a cattle.

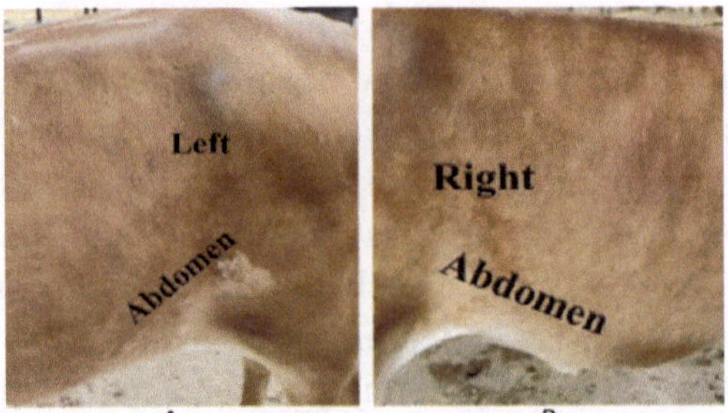

A B

Fig. 9. Showing left (A) and right (B) sides of the abdomen of a cattle .

Gross abnormal changes in external genitalia of male (enlarged preputial sheath or scrotum, tumor), and discharge from female genitalia (Fig. 10 A) can be noted on visual inspection.

Visual inspection of mammary glands and udder (Fig. 10 B) can provide information about atrophy, hypertrophy, inflammation, or injury.

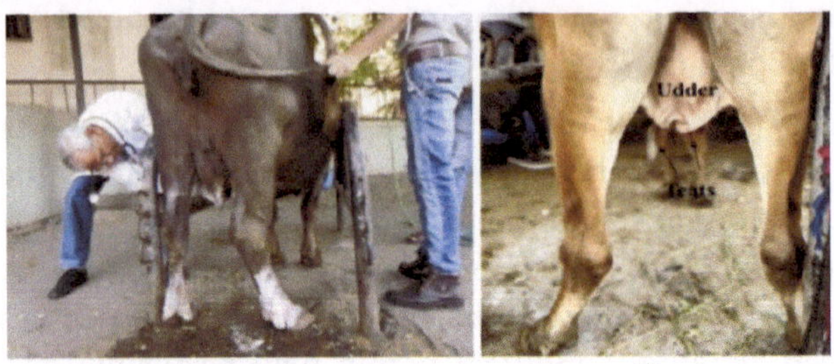

A B

Fig. 10. Showing a view of female genitalia (A) and Udder and teats (B).

Limbs (Fig.11) are to be inspected for abnormalities of posture and gait; enlargement or distortion. Symmetry of limbs is important. In case of doubt comarision can be made with counter limb.

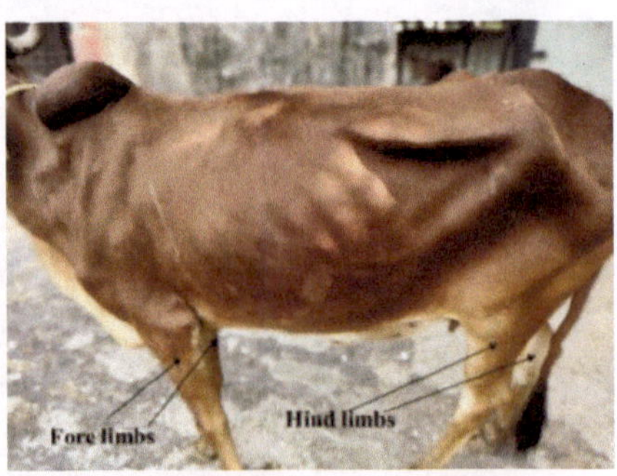

Fig. 11. Showing fore and hind limbs.

Physical Examination

The following techniques are used in physical examination of the animals

i. Palpation

- Palpation is conducted to determine the size, consistency, temperature and sensitivity of a lesion or organ.
- The lesion can be palpated directly with the fingers (Fig. 12) or indirectly with the help of a probe.

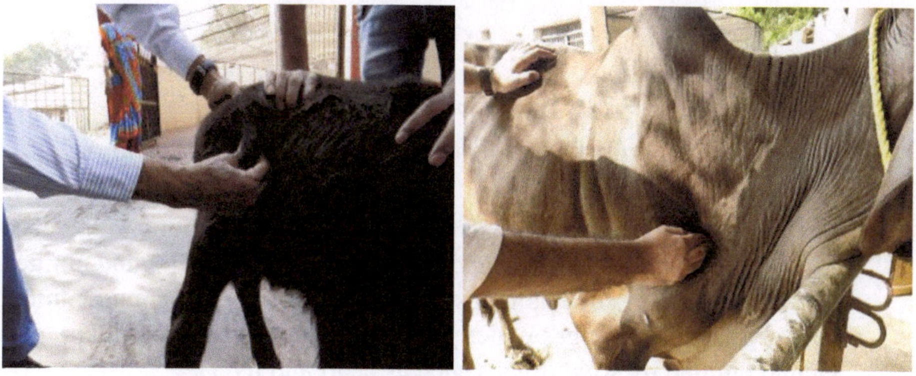

Fig. 12. Showing palpation in a buffalo calf and cattle.

- The results of palpation are described as doughy (structure pits on pressure example edema), firm (consistency of the normal liver), hard (consistency of bone), fluctuating (soft, elastic, undulates on pressure without retaining the imprint of the fingers), tense (structures giving the feeling of distended with gas or fluid), and emphysematous (swollen structures giving feeling of puffiness and crackles under pressure because of the presence of gas).

ii. Percussion

- It is the technique to set in vibrations in the deeper tissue by striking with a reasonable force using fingers (Fig.13) and interpret the emitted sound. In large animals pleximeter hammer on a pleximeter disc can be used.

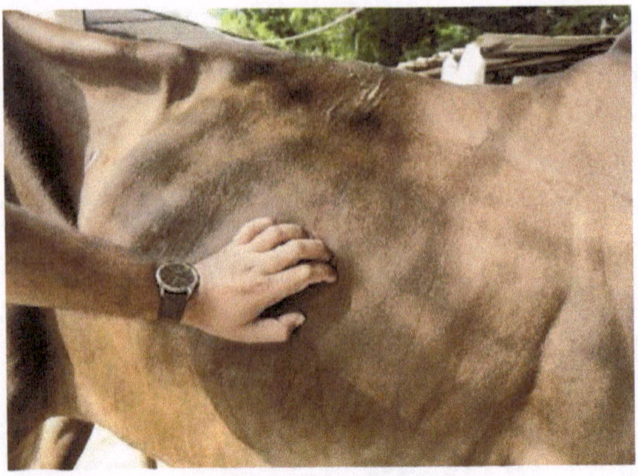

Fig.13. Showing percussion in cattle.

- It is conducted with fingers using one hand as a plexor and one as a pleximeter.
- The body surface is struck with fingers to create vibrations in the deep parts. The emitted audible sound is classed as resonant (sound emitted by air filled organ such as lungs) tympanitic (drum like sound emitted by a organ filled with gas under pressure such as tympanic rumen), dull (sound emitted by solid organs such as heart and liver).
- In fatty animals, percussion may not yield desirable results.
- In pigs and wooly sheep, percussion technique seems impracticable because of fat and wool respectively.
- In mature cattle and horses percussion of the abdominal organs has limited use because of too large organs and too thick overlying tissues.
- Increased dullness over thorax is suggestive of lung consolidation, space occupying lesion, pleural effusion.
- Increased resonance over the thorax may be due to emphysema or pneumothorax.
- Simultaneous percussion and auscultation of the abdomen (left and right) is recommended for detecting 'pings' (high-pitched tympanic sounds).

iii. Ballottement

- It is a technique used to detect floating mass or viscera in the abdomen.
- Abdominal wall is palpated with a firm push, using clenched fist/ extended fingers (Fig.14.), to displace the mass or organ away and allowing it to rebound on finger tips.

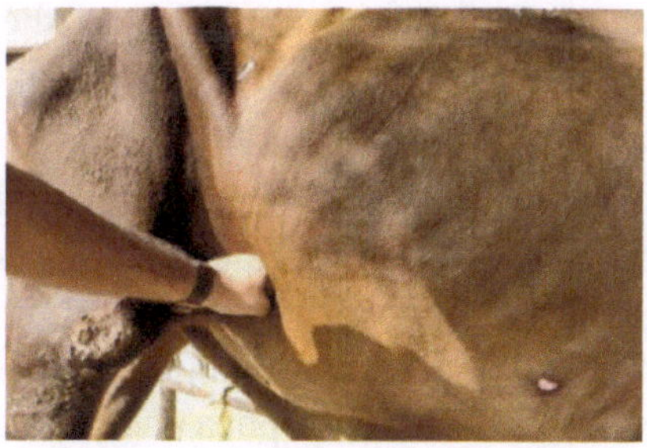

Fig.14. Showing ballottement in cattle.

- Fetus, impacted abomasum, large tumor or abscess of abdominal cavity can be detected by ballottement.
- Fluid splashing sounds in cattle can be detected by ballottement and auscultation of the flank.
- In colicky horses, detection of fluid splashing sounds on ballottement and auscultation of abdomen are suggestive of fluid filled intestines as it happens in paralytic ileus or intestinal obstruction .
- The technique is valuable in ascites as fluid thrills can be easily detected.

iv. Auscultation

(a) Lungs

- Auscultation is an important technique routinely used to listen the sounds produced by the movement of the organ such as heart, lungs, intestines.
- Sounds can be heard by direct listening (placing ears to the surface of the body over organ) or indirectly with the help of stethoscope.
- Lung sound in large ruminants can be heard in a triangle formed by the triceps cranially, the attachment of the ribs to the vertebral column dorsally and an imaginary line joining the point of the elbow with the eleventh intercostal space (Fig.15) .

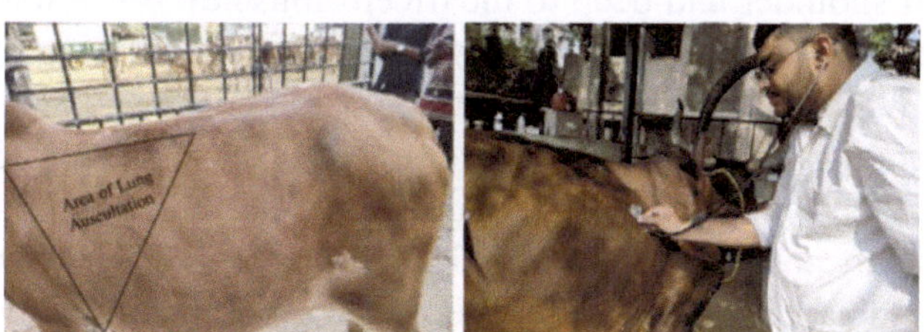

Fig.15. Area marked on the left thorax for auscultation of the lung. Similar area is auscultated on the right thorax. The triangle is formed by the triceps muscle cranially, attachment of the ribs to the vertebral column dorsally and an imaginary line joining the point of the elbow with the eleventh inter costal space.

- Lung sounds are muffled by accumulation of the fluid or fat between lungs and chest wall. Lung sounds are reduced when bronchial lumen is filled with fluid as in case of pneumonia or tissue.
- Normal sounds of trachea are like air moving through a large tube having less turbulence. Abnormal tracheal sounds are often associated with narrowing of the upper airway and/or fluid accumulation in the trachea. Normal lung sounds are soft, sipping VEE-EFF. The amplitude of breath sounds is increased in dyspnea and in early stage of pulmonary inflammation and congestion; and decreased or become inaudible in pleural effusions and in space occupying lesions in lungs or pleural cavity. Abnormal lung sounds are characterized as crackles,wheezes, and pleuritic frictional rubs.

(b) Heart

- Heart sounds are best heard under the triceps/elbow between the 3rd and 5th intercostal spaces on the left side placing the stethoscope far cranially under the elbow (Fig.16) but can also be heard on the right side. Optimum auscultation sites are the 4th and 5th intercostal spaces. Usually, heart sounds are loudest on the left side . The normal heart rate is 50 to 80 beats/minute. Only the first two heart sounds are heard. S1 the first heart sound (systolic) is the loudest and is associated with the closure of the atrioventricular valves. It is best heard over cardiac apex on both side (mitral closure on left side at 5th intercostal space just above half way point between the shoulder and the elbow, and tricuspid closure on the right side at 3rd to 4th intercostal space in ventral third of the thorax). It is loudest at the 4th intercostal space. S2 the 2nd sound (diastolic) is best heard over the base of the heart shortly after S1 and is associated with the closure of the aortic (left 4th intercostal space just below the point of shoulder and just forward of the border of the triceps muscle)and pulmonic valves (left 3rd intercostal space just below the point of shoulder and deep to the triceps muscle). It is loudest at the 3rd intercostal space. The heart is auscultated for determining the heart rate, heart rhythm, intensity and quality of heart sounds and also to ascertain whether any abnormal heart sound (murmur) is present or not. Pulse rate slower than the heart rate is suggestive of pulse deficit possibly due to weak heart contractions. Normal heart sound is three character sound described as LUBB-DUPP-Pause. LUBB sound is dull, deep, long and loud while DUPP sound is sharp and short. Irregular heart sounds are suggestive of arrhythmia.

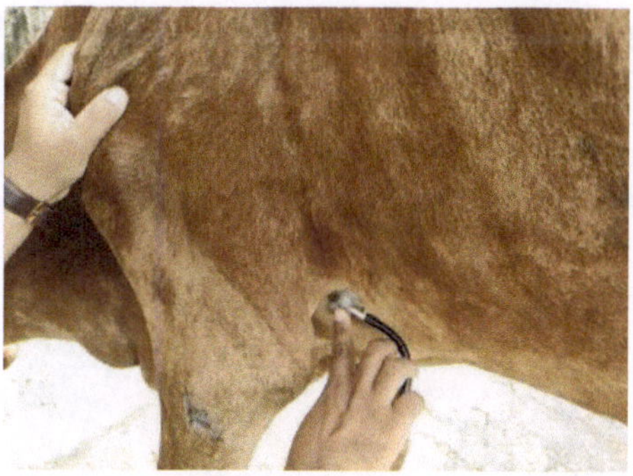

Fig. 16. Cardiac auscultation in cattle. Heart sounds are best heard under the triceps/elbow between the 3rd and 5th intercostal spaces on the left side placing the stethoscope far cranially under the elbow.

(c) Abdomen

- Abdominal auscultation is valuable in cattle, horses and sheep.
- It has limited value in pigs.
- Sounds of stomach and intestines suggest the nature of intraluminal contents and frequency and amplitude of GI tract movement.
- Sound characteristics increases just after eating feed or immediately following excitement.
- Ruminal auscultation in cattle (Fig.17) and sheep is an important part of clinical examination as ruminal movement can be easily detected. In normal animals primary contractions varies from 1-2 /minute . In rumen atony no contraction is auscultable. Secondary contraction occurs about 1 /minute. Auscultation of rumen is done in left paralumbar fossa.

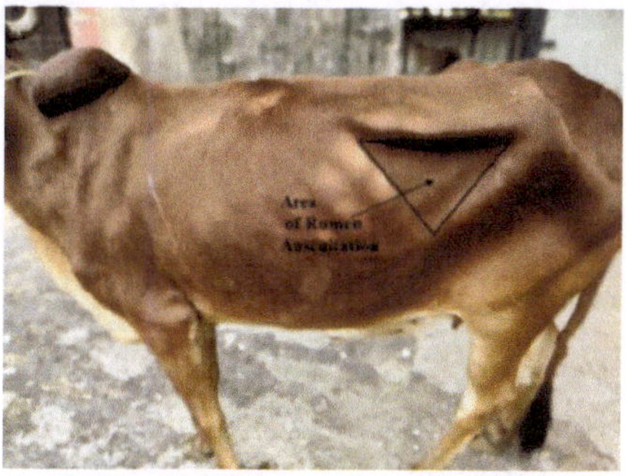

Fig. 17. Area for auscultation of the rumen in a cattle.

- Grunt can be detected over the trachea during reticular contraction phase of a primary contraction in cattle with traumatic reticuloperitonitis.

- Intestinal sounds are faint gurgling sound detected on auscultation of right ventral quadrant and flank of cattle (Fig.18) and sheep.

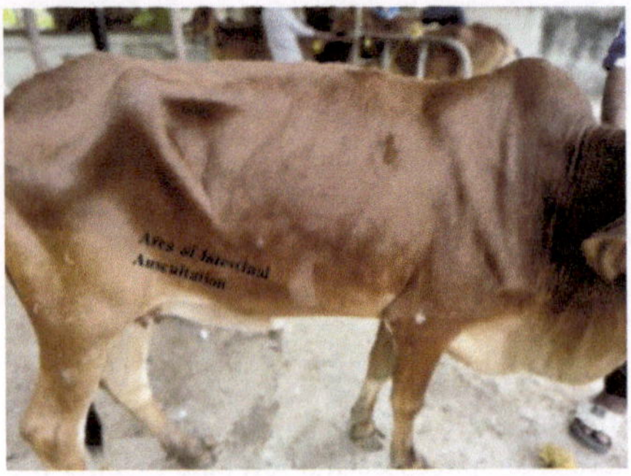

Fig. 18. Area for auscultation of intestines in a cattle.

- Contractions of abomasum and intestines leads to a mixture of sound and is difficult to distinguish.
- In horses intestinal sound are easily heard over right ventral abdomen. The character of the intestinal (colon and cecum) sounds is loud and booming occurring at the rate of 3-4 per minute. Over left abdomen faint rushing fluid sound of small intestine can be heard. The intensity and frequency of intestinal sounds is increased in enteritis ; becomes loud crackling in spasmodic colic ; and decreased in large intestine impaction and thromboembolic colic and also in intestinal obstruction.

Imaging Techniques

- Radiology (Fig. 19) is an important imaging technique. It is more frequently used in neonates and for limb defects. Its use is limited for other purpose in large ruminants. It is commonly employed for pregnancy diagnosis, abdominal survey in small ruminants.

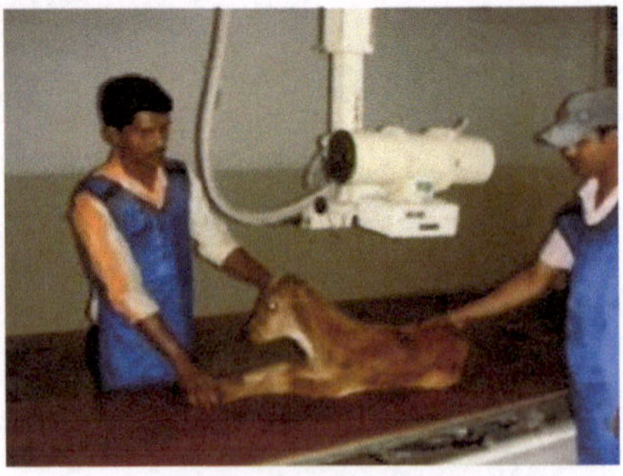

Fig. 19. Radiographic imaging of a calf.

- Ultrasound (Fig.20) is another imaging technique appears to have much more general application. It is frequently used in large ruminants for pregnancy detection and evaluation of reproductive tract.

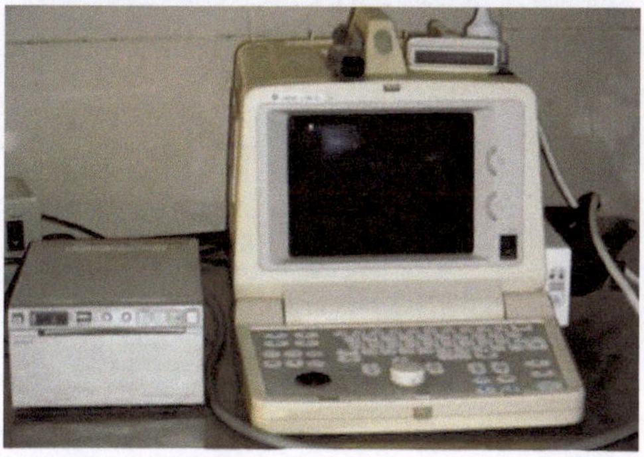

Fig. 20. Ultrasound machine for the use in ruminants for pregnancy diagnosis.

Evaluation of Vital Indices

- Monitoring of temperature, pulse and respiration is an important part of physical examination.
- Temperature is measured by inserting a lubricated thermometer in the rectum (Fig.21) . Diurnal variation up to 1^0C is common and is not an indication of any disease. Temperature of healthy animals is given in the Table 1 . Mild increase (1^0F) in temperature in late pregnancy may occur. Body temperature is subnormal before farrowing and rises with parturition in sows. In horses, temperature may remain elevated up to 2 hrs. post racing. Significant variations in body temperature indicate pathological state. Hyperthermia is simple elevation of temperature with out toxemia. While fever is associated with elevation of temperature combined with toxemia (in infectious diseases).Hypothermia is associated with subnormal temperature and occurs in shock, circulatory failure, hypothyroidism and just before death.

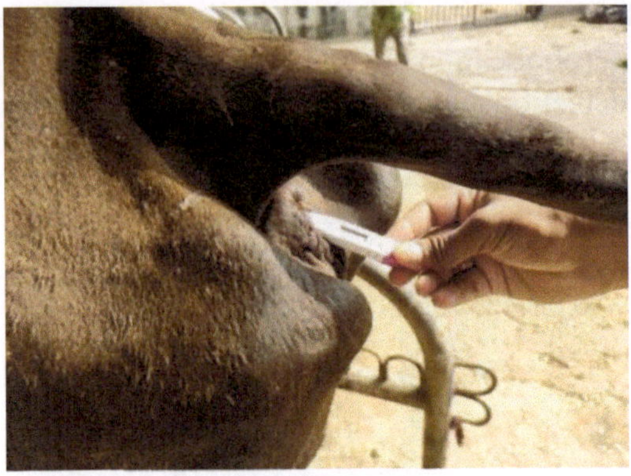

Fig. 21. Taking rectal temperature in a cattle.

Table 1: Basal Body Temperature of Healthy Animals

Sl. No.	Species of Animal	Range of Rectal Temperature °F	Mean Temperature °F
1.	Cattle	100.0 to 102.5	100.5
2.	Goat	101.5 to 104.5	103.0
3.	Sheep	102.0 to 104.0	102.0
4.	Horse	99.0 to 100.5	100.5
5.	Donkey	97.5 to 101.0	99.5
6.	Cat	100.0 to 102.5	101.0
7.	Dog	101.0 to 102.5	101.0
8.	Pig	100.0-102.0	102.0
9.	Camel	93.5 to 105.0	99.5
10.	Elephant	97.6 to 98.0	98.0
11.	Rabbit	102.0 to 104.0	103.6
12.	Poultry	105.0 to 109.0	107.0

- Pulse is taken at the middle coccygeal artery (Fig.22) or facial artery in cattle, at the facial artery in horse and donkeys, at ventral coccygeal artery in pigs, and at femoral artery in dogs, cats, goats and sheep to determine rate, rhythm, amplitude, tone pulse pressure (maximum and minimum) and form. Pulse rate is dependent on the heart rate. Pulse rate of healthy animals is given in the Table 2. Pulse deficit occurs when heart contractions are weak. Pulse rate varies from 30 to 40 /min in horses, 70 to 80 /min in colt up to one year, 60 to 80/ min. in cattle, 100 to 120 /min. in young calves and 70 to 90 /min in sheep and goats. There are variation in pulse rate in different breeds. It is difficult to palpate pulse in pigs. Its heart rate varies from 60 to 100 beats /min. Pulse rate/ heart rate lower than the specified limit is called bradycardia and higher than the specified limit is termed as tachycardia. Irregular cardiac rhythm is termed as arrhythmia. Pulse amplitude can be determined by the amount of digital pressure needed to obliterate the pulse wave. It reflect cardiac stroke volume. Pulse amplitude increases in cases of aortic semilunar valve insufficiency; and decreases in cases of myocardial weakness.

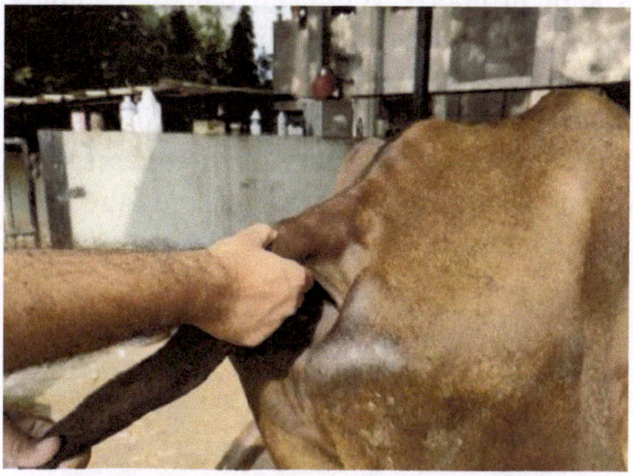

Fig. 22. Taking pulse at the middle coccygeal artery in a cattle.

Table 2: Pulse Rate of Healthy Animals

Sl. No.	Species of Animal	Range of Pulse Rate Per minute
1.	Cattle, Buffaloes Calf	60 to 80 80 to 100
2.	Goat	60 to 70
3.	Sheep	60 to 70
4.	Horse, Donkey, Mule, Asses Colt	30 to 40 70 to 80
5	Cat	100 to 130
6.	Dog (Large breed) Dog (Small breed0	60 to 90 80 to 120
7.	Pig (Adult) Pig (Young)	60 to 90 80 to 120
8.	Camel	32 to 50
9.	Elephant	20 to 30
10.	Rabbit	140 to 180
11.	Poultry	130-160

- Respiration is recorded by observing the movements of the flank, sternum and ribs, nostrils (Fig. 23), and by auscultation of thorax or trachea. Respiration rate in healthy animals is given in the Table 3. Respiration is observed not only for rate but also for type, rhythm and quality (depth). Normal respiration rate is influenced by the environmental temperature and humidity. Physiological increase in respiration is seen after exercise or work; during excitement, anxiety, stress; during pregnancy; and in animals kept in poorly ventilated places. Pathological increase in respiration is seen in cases of fever, hyperthermia, pneumonia, pleurisy, pneumothorax, peritonitis, ascites, pulmonary collapse, upper airways edema, severe anaemia, heart failure and obstruction of respiratory passage. Decrease in respiration rate is seen in cases with space occupying lesions of brain, increased intra-cranial pressure, hydrocephalus, and barbiturate poisoning.

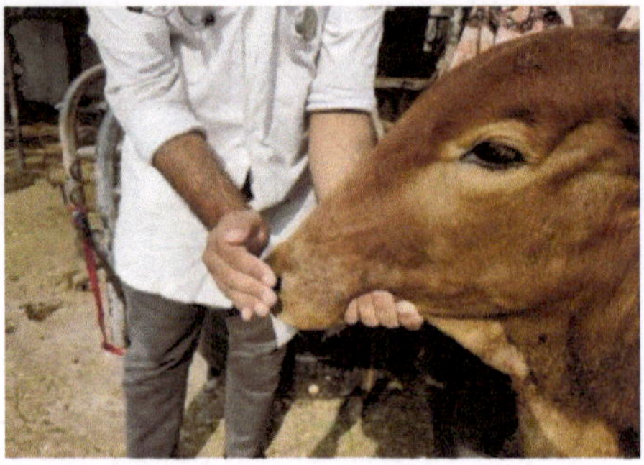

Fig. 23. Recording respiration by keeping palm in front of nostrils in a cattle.

Thoracic or costal respiration is marked in cases with diaphragmatic muscle paralysis, rupture of diaphragm, diaphragmatic tumor or abscess, ascites, ruminal tympany and peritonitis. While marked abdominal respiration is seen in cases with pulmonary edema/congestion, acute pleurisy/ tuberculous pleuritis, traumatic pericarditis, chronic alveolar emphysema, pharyngeal obstruction and paralysis of intercostal muscles.

Table 3: Respiration Rate of Healthy Animals

Sl. No.	Species of Animal	Range of Respiration Rate Per minute
1.	Cattle Cattle (Young) Buffalo	16 to 22 27 to 30 22 to 28
2.	Goat	18 to 30
3.	Sheep	18 to 30

Sl. No.	Species of Animal	Range of Respiration Rate Per minute
4.	Horse, Foal	8 to 14 14 to 16
5	Cat	20 to 30
6.	Dog	14 to 30
7.	Pig	16 to 18
8.	Camel	10 to 30
9.	Elephant	4 to 12
10.	Rabbit	30 to 60
11.	Poultry	15 to 30

Examination of Body Regions

Examination of body regions is to be conducted in orderly manner starting from head and neck to rectum and reproductive tract.

- Head - Eyes are examined for discharge, lacrimation (unilateral or bilateral), abnormalities of eyelids, membrana nictitans, conjunctiva (inflammation, pale color indicates anemia, yellow color suggests jaundice), corneal abnormalities (opacity, ulcer), eye ball (size, protrusion, retraction, abnormal movement) and deep structures (using ophthalmoscope).

  Nostrils are examined for nasal discharge (color and consistency), odor of nasal breath, volume of the breath(variation in volume between nostrils as felt on hand may indicate obstruction or stenosis), and ascertaining breathing and noting respiration rate(keeping hand in front of nostrils).

  Mouth is examined for excessive salivation (ropy saliva associated with foreign body in the mouth), tongue (actinobacillosis, Foot and Mouth Disease and mucosal disease), dryness of mouth (dehydration), local lesions on buccal mucosa (hemorrhage, discoloration), teeth, pharyngeal region and submaxillary region (lymph node enlargement, edema, local cellulitis, salivary glands or guttural pouch in horse. Thyroid glands should also be palpated for any abnormality.

- Neck- Neck is examined to ascertain the state of jugular veins in cattle and horses as bilateral engorgement may be due to right side congestive heart failure or compression/constriction of the veins. Small magnitude jugular pulse moving up the jugular vein (one third of the way up the neck) is usually normal in animals. Esophageal abnormalities can be detected by passing stomach tube. Tracheal auscultation can facilitate the diagnosis of upper respiratory tract diseases (laryngitis and tracheitis).

- Thorax- It is examined using palpation, percussion and auscultation in cardiac and lung areas.

  Cardiac auscultation is done in 5th and 6th intercostal spaces to determine the rate and rhythm of normal heart sound and to detect abnormal heart sounds if any. Systolic sounds (S1) are best heard over the cardiac apex (left side - mitral valve, right side- tricuspid valve) and diastolic sounds (S2) are best heard over the base of the heart (aortic valve closure posteriorly and pulmonary valve closure anteriorly) on the left side of the thoirax.Normal heart sound is three character sound as LUBB-DUPP-Pause. Cardiac palpation is carried out on both side of the thorax using the palm as it can assists in assessing the size of the cardiac impulse and detecting palpable thrill. Percussion of the heart has little value in large animals. Cardiac hypertrophy and dilatation of the heart is associated with an increased area of cardiac dullness.

  Lungs are examined using auscultation, percussion and palpation. Normal breath sounds are soft, sipping VEE-EFF and are audible over most of the lung area. The amplitude of breath sound is increased in dyspnea and early stage of pulmonary inflammation and congestion; and decreased in pleural effusion and space occupying lesions in lungs or pleural cavity. Crackles, wheezes and pleuritic rubs are abnormal lung sound. Palpation of the thoracic wall is valuable in detecting pleuritic thrill, bulging of intercostal spaces, narrowing of intercostal spaces and decreased rib movement over the collapsed lung. Direct percussion in intercostal spaces yields normal resonance sounds (normal lung), dullness or abnormal resonance (consolidated lungs, edematous lungs, space occupying lesion in lung or pleura or an abnormal accumulation of fluid), and definite tympanitic notes (pneumothorax, gas filled organs herniating through diaphragm.

- Abdomen- It is examined by visual inspection (gauntness, abdominal distension), auscultation (increased or decreased/ absence of gastrointestinal sounds), palpation and percussion through abdominal wall, rectal palpation, passage of nasogastric tube and paracentesis abdominis. Rumen auscultation in cattle and sheep is valuable in examining rumen movement. Similarly intestinal auscultation assists in evaluating intestinal peristalsis. Auscultation of right flank in cattle and sheep assists in hearing contraction sound of the abomasum. Decrease or absence of intestinal sounds (borborygmi) is suggestive of large intestine impaction in cattle or thromboembolic colic in horses.

Palpation of viscera and organs through abdominal wall is of limited value in cattle and horses because of thickeness and weight of abdominal wall. In horses fetus can be palpated through abdominal wall. In cattle, rumen and its contents are palpated in left paralumbar fossa.

High pitched musical pings are audible on percussion and simultaneous auscultation over upper third of the costal arch between 9th and 12th ribs on the left side in cases of left side abomasal displacement. Tactile abdominal percussion helps in detecting excess quantity of fluid in the peritoneal cavity. Abdominal palpation is difficult in pigs. Rumen, gravid uterus and impacted abomasum can be palpated through abdominal wall in sheep.

Passage of nasogastric tube in cattle assists in relieving rumen distension and to obtain rumen fluid sample for physical, chemical and microbial/ protozoal examination.

Rectal palpation (Fig.24) facilitates the examination of gastrointestinal, urinary and genital tracts and vessels on peritoneum and pelvic structures. Gastrointestinal tract abnormalities such as paralysis and ballooning of rectum, intestinal loop distension (with gas and fluid), hard mass of ingesta in cecum and colon in horse and intestinal obstruction (volvulus, intussusception or strangulation), distended abomasum (in right half of abdomen in case of abomasal torsion) can be detected on rectal palpation. In cattle left kidney can also be felt. While in horse caudal pole of left kidney is palpable. Distended bladder or its chronic thickened wall can also be palpated. Abdominal aorta is palpable on rectal palpation. Anterior mesenteric artery and some of its branches in horse can be felt. Rectal palpation is very important in cases of verminous aneurysm in horses.

Paracentesis abdominis assists in obtaining peritoneal fluid sample in cases of ascites, peritonitis, inflammation of serosa of intestines or other viscera when suspected.

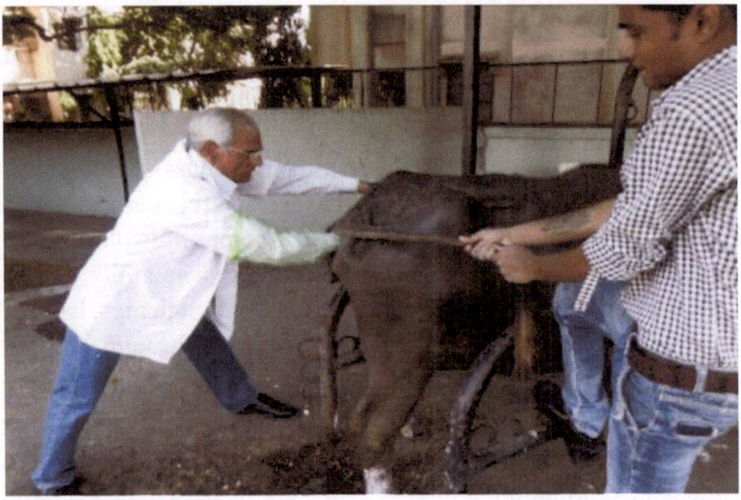

Fig.24. Rectal palpation in a buffalo.

- Reproductive tract-In post partum period, the vagina, cervix and uterus is to be examined for detecting metritis, retained placenta or ruptured uterus. Sometimes these abnormalities may be the cause of illness not detected on the examination of other body systems described earlier.
- Mammary glands (Fig.25) are palpated and examined for udder and teat swelling, hardness, erythema, temperature variations (hot or cold) and lesions.

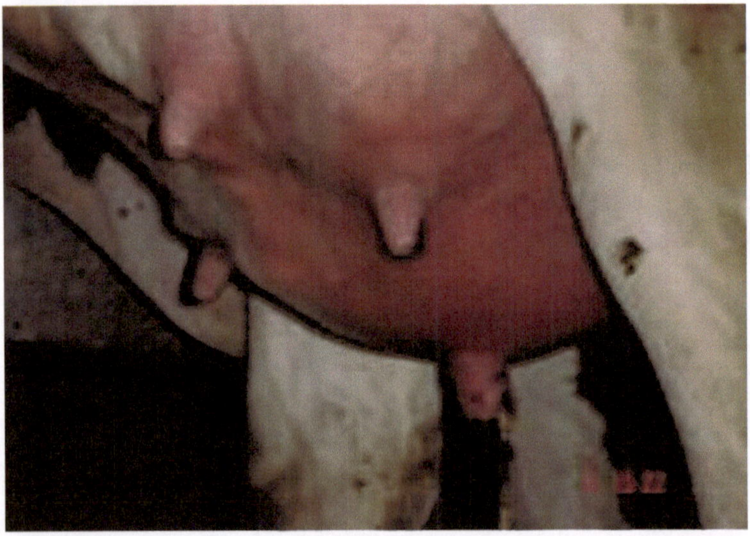

Fig.25. Udder and mammary glands of a cow

Case Record

History and Clinical Examination

Patient Identification

Case Registration No. Date:

Animal Name/number. Species:

Breed: Sex: Age: Weight:

Owner's Name and Address:

Owner's Mobile number:

History

General History

i. Primary complaint of the animal or herd
ii. Duration of signs
iii. Is there any mortality
iv. How many animals showing similar signs
v. Any treatment given and its effect
vi. Preventive and control measures (vaccination, deworming, dipping, disinfection etc.)
vii. Transportation, introduction of any animal
viii. Recent travel to other geographical region

Management History

i. Nutritional history (quality and quantity of the feed, change in feeding regimen, water supply)
ii. Hygiene
iii. Housing,ventilation and population density
iv. Climate
v. Environment

   Outdoor topography, soil type, stocking rate, feed and water supply, waste disposal

   Indoor hygiene, ventilation, flooring, floor plan,lighting etc.
vi. General management practices
vii. Opportunity to exercise
viii. Milking procedures and milking hygiene

Breeding and Reproduction History

i. Relationship between sire, dams
ii. Bull /cow ratio
iii. Natural service/ A.I.
iv. Conception rate
v. Length of breeding season
vi. Percentage of females pregnant at a specific time
vii. Percentage of abortions

Disease History

a. Previous disease history (previous disease (s), treatment given and outcome
b. Present disease history (signs, lesions, sequence of illness, duration)

Clinical Examination

Temperature: __________ Pulse: __________ Respiration: __________
Heart rate: _________________ Body condition: ______________
Hydration status: ________________

General Inspection from distance

1. *Behavior and general appearance*
   - Bright (responding to external stimuli such as sound and movement)
   - Dull (poor response or indifference to external stimuli)
   - Dummy (pronounced state of indifference, not responding to external stimuli)
   - Coma (stage of apathy and depression, unconsciousness and animal can not be aroused)
2. *Excitation state*
   - A state of anxiety or apprehension
   - Severe restlessness (moving, lying down and getting up, looking at flank, kicking at belly, rolling and bellowing)
   - Mania (abnormal movement with vigor, violent licking its own body, chewing, and licking inanimate objects, pressing forward with the head)
   - Frenzy (wild and uncontrolled action)

3. Voice
4. Eating pattern
5. Defecation pattern
6. Urination
7. Posture
8. Gait
9. Body condition
10. Body conformation
11. Skin

Body Region Inspection from Distance

1. Head
2. Neck
3. Thorax
4. External Genitalia
5. Mammary Glands
6. Limbs

Close Clinical/Physical Examination Using Following Techniques

Techniques

1. Palpation (findings as doughy, firm, hard, fluctuating, tense or emphysematous)
2. Percussion (resonant. tympanitic or dull)
3. Ballottement (detection of floating viscera or mass)
4. Auscultation of heart, lungs intestines
5. Percussion and simultaneous auscultation of abdomen
6. Succussion (moving body from side to side to detect the presence of the fluid)
7. Special techniques using biopsy and paracentesis

Sequence of Examination

1. Thorax (heart for rate, rhythm, intensity, murmurs; lungs for normal or abnormal sounds)
2. Abdomen
3. Head and neck including eyes, ears, nose, oral cavity, facial structure, jugular veins

4. Rectal examination
5. Urinary tract examination
6. Reproductive tract examination
7. Examination of mammary glands
8. Examination of musculoskeletal system
9. Examination of nervous system
10. Examination of skin, hoof and horn

Special Laboratory Examination

1. Blood
2. Feces
3. Urine
4. Skin Scrapings
5. Impression Smears
6. Serum Investigation
7. Culture Examination
8. Biopsy Examination

Date: **Signature of Attending Veterinarian**

3

History Taking and Physical Examination: Small Ruminants

Small ruminants are grazing mammals and this group consists of sheep (*Ovis aries*) and goats (*Capra hircus*).Advantage of these animals is their low cost, small size and marginal land use, which convert into high quality animal products. Sheep is a grazing animal and goat is a browsing animal. They are generally kept in flock. Therefore flock management is more important than the individual animal. Though, procedure of history taking and physical examination is almost similar to other animals, some peculiarities warrant a separate discussion.

Sheep

The general methods of history taking and physical examination of sheep are similar to cattle and goat starting from owner's complaint (history), signalment of the patient, history of the ailing animal (s), history of the farm, examination of the environment, observing the animal from distance, detailed observations of the sick animal, physical examination of the animal to laboratory investigations.

Owner's Complaint (History)

Sheep are generally kept in small or large groups. Perception of magnitude of real problem varies with knowledge and experience of the owner.

Signalment

Age, sex, breed, color and genotype of the sheep do have bearing on occurrence of diseases. For example watery mouth and lamb dysentery are only seen in young lambs; mastitis is seen in ewes after lambing / weaning; sunburn is seen in pale faced sheep (Charollais breed); Texels breed are predisposed to chronic chondritis of the larynx ;and susceptibility to scrapie is related to the genotype.

History of the Farm and Patient

Information related to the presence of disease(s), deficiency or excess of mineral in the soil in the area, presence of other animal species on the farm and number of the sheep on the farm and how they are being kept is valuable in ascertaining the cause of illness.

Recent management and husbandry practices such as shearing, tagging, castration, foot trimming and dipping may sometimes predispose to illness. High incidence of lameness after dipping (due to erysipelas infection gaining access to the joints through superficial injuries sustained during dipping and clipping) or feet trimming have been reported in sheep. Increase in cases of caseous lymphadenitis is noticed after shearing in already infected flock.

Preventive Health Program

Information regarding vaccines, fly control products, antimicrobials, anticoccidials, mineral/ vitamin supplements, and anthelmintic given is valuable in providing lead to diagnosis. Endoparasites may be a serious problem in some flocks. History of anthelmintic and its dosages may reveal inappropriate usage or possible anthelmintic resistance if any. Biosecurity measures taken in the light of the current problems should be reviewed.

Practice and Farm Records

Details of current and previous health problems are relevant to the problem under investigation. Practice records provide an overview of the disease initially diagnosed and drugs used. Farm records are valuable in providing information related to lambing percentages, lamb mortality, pregnancy rates, ewe culling rates, ewe mortalities, growth rates, condition scores in relation to the production cycle, prevalence of hypocalcaemia, hypomagnesaemia and pregnancy toxemia in the flock. Recently introduced animals may spread new diseases. Diet, stocking density and pasture management are relevant to endoparasite control. Lack of magnesium supplementation to postpartum lactating ewes may increase the risk of hypomagnesaemia. Urolithiasis in male fat lambs can be associated with increased and incorrectly balanced mineral content of the fed concentrate. Warm wet wooded pastures are associated with an increased risk of fly-strike.

Observation of the Environment

Place where sheep are being kept or have been kept in recent past, and pasture where flocks are grazing make the environment that needs to be examined. Sheep grazing close to grass roots may be predisposed to ‘rye grass staggers’. Water deprivation increases the risk of urolithiasis. Damp conditions predispose sheep to foot-rot.

Condition Score

Condition score is assessed by palpating the transverse and spinous process of the lumbar vertebrae and observing the amount of muscle and fat. All parts of the lumbar vertebrae are readily palpable in animals with condition score 1 having very little muscle and difficult to palpate at a score of 5.

Examination of Vitals

Temperature, pulse and respiratory rate are recorded; and mucous membranes and lymph nodes are examined. Heart and respiration rates increases in warm weather in fully fleeced sheep. Respiration rate in resting sheep varies from 20 to 30 breaths/minute. The pulse is taken at the femoral artery or by auscultation of the heart. It varies from 70 to 90 beats/minute in resting sheep. Temperature in healthy sheep remains in the range of 38.5 to 40.0°C.

Clinical Examination

Mucous membranes- Mucus membranes in healthy sheep is salmon pink and is conveniently assessed at conjunctiva, buccal cavity or vulva (in females). Mucus membranes of pale, icteric, hyperemic/congested and cyanotic color are suggestive of anemia, intravascular hemolysis or hepatic dysfunction, febrile / toxemic states, and severe respiratory and cardiovascular compromise respectively. Capillary refill time (CRT) can be assessed on gums, lips or vulva (in females). Prolonged CRT indicates poor peripheral perfusion.

Superficial lymph nodes- Superficial lymph nodes (submandibular, parotid, retropharyngeal, prescapular, precrural and supramammary glands) are palpated for enlargement and compared with the opposite one. The parotid lymph nodes, the submandibular lymph nodes and the cervical chains in the neck have been found to be an external predilection site for caseous lymphadenitis.

Skin lesions and/or loss of hairs Skin lesion and /or hair loss is suggestive of mycotic, bacterial and/or viral infections or pruritus caused by ectoparasites, head fly or scrapie infection. Entire skin surface is looked for areas of non-functional hair follicle, damaged or thickened skin; and for the presence of specific diseases. Subcutaneous abscess on the poll of rams are often caused by fighting and hair loss over the brisket in rams may be caused by mounting and serving ewes. Greenish discharging abscess in the proximity of a superficial lymph node is suggestive of caseous lymphadenitis. Yellow crusty and pruritic lesions on dorsum of the sheep is suggestive of sheep scab and other ectoparasites. Confirmation of the disease requires microscopic examination of skin scrapings. Keds can be easily seen with the naked eye; they are brown in color. Lice and their eggs attached to wool are observed using a bright light and a magnifying glass. Engorged female ticks are easily seen. Blow-fly can be seen around the perineum. Head fly may provoke self trauma of the skin of the head, particularly in horned breeds. Hypersensitivity (atopy) to culicoides may present with pruritus and loss of hair around the face and ears with scabby lesions in summer and autumn. Orf, a zoonotic disease, causes crusty lesions on the lips, udders and teats of ewes. Matting of the wool may indicate infection by *Dermatophilus* or if stained yellow/green by *Pseudomonas*. Staphylococcal dermatitis and actinobacillosis can produce scabby, pustular, necrotic granulomatous lesions at various sites.

Head Conformation- It is examined for symmetry. Unilateral facial paralysis result in flaccid paralysis of the lip, cheek, eyelid, nostril and ear. Intermandibular space edema is seen in cases of endoparasites causing hypoproteinaemia. Generalized facial edema may be due to photosensitization. Lower jaw bony swellings may be due to dentigerous cysts and tooth root abscesses. *Coenurus cerebralis* cyst may lead to focal thinning of skull.

Examination of Eyes- Keratitis and epiphora with hyperaemia of the conjunctiva may be detected on external examination of eyes. A few drops of topical ophthalmic anesthetic can facilitate the examination of conjunctival fornices. Iritis may be associated with superficial infection with listeria in sheep on silage feeding. The menace response is absent in blind sheep. Internal examination of the eye using ophthalmoscope is rarely done. Narrowed retinal blood vessels and hyper-reflective tapetum lucidum has been observed in sheep with bright blindness associated with bracken poisoning. Increased intracranial pressure may cause papilloedema (a swelling of the optic disc).

Examination of Ears - Ear tears and hematomata of the pinnae are not uncommon in sheep. Photosensitization may lead to crusting and peeling of the pinnae skin. Some cases of listeriosis may show abnormal unilateral ear carriage.

Examination of Nares- Nasal discharge is seen in cases of some upper and lower respiratory infections, in some cases of chronic sinusitis in older ewes, and infestation with *Oestrus ovis*.

Mouth Examination- Sheep generally resent mouth examination. Sedation can facilitate mouth examination. Vesicles on dental pad or tongue can be seen in FMD affected sheep. Puffing of the cheeks during respiration usually indicates respiratory distress. Retropharyngeal lymph-node abscessation can sometimes be felt by external deep palpation or delineated by ultrasonography or radiography. Endoscopic examination of the pharynx and larynx can also be done. The jugular vein, can be examined for signs of thrombophlebitis.

Esophageal palpation is to be done if foreign body obstruction is suspected in sheep with excessive salivation, free gas bloat, teeth grinding and apparent quidding.

Examination of Thorax- Lungs are auscultated in sheep as is done in other species. Percussion is restricted by the presence of fleece in sheep. Auscultation in sheep requires parting of fleece. The normal heart sounds are heard at the 4th or 5th intercostal space behind the foreleg on both sides. Cardiac diseases are not common in sheep. Radiography is valuable in assessing the thorax, though presence of wool reduces the clarity.

Examination of Abdomen-Abdomen is examined in a way similar to cattle. Ruminal auscultation can assist in assessing its function. In sheep, rumen has an appearance of 'cottage loaf '(dorsal sac smaller than ventral sac). Rumen contractions are @1-2 per minute in sheep. Direct palpation through left paralumbar fossa can ascertain whether the rumen is full or not. Bloat is less common in small ruminants. Abomasal disorders are also rarely diagnosed in adult sheep. Ballottement may indicate the presence of large amounts of fluid in the abdomen. The caudal abdomen can be readily examined through the flank by ultrasound

Rectal examination- It is limited to digital palpation. Feces of normal sheep are pelleted.

Examination of Urinary System - Catheterization is not done in male sheep. Urine sample can be collected by gentle 'tickling' of the preputial opening with the gloved hand or a piece of straw. Gross examination of a urine sample may reveal red urine, indicating haematuria or haemoglobinuria; a brown urine indicates a myoglobinuria. Ultrasonography and abdominocentesis may provide diagnostic help.

Examination of External Genitalia - In male, the scrotum is examined in standing position by palpation, for their size, and consistency and mobility is checked. Painful, enlarged and hard testicles suggests orchitis. Inguinal hernias may be detected by visual inspection, palpation or ultrasonography. Penis and prepuce are also inspected for abnormality if any. A digital rectal examination may reveal pulsation of the urethra in urethral obstruction.

Visual inspection of the vulva may reveal a vaginal eversion or prolapse, swelling or erosions, uterine prolapse, or discharge. Normal lochia following an uncomplicated birth is reddish brown and odorless,

Examination of the Udder- Visual inspection and palpation of the mammary glands may reveal abnormalities of the suspensory apparatus, abnormal swellings and discoloration of the skin and identify edema, inflammation, fibrosis, scarring or abscessation. Udder pain and swelling may initially cause a stiff gait or lameness.

Examination of Locomotor System- Lameness and abnormalities of gait are common in sheep. Affected limb can be identified by observing the weight bearing while standing and at the walk. Severe lameness of a single limb may be due to bone fractures. Gentle manipulation of the limb may reveal abnormal movement, pain and crepitus. The foot inspection is done after washing the foot. Hoof testers may facilitate in identifying the seat of pain. The interdigital area is inspected from above and below for the presence of growths, integrity of the skin and inflammation. Foot rot is associated with necrosis and an offensive odor. Coronary band and heel are inspected for swellings and eruptions.

Examination of Nervous System- It is done in manner similar to cattle. Observations are made on behavior (wandering, apprehension, circling, head pressing and fits), mental state (dullness and depression, hyper excitability), head position (aversion, tilted, high or low head carriage), head coordination (intention tremors) and movement (recumbency, incoordination, ataxia). Exaggerated attentiveness may be caused by hyperaethesia or defective vision. Head tilt may indicate inner- or middle-ear infection or a lesion in the upper medulla. Lateral deviation of the head may be seen in unilateral blindness. Head tremors usually seen in cerebellum or diencephalon lesion. Compulsive circling movements may indicate a brainstem or cerebellar lesion. Facial paralysis may result in to unilateral bulging of the cheeks by retention of the cud, lingual paralysis, drooping of the ear and/or eyelid. Lesions in the vestibular apparatus or cerebellum may cause nystagmus. The pupils should be checked for symmetry. Loss of prehension and drooling of saliva may be due to flaccid paralysis. Spinal lesions may result in tetraparesis or paraparesis.Localization and characterization of the lesion is done by testing proprioception, skin sensation, the panniculus response, deep pain, pedal reflexes, triceps reflex, the patellar reflex, tail tone, the anal reflex, and observing bladder control. The wheelbarrow and hemi walking test can be performed.

Goats

Goats are kept in small groups or individually also. Though goat is classified with sheep as small ruminant, there are some differences. Goat are browser and sheep are grazer. Illness in goat is recognized only when it is in advanced stage. Clinical examination of goat consists of owners complaint and history of illness, signalment details (breed, age, sex), examination of goat environment, observation and inspection (of the goat or group of goats), clinical examination of the patient (temperature, pulse, respiration, color of mucus membranes, capillary refill time, palpation, inspection of lymph nodes; examination of head and neck, cardiovascular system, respiratory system, gastrointestinal system, udder and perineum, external genital of the female or male, locomotor system, neurological examination), clinical pathology and post-mortem examination.

History

It should consist of the long-term and short-term history of the herd and patient,the background health status of the herd, parasite control and disease prophylaxis.

Signalment

The breed, age and sex of the goat can help in arriving at diagnosis as the sticky kid syndrome is seen only in male kids of the Golden Guernsey breed; Saanen goats more prone to ketosis and hypocalcemia and gynaecomastia is common in British Saanen.

Environment

Goats are predisposed to inclement weather. Therefore shelter and housing conditions should be examined. Goats are prone to photosensitization.

Clinical Manifestations and Diseases

The goat should be observed discreetly. Star gazing is seen in some goats. Heart failure in not common in goats. Weight loss in goats is generally associated with malnutrition, a high parasite burden and chronic diseases (Johne's disease). Distension of the left upper quadrant of the abdomen is commonly seen in cases of bloat. Vagal indigestion is associated with right-sided distension in the ventral quadrant. Enlargement of abdomen is also seen in pregnancy. Pendulous abdomen may be seen in some elderly healthy doe goats. Sarcoptic mange is common in goats and is associated with intense itching. Caprine arthritis encephalitis virus (CAEV) syndrome is associated with joint swelling in goats. Foot lameness is common. Chronic mastitis of one sided udder causes disparity of udder. Disparity in the size of testicles may be due to unilateral orchitis. Intersex in goats is associated with the polled (hornless) state. Vaginal discharge in goats is indicative of a uterine infection or a genital tumor. Scrapie, listeriosis and gid are neurological diseases affecting behavior, gait and balance.

Close Observations

Goats are closely observed with or without restrain. Inspection of goat begins at the head. Healthy coat is sleek and shiny. If there is focal or generalized hair loss or any skin lesion, detailed examination needs to be done. Normal eyes are bright and free from any ocular discharge. Goat carries the ears erect except Anglo-Nubian breed. Drooping ears are seen in goats with psoroptic ear mange. Goat's head and neck are mobile. Healthy goats breath with minimal efforts. Abdomen of healthy goats looks full but not distended. Normal weight bearing is on all four legs. Feces of healthy goats are dry pelleted and passed frequently with little efforts. Goats urinate effortlessly when restrained.,

Abnormalities

Behavioral abnormalities are seen in diseases such as scrapie, hypocalcaemia, hypomagnesaemia, ketosis and terminal illness. Scrapie is also associated with neurological signs.

Clinical Examination

Clinical examination of the goat is conducted after proper restrain. Respiratory rate and its character is observed before restraint. Temperature (rectal) and pulse (femoral artery) is taken immediately after restraint. The color of mucus membranes (conjunctiva and vulva) is salmon pink in health. It turns pallor in anemia (copper deficiency or heavy parasitic load), yellow in jaundice and congested in infectious diseases with pyrexia. The capillary refill time (can be assessed on vulva, dental pad or gum) in healthy goats is < 2 seconds. Hydration status of the goats can be assessed by raising skin fold on the neck or chest wall. It falls back immediately in well hydrated goats. Enlarged lymph node are easily palpated. The parotid and submandibular lymph nodes are usually involved in cases of caseous lymphadenitis. Enlargement of other lymph nodes may indicate active or recent local infection. Skin diseases require differential diagnosis. Chronically ill goats may have dry scaly skin with reduced elasticity and hair growth. Dermatophytosis and orf in goats are of zoonotic importance. Perineal lesions are easily accessible to the goat, and any irritation in this area may be the subject of biting and severe secondary damage. Dermatitis on caudal aspect of fore limb in male goats is seen during breeding season owing to frequent urination. A number of diseases cause dry crusty lesions with a varying degree of pruritus requiring scrapings, culture and biopsy for differentiation. Louse infestation (chewing and sucking lice) is common in goats. Sarcoptic mange lesions are dry, crusty and intensely pruritic and found on the head (especially on the outer surface of the pinnae, around the eyes and the lips and chin), the legs, perineum and tail. Chorioptic mange lesions (crusty superficial and pruritic) are found chiefly on the caudal surface of the lower parts of the legs. Psoroptic mange lesions are found on the inner surface of the pinnae. Demodectic mange lesions are in a pustular form and found often in the skin covering the ribs. Blow-fly strike is less common in goats than in sheep. *Staphylococcus aureus* can produce a pustular dermatitis. *Arcanobacterium pyogenes* is an opportunist pathogen causing subcutaneous abscesses in various parts of the body. Contagious pustular dermatitis (orf), a viral disease, is characterized by dry proliferative crusty lesions on lips udder, feet and tail. Foot-and-mouth disease vesicles and ulcers are seen on the lips, gums, teats, coronary band and the interdigital region of the feet. Warts (paillomatosis) are seen on the skin of various parts of the body, udder and teats. Ringworm is not common in goats. Lesions are dry, slightly raised white or grey Photosensitisation following consumption of plants can lead to skin lesions.

Head and neck - Both sides of the head are compared for any disparity. Facial muscle paralysis is seen in listeriosis. Ketone body smell in breath is suggestive of ketosis. Foul smelling breath may suggest oral or respiratory disease. Excessive salivation is not common in goat. However, it can be seen in cases, with oral disease, esophageal obstruction (choke) and enlargement of the retropharyngeal lymph nodes. Salivary cysts occur chiefly in Anglo-Nubian goats. Facial abscesses may be due to injury or tooth root infection. Goats with tetanus may show trismus, hyperaesthesia and uncontrolled tetany of the muscles. Nasal discharge is seen in sinusitis and upper respiratory infection. Tongue control and swallowing, are observed. The eyes are examined for any signs of disease as per standard procedures. Mouth is carefully examined.

Cardiovascular system- Heart disease is not common in goats. Kids are occasionally born with cardiac anomalies. Severe vitamin E and/or selenium deficiency may cause cardiomyopathy in kids. Congestive heart failure (jugular vein distension – Fig. 26 and brisket edema) may be an occasional finding in old goats.

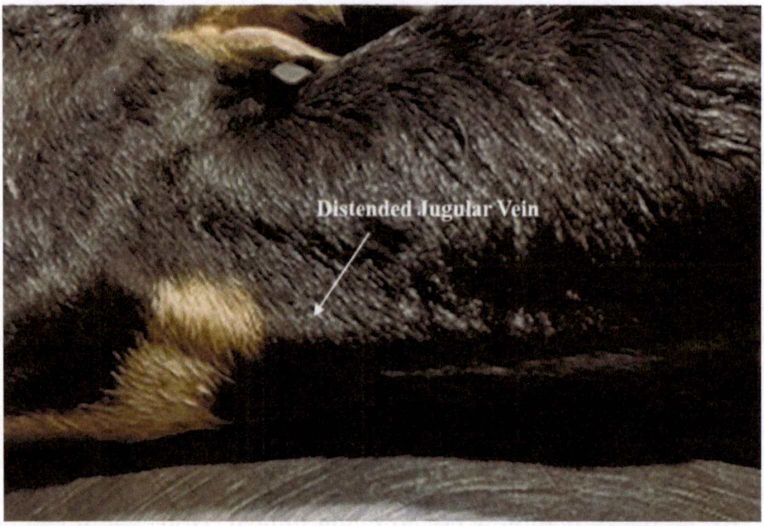

Fig. 26. Jugular vein distension in a pregnant recumbent goat. Distended jugular vein has been marked with white arrow.

Respiratory system- Respiratory system should be examined in goats with difficult breathing and coughing. Hyperthermia is also associated with rapid breathing. Tracheal collapse and emphysema may cause inspiratory and expiratory dyspnoea respectively. The thorax can be examined by auscultation, percussion, radiography and ultrasonography. Auscultation may reveal increased and abnormal sounds (squeaks, bubbling sounds) in pneumonia, rubbing sounds in pleurisy, and crackling sounds in lung emphysema. Percussion may provide evidence of ventral consolidation of diseased lungs. Diagnosis of lungworm infestation may involve larval counts in feces.

Gastrointestinal system - Dietary changes and exposure to toxic material may cause diarrhea and colic. Information about internal parasite control in the herd is important in evaluating gastrointestinal tract. Carbohydrate engorgement is the cause of rumen acidosis. Ruminal bloat (Fig.27A) is accompanied by left abdomen distension and tympanic note on percussion. Ruminal auscultation can assist in the examination of rumen movements. On palpation normal rumen gives a doughy consistency. It changes to non-compressible hard in rumen impaction. Vagal indigestion may reduce or increase ruminal movements and interfere with abomasal emptying. Borborygmi can be heard on auscultation in the lower right flank (Fig.27B). Radiography (Fig.28) is valuable diagnostic aid in detecting penetrating / non penetrating foreign bodies in the reticulo-rumen, intestinal distension (functional or physical) of the small ruminants.

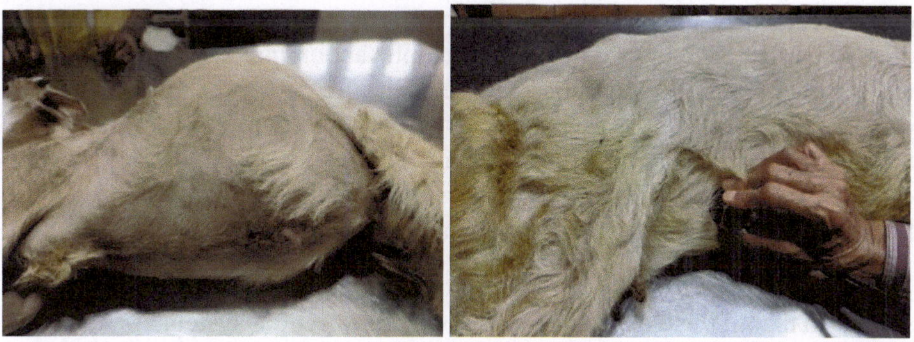

A B

Fig. 27 A. Bloated goat showing distension of left abdomen.

Fig.27 B. Auscultation of lower right flank of a goat for borborygmi.

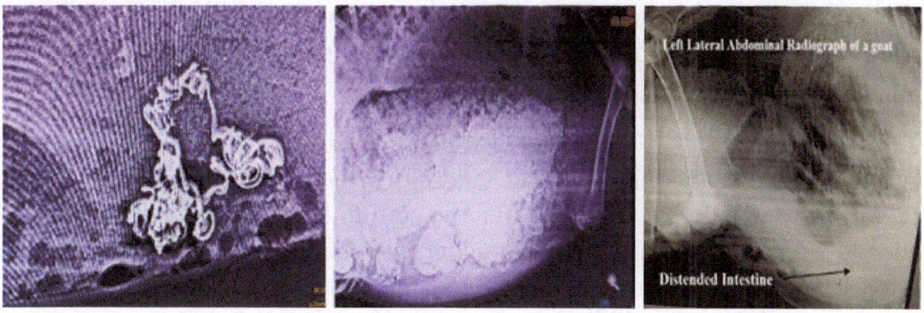

A B C

Fig. 28. Barium meal radiographs of goats are showing coiled ropy material (A) and plastic material (B) in the reticulo-rumen. and distended intestine at the floor of right abdomen (C). Only mucus and no fecal pellets were detected on rectal palpation of the goat at C showing distended loop of the intestine. It appears that the goat at C was having intestinal obstruction.

Urinary system- Changes in the pattern and frequency of urination, volume and composition of urine and pain at urination, are suggestive of urinary tract involvement. Radiography, ultrasonography and urine analysis play very important role in the diagnosis of urinary tract diseases.

Female genital system- Anestrus in an emaciated and weak doe may be due to nutritional deficiency rather than specific reproductive abnormalities. Vulva and perineum are inspected for vulvar seal or an abnormal discharge. Intersexes are quite common in goats. Most affected animals are male pseudo hermaphrodites. They are genetically female with an XX karyotype. Pseudopregnancy is quite common in goats. It is characterized by progressive abdominal enlargement, mammary development and lactation at end of pseudopregnancy. A foul vaginal discharge may be seen and the presence of a fetus detected by vaginal examination, abdominal ultrasonography or radiography. Necrotic vaginitis may be seen after dystocia. Endometritis is usually seen with a whitish-yellow vulvar discharge two weeks or more after kidding. Chronic vaginal discharge may indicate the possibility of vaginal, cervical or uterine tumors. Ovarian dysfunction occurs in does and require hormonal investigations. Rectal examination is not possible in does.

Male Genital System- The approach to examination of the buck is almost similar to that in the bull. Problems involving the genital system are relatively uncommon in bucks. Poor husbandry may lead to poor fertility in males and a complete clinical and genital system examination should be conducted. Ultrasonography is valuable in establishing a diagnosis. Accessory sexual glands may be palpated using a gloved and lubricated forefinger. The anterior part of the penis including the glans can be palpated within the prepuce. Urethral process is a common site for urolithiasis in male goats. Infertile bucks are investigated on the pattern similar to bulls.

*The Udder-*The udder (Fig.29)is examined initially from behind the doe. It is slightly edematous shortly after kidding, In per-acute mastitis, the udder is warm and swollen, some areas may be cold to the touch and subcutaneous edema extending forward from the udder into the muscles of the abdominal floor. Udder may be hard and painful in acute mastitis. Chronic mastitis may have areas of fibrosis within the affected half of the udder. Crusty lesion on udder may be due to pustular dermatitis caused by *Staphylococcus aureus*. Teat should be palpated to detect any small round structure within the teat in cases of milk flow obstruction.

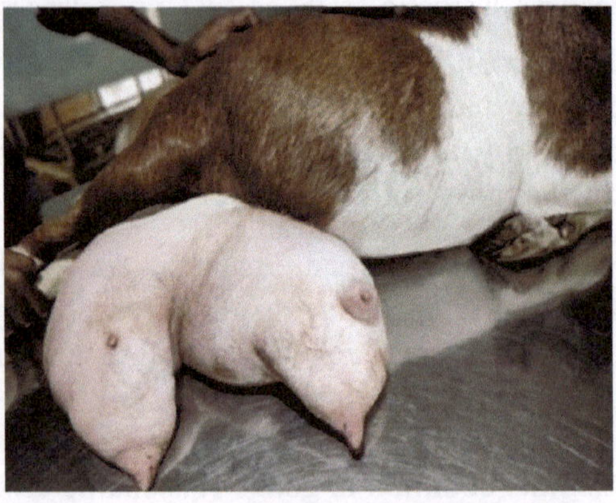

Fig.29. The udder of the goat is examined for edema, inflammation, softness/hardness, cold/hot etc.

Locomotor system- Lameness is not uncommon in goats and requires examination of feet as well as of whole animal. Septic arthritis (joint ill) caused by bacterial infection is much more common in kids than in adult goats. Osteoarthritis and infections of the feet are common cause of lameness in in older goats. Laminitis is associated with reluctancy to move at all. Bending of the long bones and ricket rosary in kids is suggestive of rickets in kids. In older animals softening of the bones may be osteomalacia or osteodystrophia fibrosa (Fig.30). Foot-rot may lead to lameness in one or more limbs and is associated with prolonged exposure to wet conditions. Radiography is useful to identify the degree of damage sustained by the articular surfaces of the joint. Distension of the joints may indicate septic arthritis ('joint ill') in kids.

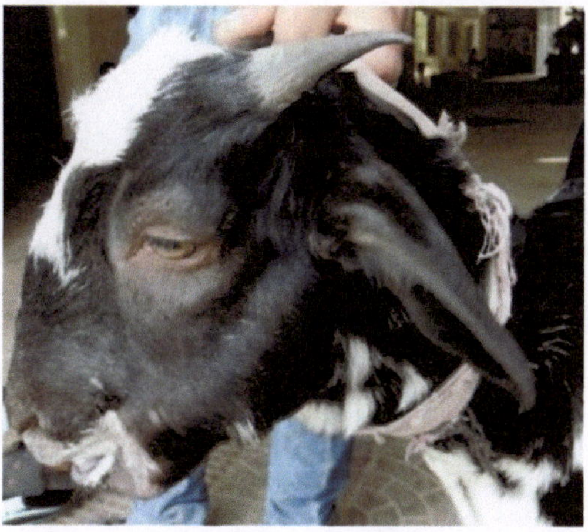

Fig.30. A goat with osteodystrophia fibrosa. (Varshney et al. 2018.) showing maxillary and mandibular enlargement.

*Neurological Examination-*Specific signs of neurological diseases in goats include abnormalities of mental state (obvious or subtle), involuntary movements and muscular tremors, abnormal posture and gait,paralysis (flaccid or spastic), disturbance in sensory perception, blindness and abnormalities of the autonomic nervous system. The presence of upper and lower motor neuron deficiencies are among specific signs that may be seen. Caprine arthritis encephalitis virus encephalitis can occur in both young and adult goats. The congenital form of the swayback is uncommon in kids. Delayed swayback is seen in older kids which are normal at birth. Progressive hind-limb ataxia is seen in copper deficiency. Scrapie is seen chiefly in 3-4 years old animals. Affected goats may nibble or scratch any accessible part of the skin and show hyperesthesia, muscular tremors, incoordination and a high stepping gait.

Further Readings

Jackson, Peter, G.G. and Cockcroft, Peter D. (2002). Clinical Examination of Goats. In: Clinical Examination of Farm Animals. Blackwell Science.

Varshney, J.P. Kumar, G. and Singh, S.K. (2018). Evaluation of promising biochemical markers of nutritional osteodystrophy in goats. Small Ruminant Resch. 169:86-89.

4

History and Physical Examination: Swine

In pigs, making a diagnosis of disease is no way different from that of other species. It is based on assessing the owner's complaint, history of the farm and the ailing pig, signalment factors, detailed observations of the ailing pig, the other pigs in the group and their environment, detailed physical examination. These observation may lead to a provisional diagnosis. Further clinical tests, clinical pathology and a post-mortem examination of the dead pigs are to be taken up to confirm or refute the provisional diagnosis. Because pigs are kept in groups and owing to infectious nature of many diseases, the health of other pigs in the group should also be taken into account.

History and General Examination

General examination of swine is conducted without handling them. The physical examination of swine includes history taking, examination of environment, examination of group of swine, post mortem examination of dead swine and ancillary diagnostic testing. Examination of history, herd records, environment, and herd is performed before an individual swine is physically examined. Post-mortem examination is a very important part of a health examination on swine in case of mortality in the herd. History is the most valuable part of the examination of swine as in other animals as it provides the lead for further observations. The breed, age and sex should be noted. As enteritis in piglets is commonly due to *Escherichia coli* and rotavirus.These organisms rarely cause disease in adults. Mastitis is confined to females.

Examination of Environment- Examination of the environment is also an important aspect of the history.The general appearance of the farm may reflect the standard of management and level of cleanness. The housing of the pigs should be assessed for ventilation. An intense smell of ammonia is suggestive of poor drainage and ventilation. Stocking density, ambient temperature, light and bedding should be examined. Feces are normally passed in one corner of the pen. Fecal assessment may be done. Loose feces, passed anywhere in the pen, is suggestive of diarrhea. Blood in the faeces may be seen in swine dysentery. Blood may indicate haemorrhage through injury or internal bleeding. Vomitus may occur in overeating or in gastritis. Floors should free from sharp corners and other hazards. Water availability, number of drinkers and feeding facilities should also be examined. Outdoor environments should be looked for water supply, fencing, wallows and shade.

Though, physical examination of an individual swine is difficult and less common because of its characteristic nature (squeal and runaway if frightened), it is conducted after proper restraining in the manner similar to any companion animal.

Examination from a Distance

Any departure from normal should be noted. Pigs naturally spend much of their time sleeping, and young ones lie comfortably huddled together. Usually one member of the group is keeping an eye open. Huddling also indicate that pigs are cold. Signs of tail biting and other vices may be seen. Any pig standing or lying alone away from the group is suggestive of illness.

Pigs are often initially startled by the arrival of person or any unusual activity in their environment. Ignoring human presence by the pigs may be due to severe sickness. Sick animals may show clinical signs such as poor body condition, respiratory abnormalities including coughing, sneezing or laboured breathing, neurological signs, lameness, diarrhoea, pruritus and abnormal body shape.

The sow and her litter is considered as one unit, as sow illness may affect the piglets and vice versa. The piglets, when not feeding, are mostly lie and sleep in the creep area under their lamp. If hungry, the piglets will stay close to the sow making a high pitched rather disgruntled sound. The sow, when not feeding her piglets, usually lies in sternal or less commonly in lateral recumbency.

Initially both the individual sick animal as well the whole group should be observed. After having observed the pigs in their natural environment without disturbance, close examination should be done. Breathing pattern, coughing, sneezing should be taken in to account. Pigs may be checked for nervous signs such as incoordination, ataxia, tremor, head tilt or fitting. Sleeping pigs may occasionally show muscle tremors that disappear on waking.

Restrain

Swine's habit of escaping from any restrain or confinement and tendency to squeal makes chest auscultation difficult. Sows with litters and breeding boars are potentially aggressive and dangerous. Therefore examiner should not enter in the pens without their stockman. Individual pigs should be taken out for a brief period for their physical examination. Patience, quietness and a knowledge of pig behavior are very important. For blood sampling and X-ray, restraint and sedation are needed

For an individual examination restrain of the animal is a pre-requisite. The swine is approached from the front and snout loop is quickly applied over the maxillae. The hog should pull back slightly on the loop. While making the close contact with the pig, it should be scratched or rub behind the ears with a blunt object like finger or stick.

Individual Examination

Fitting with nystagmus and opisthotonus is seen in cases of meningitis (caused by *Streptococcus* and *Haemophilus* species), water deprivation and classical swine fever. Incoordination may be seen in bowel edema. Ataxia may occur with spinal injury or abscessation. Various forms of congenital tremor are seen in piglets. Head tilt is often seen in cases of middle ear or vestibular disease. Animals should also be looked for lameness.

Before entering the pen, eating pattern and behavior of the individual ill pig; as well as lying pattern of the group and individual being different (separation from the group) should be observed. Also note the noises (pig make sound when some one enter to the pen; pigs with influenza are very quite and reluctant to get up), coughing, sneezing and smell (malodorous smell in swine dysentery and swine fever).After entering the pen make the pigs walk and note their movement (getting up and moving, unable to get up, stiffness), urination (on rising in the morning pigs urinate with in 5 minutes. Urinating posture and color of urine should be noted), blood (look for any blood on floor or wall) , floor (examined for the consistency of feces), noise and smell. Any sick pig should be examined in detail as per procedure adopted in other animals. Auscultation in pigs is generally unrewarding.

The animal is examined for behavior (demeanor such as dull, excitable or normal), posture (lying down posture as sternal / lateral recumbency, or standing posture such as normal, lameness, arched back or abducted elbow) and gait, body scoring, breathing rate (thoracic, abdominal or both), pulse rate and temperature. Then body systems should be examined.

After restraining, individual examination can be taken up. Rectal temperature is taken and anus is examined for trauma or prolapse, if any. Routinely herd examination, without disturbing, is more valuable in swine. Sampling or any minor surgical procedure in swine requires chemical restraint. Examination of the environment of the pig can provide valuable information. Type of water drinker, their height, and flow; food feeder space, feeder type, quantity of feed and composition of the feed; floor space and stocking density; air ventilation patterns, temperature, humidity and gas concentration needs to be examined. Examination of the farm performance requires a systematic approach.

Signs of Illness

- Abortion
- Ataxia
- Huddling
- Nasal discharge
- Eye discharge
- Mouth discharge
- Food at mouth
- Dropped ears or ear discharge
- Raised, rough and irregular hairs
- Signs of fight on neck
- Thin, wasting back
- Uneven hind quarters, lameness, swelling
- Tail biting, twitching
- Constipation or diarrhoea
- Discharge from vulva
- Swelling on testicle
- Swelling on joints, pain, lameness
- Wound, ulcer or swelling on feet
- Swelling on mammary gland or any discharge
- Belly (empty or distended, kicking)
- Heave line on chest, abnormal breathing or coughing
- Tremors
- Sunken flanks
- Dirty rump
- Rough, messy standing hairs

5

History Taking and Physical Examination: Equine

History and thorough physical examination is crucial procedures in making diagnosis in horses. Mental attitude of a horse may help in understanding whether it has a localized problem or is systemically ill. Horse is better approached by its left side, keeping himself/herself next to the shoulder. Examination of the horse should be started in a non-threatening, easy, and quiet way. Generally physical examination starts at the head to prevent possible injury to unfamiliar examiner. Abnormalities in demeanor and mentation can be assessed by watching the horse's initial response to the environment. The locomotor and neurologic systems are examined by observing the horse's movements and gait at walk and trot, both on straight line and in turns, and backing. Horse weight should be taken. General inspection of the horse should be conducted by standing back keeping a safe distance and looking at the horse for it condition, abdominal shape, muscle symmetry, bloating and the manner in which it is standing. Attitude and behavior of the horse is also noted.

External Physical Examination (Fig. 31)

Head

Nares are checked for symmetry, air flow and smell of the exhaled air. Mouth is examined by applying light pressure to the upper corner incisor teeth for capillary refill time. Lips are rolled away, and incisors are examined. The mouth is opened by applying pressure to the top of the hard palate, tongue is grabbed and pulled to the side. Maxillary and frontal sinuses are percussed. Dull sounds indicate fluid or soft tissue in the sinuses. Eyes are examined for corneal scars, conjunctivitis, cataracts, etc. Quick menace and pupillary light response tests may be performed. Color and scleral vasculature is also evaluated. Pulse is taken on facial artery as it passes along the ventral border of the ramus of the mandible. Submandibular lymph nodes should be palpated.

Neck

The jugular grooves, the tracheal rings and the crest of the neck is palpated for any heat or swellings. The neck is palpated for muscular or bony protrusions, indentations, or pain.

Limbs

The fore and hind limbs and all the joints (knee, fetlock, ankle, pastern and coffin of fore limbs; stifle, hock, ankle, pastern and coffin of hind limbs) are examined for any signs of swelling, heat or pain.

Udder, the inguinal area, penis/sheath, rectum, and perineum

Examination of these organs is conducted to find any abnormality if any.

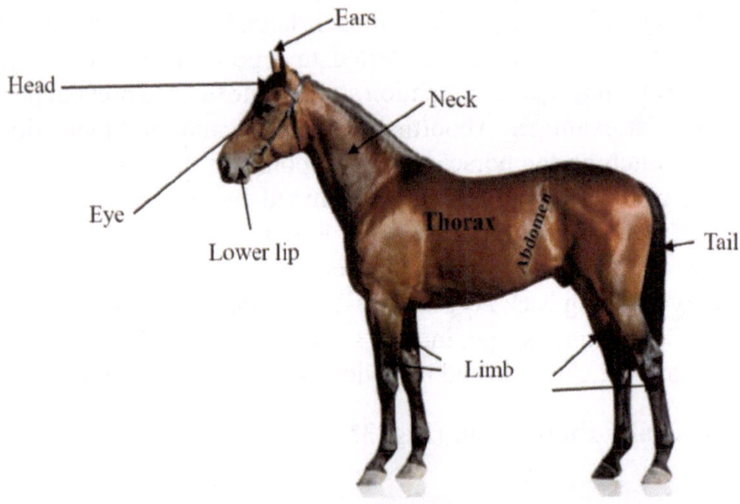

Fig. 31. External appearance of a horse

Vital Indices

Temperature is recorded by inserting a clinical thermometer in the rectum. Normal temperature of an adult horse varies from 99.0 to 101.5 ^{0}F. Heart rate can be determined by heart auscultation. Heart rate in adult horses varies from 32 to 40 beats per minute. Pulse is taken on facial artery (Fig.32) as it passes along the ventral border of the ramus of the mandible. Respiration rate can be counted by watching flaring of nostrils. Each flare is one breath. Respiration rate can also be counted by watching going in and going out of the flank. Respiratory rate in healthy horses varies from 12 to 40 breaths per minute. Capillary refill time is determined by pressing gently on the gums and count the time it takes for the blood to return. Normal capillary refill time is 1 to 2 seconds. Gums which are pale, or dark red, or muddy, or which have a slow refill time indicate a serious problem which requires immediate veterinary assessment.

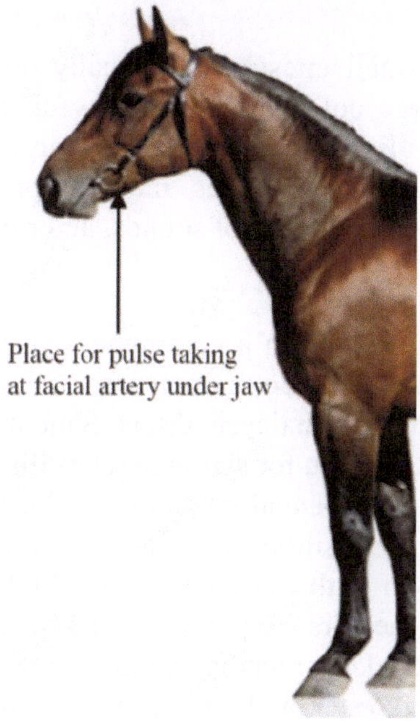

Fig. 32. Showing site for taking pulse on facial artery as it passes along the ventral border of the mandible.

Auscultation of the Thorax

Auscultation of the heart- The heart is auscultated to monitor heart rate and rhythm. If a murmur is detected, it needs to be characterized (timing in the cycle, point of maximum intensity, and intonation). Sounds over each valve should be heard. Pulmonary valve is auscultated on left side at 3rd intercostal space at costochondral junction. Aortic valve is auscultated on the left side at 4th intercostal space just below level of shoulder. Mitral valve is auscultated on the left side at 4th intercostal space at level of olecranon. Tricuspid valve is auscultated on the right thorax at 3rd to 4th intercostal space between olecranon and costochondral junction.

Auscultation of lungs- Lung sounds of the horse are of low intensity. Breath sounds can be increased by the use of a rebreathing bag. Essentially the horse's nostrils are enclosed in a large plastic bag so as to increase the CO_2 inside the bag making the horse to breath deep. Lungs of a healthy horse are without any crackles or wheezes. Lung auscultation area in horses is marked cranially by shoulder, dorsally by epaxial muscles, and caudoventral border by a line extending from the 17th intercostal space at the level of the tuber coxae to the 11th intercostal space the point of the shoulder.

Auscultation of the Abdomen

Contractions of stomach and small intestine are usually not detected. The contractions of cecum and large colon are mixing (weak and moderately pitched lasting for 10 to 20 seconds) as well as propulsive (large , deep sounding occurring less frequently). Abnormal sounds include `pings', increased, frequently high pitch sounds, complete lack of sounds, and rubbing sounds.

Examination for Lameness

The examination for lameness starts with a comprehensive history, training regimen, period of lameness, improvement in lameness with rest or exercise, and response to anti-inflammatory or analgesic drugs. Shoeing history is also important. A thorough visual inspection for symmetry, swellings, muscle loss, abnormal stance, and injuries; and manual palpation of the limbs in weight bearing and non-weight bearing positions is necessary. There is variation between horses, and comparison with the contralateral limb can provide a useful control. The reaction of the horse to palpation and the range of flexion and extension of all joints is noted. The feet is examined by compressing the walls, sole with the help of hoof tester, and noting the wearing pattern of the shoes and feet for revealing any abnormality of hoof pastern axis, hoof angle, under run, contracted and sheared heels, and disproportionate hoof size.

The back and neck is thoroughly examined with the horse restrained and standing square on a level surface. Examination during exercise is also required to localize the lameness to a specific limb or site and also to see the response to diagnostic regional anesthesia. If fracture is suspected in cases of acute lameness, exercise and diagnostic regional anesthesia is not recommended.

Head nod is a consistent sign of a unilateral forelimb lameness and the sacral rise (pelvic rise or hip hike) is a consistent and easily observable sign of hind limb lameness. Both head nod and sacral rise serve to reduce concussion on the lame limb.

Flexion tests are performed during the evaluation on the straight line followed by the range of movement and response to passive flexion. The distal phalanges in both forelimbs and hind limbs are examined by flexing independently of the carpus and hock. All tests are performed on both sound and lame limbs for comparison. For detecting subtle lameness a ridden assessment is done. Diagnostic regional anesthesia is used to determine the area of pain in the affected limb. A complete neurological examination (cranial nerve, and upper and lower motor neuron function) should also be conducted as neuromuscular disorders may also lead to lameness. Hind limb lameness is better assessed from the side and from behind while fore limb lameness is better assessed from the front and side.

Examination of the eye is conducted by a rapid visual assessment using a penlight and menace response tests. Evaluation of the cornea and cranial and caudal lens capsules can be done by moving the penlight horizontally and noting the crossing light reflexes. In cases of primary eye complaints, an ophthalmoscope should be performed. Fluorescein dye is of diagnostic value in detecting corneal ulcers. Blind folding of one eye at a time can assist in detecting unilateral blindness.

Evaluation of the Circulatory System

It starts with assessment of heart rate, rhythm, and any presence of murmurs (done by bilateral auscultation of the heart); color of mucous membrane; capillary refill time; scleral injection, palpable changes in the temperature of the ears and extremities; jugular pulsation, and pitting subcutaneous edema. Characterization of cardiac arrhythmias requires electrocardiography. Echocardiography is used to evaluate cardiac structures and valves. Sphygmomanometer with Doppler ultrasound is used on the base of the tail to measure indirect blood pressure.

Examination of Respiratory System

Breathing rate and mucous membrane color are important assessment factors of respiratory system. A quick visual assessment of the alar folds may detect conformational abnormalities or severe compromise (e.g., nostril flaring). Respiratory rate can be determined by holding the wet hand gently over the nostrils and assessing intensity and equality of air movement and counting numbers in one minute. Nasal discharge, its color, amount and character; and smell of breath should be noted. Nasal septal mucosa can be visualized using a penlight. Sinus percussion may be helpful in suggesting a possible mass (e.g. ethmoid hematoma), sinusitis, or the presence of fluid within the sinus cavity. Lymph nodes should also be examined for lymphadenopathy. The upper (larynx and trachea) and lower airways should be auscultated.

Percussion of the thorax can assist in cases of suspected abscess, tumor, or pleural effusion. In foals, percussion can be performed by placing a stethoscope on one side of the chest and manually percussing the other side. Horses with pleuropneumonia may show a reluctance to move.

Thoracic ultrasonography is valuable in the diagnosis of subtle pneumonia, abscess, and pleural effusion as radiographic changes often lag behind clinical disease.

Examination of Gastro-intestinal Tract

In equines, colic or acute intestinal disease are the emergencies requiring the systematic approach. Bilateral auscultation of intestinal sounds should initially be performed for hearing the borborygmi. In general, small intestinal sounds are the least dramatic and heard in the area of the left flank. Abdomen should also be auscultated ventrally for sounds similar to "ocean waves" or sand pouring on itself, indicative of the presence of sand within the gastrointestinal tract. Horses should be observed for the presence of normal prehension, eating, and drinking whenever dysphagia is present or neurologic dysfunction is suspected. Tongue strength and tone can be assessed by observing the response of the horse to careful grasp and release of the tongue. Dental examination should also be conducted for detecting malocclusion and dental abnormalities. Nasogastric intubation can assist in evaluating dysphagia and esophageal blockage (choke) and determining the presence or absence of gastric reflux. Esophageal stricture and gastric ulceration can be detected by endoscopy. Diagnosis of peritonitis, uro-peritoneum, hemorrhages, ascites, visceral rupture or abdominal mass is facilitated by abdominal ultrasonography.

Examination of Urogenital System

Manual palpation, rectal palpation, vaginoscopic (speculum examination) viewing, endoscopy, and ultrasonography are of great diagnostic assistance in examining the urogenital system in equines. The caudal portion of the left kidney is easily palpable in horses. Sphincter tone assessment can be done by urethral pressure profile in horses with urinary incontinence or stranguria. Scalding of perineum and hind limbs is a common manifestation in horses with incontinence.

Neurological Evaluation

A horse should be assessed from a distance for its attitude, posture and head carriage. Functioning of cranial nerves, spinal reflexes and tone of the tail should be evaluated and sensory deficits, if any, should be noted. Conscious proprioception and postural responses (e.g., placement, hemi-hopping, sway) are assessed before observing the patient in locomotion. Observation should be made on symmetry, ambulation, paresis, muscle atrophy, and upper and lower motor neuron deficits to aid in determining the sites for any additional ancillary tests such as radiographs, cerebral spinal fluid collection, or myelogram. Horses with vestibular disease should be subjected to blindfolding in a safe area.

Examination of Lymphatic System

The lymphatic system is usually examined by palpating the lymph nodes and detecting lymphadenopathy, if any. Local or regional lymphadenopathy is seen in strangle. Lymphangitis or the presence of edema is also taken in to consideration.

6

History Taking and Physical Examination: Dogs and Cats

Making a correct and an early diagnosis is very much important for undertaking rational and effective treatment for fast recovery, avoiding unnecessary medication, reducing the load of so many drugs on body metabolism, reducing environmental pollution and the cost of treatment. There are four important pillars for making a correct diagnosis viz. the history, the systematic clinical examination, the correct clinical diagnostic techniques and the laboratory support.

History Taking

Clinician's first step in making a diagnosis is through asking the owner's complaint. Accurate and complete history may be obtained from focusing on the details of the patient (name, number, species, breed, age, sex and weight), vaccination status, deworming schedule, nutritional history, past and present medical history and environmental history. Physical examination of the patient involves general inspection, palpation, percussion, and auscultation to detect manifestations of abnormalities.

Physical Examination

Physical/clinical examination is a fundamental part of the process of diagnosis.It is routinely done to diagnose diseases in dogs and cats presented in the clinics.Physical examination provides important information needed for determining the disease or diseases producing the clinical abnormalities. Information gathered during physical/clinical examination of the ailing dogs /cats assist in determining the organs/system involved, type and location of the lesion, pathophysiological process and severity of the disease. Results of physical examination depends heavily on the knowledge of the clinician regarding anatomy, physiology, pathology and animal behavior as well as his skills in the methods and techniques of clinical examination, recognizing clinical sign and understanding of disease pathogenesis. Without a proficient physical examination, an accurate diagnosis is unlikely leading to treatment failure.

Physical Examination Techniques

- *General observations*- General observations start as soon as clinician sees his/her patient. Patient's general condition, gait, posture, mentation and behavior should be noted without disturbing the patient.
- *Palpation*- Palpation (Fig.33) is a routine technique that is used for ascertaining the size, shape, consistency, and temperature of the area/ organ under examination. Palpation of the area or structure is done using the fingers (direct palpation) or a probe (indirect palpation). Palpation may be light or deep. Tissue/organ is felt gently in light palpation. While more pressure is applied in deep palpation, for obtaining information about the consistency of the tissue or to detect pain or to feel abdominal organs.

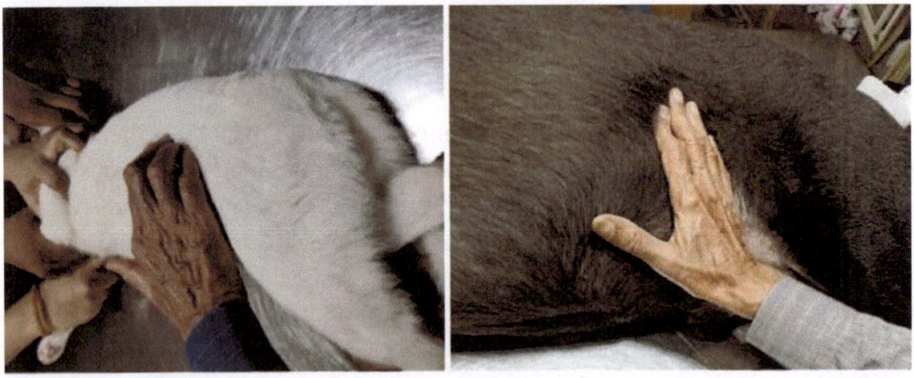

A. **B.**

Fig.33. Showing the technique of palpation using finger tips in cat (A) and in dog (B).Palpation assists in ascertain the size, shape, consistency and temperature of the area under examination.

- *Percussion*- In this technique, the area is tapped with fingers (Fig. 34), hands or small instruments to determine the size, consistency and borders of the organs; and to ascertain the presence or absence of the fluid in the body. A sound is elicited on tapping the area / organ. The elicited sound has a potential to differentiate whether the organ is air filled (as lungs), dense (as liver) or fluid filled (as urinary bladder). The air filled organs (such as lungs) produce hollow low pitched resonance. Solid organs, bones and joints produce dull sound as is produced on percussion of liver. Tympanic high pitched drum like resonance of longer duration is produced on percussion of fluid (urinary bladder), air filled (lungs) or air and fluid filled organs (stomach, intestines).

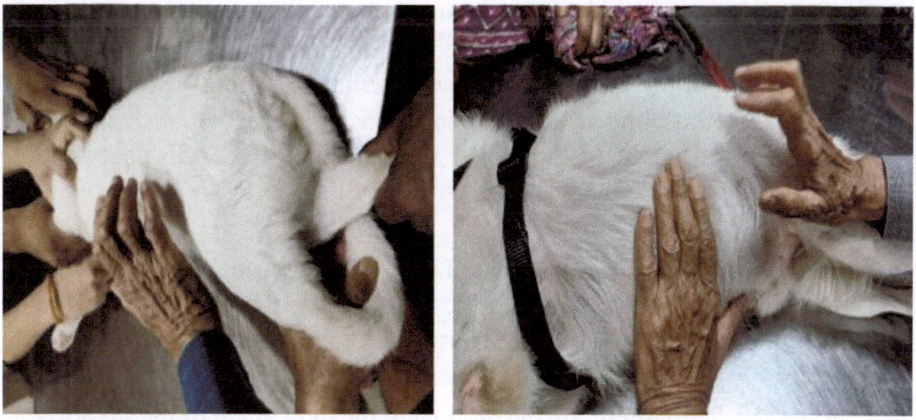

A. B.

Fig.34. Showing the technique of percussion (tapping with finger tips) in cat **(A)** an dog **(B)**. Tapping of the area elicits sound depending on the contents of the organ.

- *Auscultation*: Auscultation is widely used most common clinical examination technique of listening sounds of internal body organs (heart, lungs and intestines) either directly (putting the ear directly over the area of organ to be listened) or indirectly using an instrument called "Stethoscope". The stethoscope was invented in France in 1816 by Rene Theophile Hyacinthe Laennec. It is composed of chest piece, diaphragm, bell, tubing, stem, head set, ear tubes and ear pieces (Fig. 35).

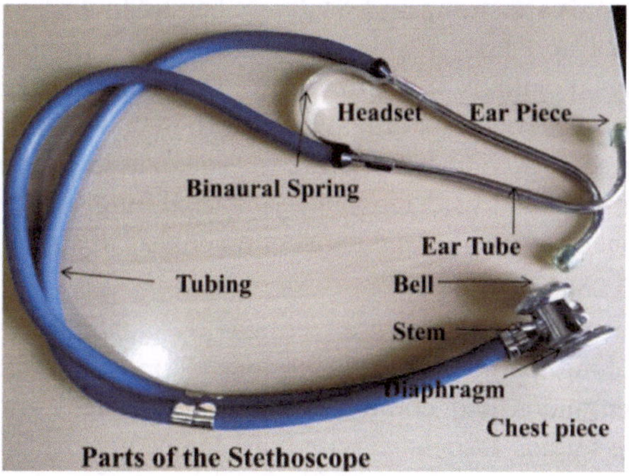

Fig. 35. Showing different parts of the stethoscope (Courtesy Varshney, 20 25).

The pet should be made comfortable and relaxed in a room before it is auscultated to avoid disturbing humming sounds. Auscultation can be performed in a standing position or in recumbent animals also. The stethoscope should be pressed firmly (not too hard or not too loose) on the area to be auscultated to avoid sound of hair scratching on the diaphragm of the stethoscope. The auscultation should be performed by placing stethoscope at least five locations (4th, 6th, 8th intercostals space at two third height of the thorax ; 4th and 6th intercostals space at one third height of the thorax) on each side of the thorax.

(i) Lung Auscultation: It is of much significance in cases suspected with cardiac or pulmonary disease. The lungs are usually auscultated on the both side of the thorax in the area shown in the Fig. 36.On thorax auscultation, extra thoracic, pleural and broncho-pulmonary sounds are heard

Extra thoracic sounds are the product of the movement of the stethoscope tubing on the hairs or muscle trembling. Pleural rubs are audible owing to pleural movements in cases of pleuritis. Because of exudative pleuritis in dogs and cats, sounds of pleural rubbing are generally not audible. Normal bronchopulmonary sounds vary in intensity from inaudible, weaker, of normal loudness only during inspiration, to louder. The character of respiratory sound is almost similar to those heard over the trachea (inspiration and expiration are alike). In general, respiratory pattern and lung sounds are subtle in cats as compared to dogs.

The normal breath sounds can be categorized as bronchial breath sounds, bronchiovascular breath sound and vesicular breath sound. Bronchial breath sounds are strong and relatively harsh blowing sound heard over the trachea and hilar region. Inspiratory and expiratory components of bronchial sounds are prominent. Bronchiovascular sounds have both inspiratory (soft) and expiratory (short but more prominent than vesicular sound) character and are intermediate between bronchial and vesicular sounds. Vesicular sound is normal lung sound and is heard in the periphery of lung. Vesicular sound consists of soft inspiratory and a soft short expiratory sounds.

Abnormal lung sounds are termed as adventitious sounds that indicate lung pathology. Adventitious sounds are further classified as discontinuous and continuous. Crackles (rales) are discontinuous explosive sounds associated with lung edema and are heard throughout inspiration indicating disease of large airways. Whereas crackles heard at the end of inspiration are suggestive of small airways disease. Crackles can be of low pitch (associated with airways secretions), high pitch (associated with opening of collapsed peripheral airways) or pleural friction rub (low pitched associated with roughened pleural surfaces).

Wheezes are continuous sounds with musical character associated with asthma or bronchitis. Wheezes heard during expiration are due to dynamic compression of airways. Wheezes at the end of inspiration are suggestive of opening of smaller inflamed airways. Stridor is another type of abnormal continuous monophonic musical sound primarily heard during inspiration and is associated with laryngeal or tracheal disease. Stertor is snoring sound without musical character. It is harsh continuous crackling sound in trachea or larynx due to edema and accumulation of secretions in upper airways.

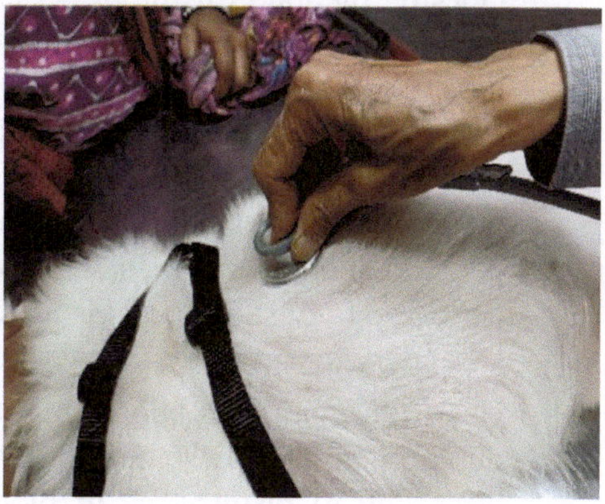

Fig. 36. Area marked for auscultation of lungs. Auscultation of lungs is done on both side of the thorax (Courtesy Varshney, 2025).

(ii) Cardiac Auscultation- Cardiac auscultation is necessary for obtaining information about heart rate and rhythm; detecting cardiac murmurs and muffling of heart sound. Heart sound is characterized by LUBB-DUBB PAUSE. Of these, LUBB (S1) is associated with closure of mitral and tricuspid valves; and DUBB (S2) is associated with closure of aortic and pulmonary valves. Cardiac auscultation should be carried out on a relaxed standing dog (Fig.37A and Fig.37B) /cat at apex, at base and at the level of all valves' (mitral, tricuspid, aortic and pulmonary) in a quiet room.The pet should be gently restrained and the clinician should also be relaxed for a better auscultation result. A cat can be restrained gently with one hand under the abdomen. Ventilation artifacts are reduced by gently holding the mouth close. Showing the cat water in sink or gently pressing over the larynx reduces the degree of purring in cat. Mitral, aortic and pulmonary valves are auscultated on the left side of the thorax in standing position at 5th intercostal space around costo-chondral junction (opposite to elbow point in standing position), at 4th intercostal space dorsal to mitral valve (at the level of the point of shoulder), and at

3rd intercostal space at sternal border respectively (Fig.37 C). Tricuspid valve (Fig.37 D) is auscultated on the right side of the thorax at 3rd to 4th intercostal space at costo-chondral junction. In cats, mitral valve and pulmonic valves are auscultated on the left side at 5th/6th intercostal space above the sternum, and at 2nd/3rd intercostal space one third of the way up from sternum respectively. The tricuspid valve in the cat is auscultated on the right side of the thorax at 4th/5th intercostal space near sternum. Heart sounds are more intense in young dogs, thin/emaciated dogs as well as in disease conditions such as fever, anemia, or hyperthyroidism. Obesity, pleural effusion or pericardial effusion decreases the intensity of the normal heart sounds. Muffled heart sounds (decrease intensity of heart sounds) may be due to pericardial effusion, pericarditis, pleural effusion, diaphragmatic hernia, thoracic neoplasia, hypothyroidism or cardiac temponade. Obesity can also muffle the heart sounds because fat acts as an insulting layer dampening the heart sounds. In dogs, pericardial effusion results from neoplastic or idiopathic diseases. While pericardial effusion in cats is most often due to congestive heart failure secondary to cardiomyopathy.

Heart murmurs are abnormal, extra, unusual sounds of relatively long duration in the heart beat, produced as result of turbulence within heart due to disturbance in blood flow. Murmurs are also heard in physiological conditions (in young growing animals even without cardiac abnormality. Murmurs in growing animals generally resolve as the animal grow); anemia and hypoproteinemia; as well as in cardiac diseases (valvular insufficiency, valvular stenosis, interatrial or intraventricular septal defects, patent ductus arteiosus or defect of great vessels). Murmurs are classified according to timing of their occurrence at the point of maximum intensity as systolic, diastolic or continuous (occurring all time during systole and diastole). Systolic murmurs are of common occurrence and are of soft nature that occurs during an early systole. While diastolic murmurs are of less common occurrence and are of low frequency. Continuous murmurs vary in intensity and are associated with patent ductus arteriosus (PDA). Murmurs have been graded into 6 grades (grade 1 to grade 6) depending on their intensity. Murmurs of grade 1 are soft and localized barely audible. Murmurs of grade 2 are similar to grade 1 except these are easily audible. Murmurs of grade 3 are moderately intense and are detectable at more than one location. Grade 4 murmurs are similar to grade 3 but are detected at many places in left and right thorax. Murmurs of grade 5 are loud at point of maximum intensity and are heard in association with precordial thrill. Grade 6 murmurs are loudest and are associated with precordial thrill.

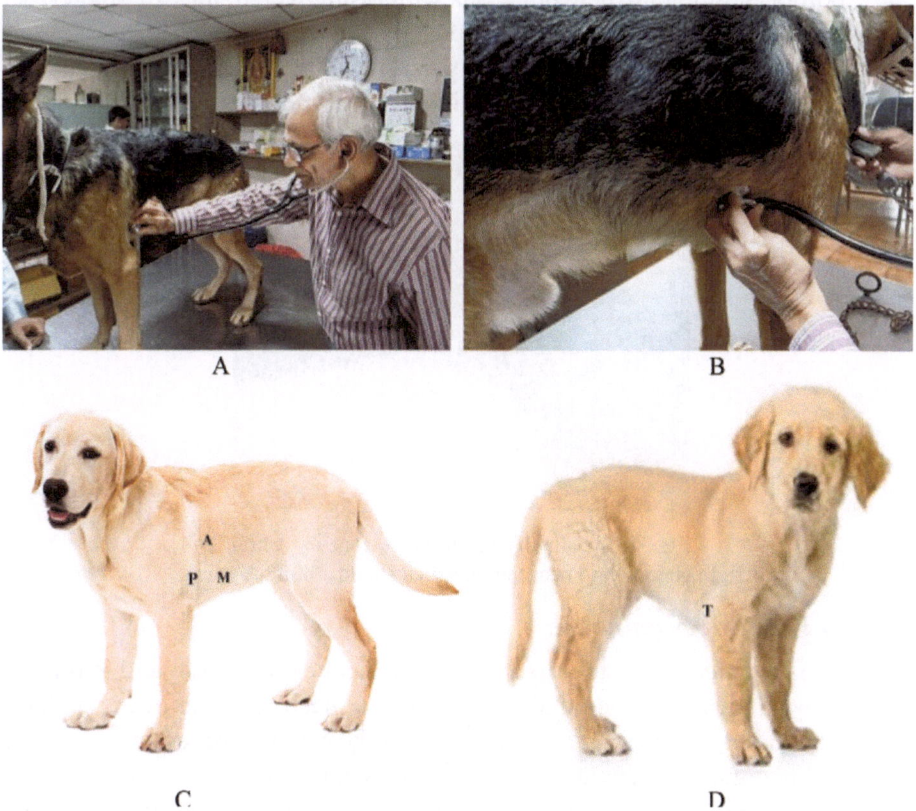

Fig. 37. Cardiac auscultation in a standing dog on the left (37 A) and right (37 B) thorax. Sites for auscultation of heart valves in a dog are also shown (Fig.37 C and Fig. 37 D). On the left thorax P, A and M denotes pulmonary valve, aortic valve and mitral valve respectively Fig. 37 C). Pulmonary valve (P) is auscultated at 3rd intercostal space near the elbow just above sternum. Aortic valve (A) is auscultated slightly dorsal and caudal to Pulmonary valve at 4th intercostal space. Mitral valve (M) is auscultated slightly caudal and ventral to Aortic valve at 5th intercostal space. On the right side of the thorax tricuspid valve (T) is auscultated (Fig.37 D) at 3th /4th intercostal space near costochondral junction. Figures at A,C,D are with the courtesy Varshney,2025)

(iii) Intestinal Auscultation - Borborygmi are the normal intestinal sounds audible on intestinal auscultation (Fig.38) in healthy dogs and cats. These sounds are generated owing to intestinal peristalsis propelling the digesta through gastrointestinal tract. The intestinal sounds (borborygmi) are classed as hypoactive, normoactive or hyperactive. Borborygmi's characterization in isolation is of little significance. Therefore character of borborygmi sounds should always be considered along with the history and results of other physical examinations. Hyperactive borborygmi in diarrheic dogs/cats with losing weight may be suggestive of malabsorption syndrome. Hypoactive borborygmi in dogs/cats with vomiting may be suggestive of gastrointestinal obstruction. Complete absence of borborygmi is indicative of paralytic ileus.

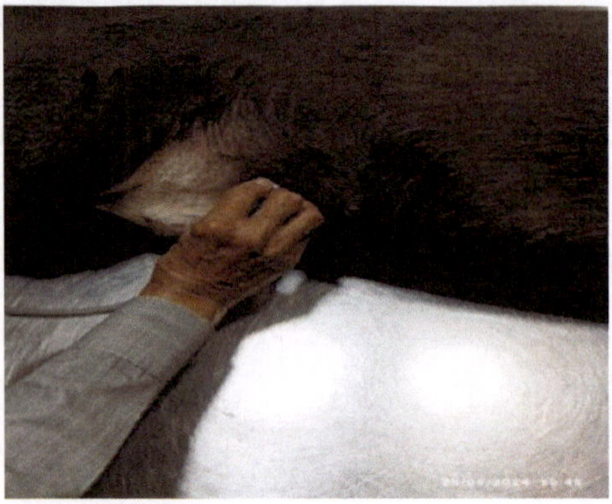

Fig.38. Intestinal auscultation is done in ventral right abdomen to hear the intestinal sounds (borborygmis).

Steps in Physical Examination

1. *Signalment*- It is the first step in physical examination and is related with identification details of the patient such as name of the patient, number, species, breed, age, sex, body weight, reproductive status, distinguishing characteristics, owner's name and address with mobile number. Signalment data of the patient is a pre-requisite for its identification.
2. *History Taking*- History taking is a process of getting valuable detailed information about the illness of the patient through careful questioning the owner. In canine/feline practice, disease is presented indirectly in the form of complaint by the owner or attendant of the pet. Many times, owner or attendant fails to provide relevant and adequate information leading to misdiagnosis. It is imperative on the part of the clinician to ask rational questions utilizing his professional knowledge and experience.

   History includes information about environment, diet, past medical history including drugs being given, allergies, reproductive history, vaccination status, deworming status, life style of the pet, and main presenting complaint, its genesis and duration. Direct question such as "Whether the dog had diarrhea?" is to be avoided. The owner may be asked whether he has seen the feces, its color, consistency and frequency to ascertain whether the patient had diarrhea or not.

3. *General Appearance*- The dog/cat is observed from a distance as well as closely for symmetry, size and shape of extremities, gait/posture, general looks, body condition (obese or emaciated), hydration and mentation (level of consciousness, alertness, responsiveness), before it is handled. Hydration status may be classed as adequate, marginal or inadequate. Loss of skin elasticity (skin turgor) reflects dehydration. Its level is measured as the time skin tenting persists. Skin tenting test can be performed on the skin of upper eye lid (Fig. 39A) or neck (Fig.39 B) by gently pulling up the skin and then leaving. Quick return of skin fold to its normal flattened position indicates normal hydration status and persistence of the tenting for more time is suggestive of dehydration. Skin may tent more in emaciated animals.Performing skin tenting test in obese animals is questionable as they may not show tenting even when dehydrated.

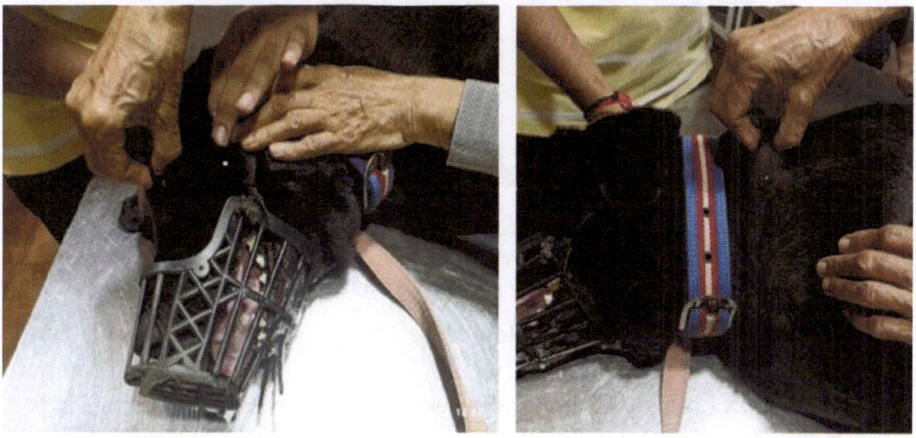

A B

Fig. 39. Showing skin tenting in dogs on the skin of upper eye lid **(A)** and on the skin of neck **(B)**. The test is performed by gently pulling up the skin and then leaving. The time of persistence of the skin tent is noted.

4. *Evaluation of Vitals*- Body weight, rectal temperature, pulse rate and its character, respiratory rate and its character, mucus membrane color, skin tenting and capillary refill time are to be evaluated in each case.

   Rectal temperature is recorded by inserting the lubricated digital thermometer into the rectum in a manner that it touches mucus membrane and keeping it for specified time (Fig.40). Digital thermometers are programmed in a manner to sound when an accurate temperature has been obtained. The thermometer is then taken out, wiped off and the temperature is read. The thermometer is cleaned with disinfectant (alcohol). Mercury thermometers are to be avoided as their breaking may expose the pet to injury and toxic mercury. Rectal temperature of healthy dogs (100.5 to 102.5^{0}F) and cats (98.1 to 102.1^{0}F) is variable.

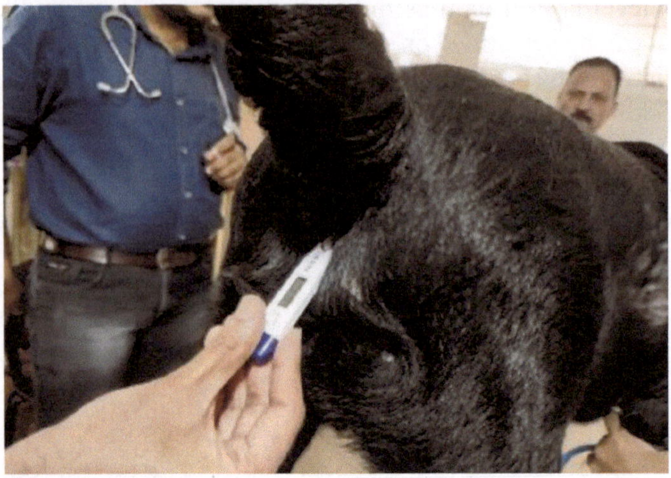

Fig.40. Showing the process of taking temperature in a dog. The lubricated thermometer is inserted into the rectum and is tilted slightly upward so that it is in contact with mucus membrane and then temperature is recordrd by keeping it in position for 30 to 60 seconds.

Pulse is generally taken at femoral artery (Fig.41) and it is evaluated for rate, strength and quality (weak, thread, strong, bounding). Compare pulse of both sides' femoral artery and heart rate. If pulse rate is less than heart rate, it indicates pulse deficit and further investigations are warranted. Pulse rate varies from 60 to 160 per minute in healthy dogs (60-120/minute in large dogs and 120 to 160/ minute in small dogs) ; and from 160 to 200 /minute in healthy adult cats. Irregular pulse is indicative of arrhythmias. Bounding pulse (increased pulse pressure) is seen in hyper dynamic (compensatory) state of shock. Weak pulse in seen in patients with decreased cardiac output, peripheral vasoconstriction, thrombosis or decreased pulse pressure. Decreased pulse rate or bradycardia (decreased heart rate) is due to decreased cardiac output and subsequent poor perfusion. Cats with shock may develop bradycardia. Tachycardia (> 180 bpm dogs, > 220 bpm cats) is associated with compromised diastolic filling, hypovolemic shock, pain, hyperthermia or primary cardiac disease.

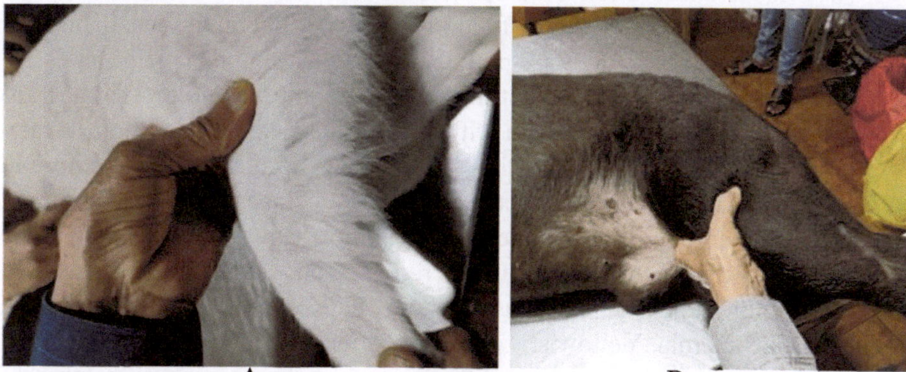

A B

Fig. 41. Pulse taking in a cat **(A)** and in a dog **(B)** at femoral artery. Arterial pulsation is recorded for one minute.

Resting respiration rate is taken by watching a resting dog or cat taking a breath in (inhaling) and then out (exhaling). Count the breaths in one minute. If number of breaths are counted for 30 seconds then multiply the breath number by 2 for getting the respiratory rate in one minute. Respiration rate can also be determined by lung auscultation (Fig.42). In general normal dogs and cats have respiration rate of 15 to 30 per minutes.

Fig. 42. Counting respiration during lung auscultation. Number of respiration (either inspiration or expiration as inspiration and expiration makes one respiration) can be counted for one minute during lung auscultation.

Perfusion indicators includes color of the mucus membranes and capillary refill time. Color of the mucus membrane is determined by visual observation of the mucus membrane. Color of the mucus membrane is the indicative of blood flow to peripheral tissues. In healthy pets, mucus membranes are pink and moist. Cats mucus membranes are little paler than dogs. Sometimes mucus membrane color changes are subtle and may not correlate with significant medical condition. It may be kept in mind that sometimes seemingly normal mucus membrane color do not rule out significant disease. Petechiae (red/purple spots on mucus membrane) are suggestive of hemorrhages. Mucus membranes of pale white, yellowish, bluish, reddish, chocolate brown or cherry red color indicates pathological state. Pale white mucus membranes indicate anemia. Yellowish color of the mucus membranes is suggestive of jaundice. Cyanotic or muddy mucus membranes are suggestive of severe hypoxemia or decompensatory shock. Capillary refill time (CRT) reflects the perfusion of peripheral tissues. It is conducted on gums. An area of gums is pressed (Fig. 43). On pressing the area is blanched white and turns pink again within 2 seconds when pressure is released in normal cases. Prolonged CRT (> 2 seconds) indicates compromised circulation leading to poor perfusion. It may be due to dehydration, anemia, shock, cardiac disease, hypothermia. An early CRT (<1 second) is suggestive of hyper dynamic states as seen in cases of fever, heat stroke (hyperthermia), distributive shock or an early compensatory stage of hypovolemic shock.

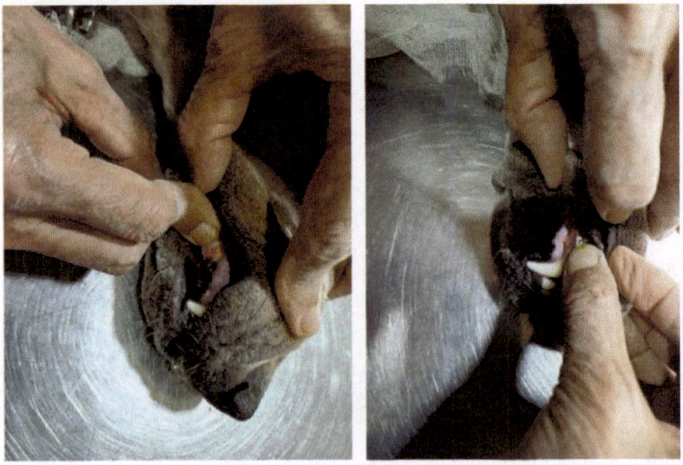

Fig. 43. Performing capillary refill time test on gum. The test is conducted by pressing an area of the gum. On pressing the area becomes blanched or white. Time taken in turning white area to pink is noted. Normally it should be within two second.

5. *Systemic Body Examination (from Head to Toe)*- Systematic body examination is initiated with the examination of head and neck (including ears, eyes, nose, mouth), integument, thoracic limbs, thorax, abdomen, pelvic limbs and tail, external uro-genitals, perineal region, and rectum (Fig. 44).

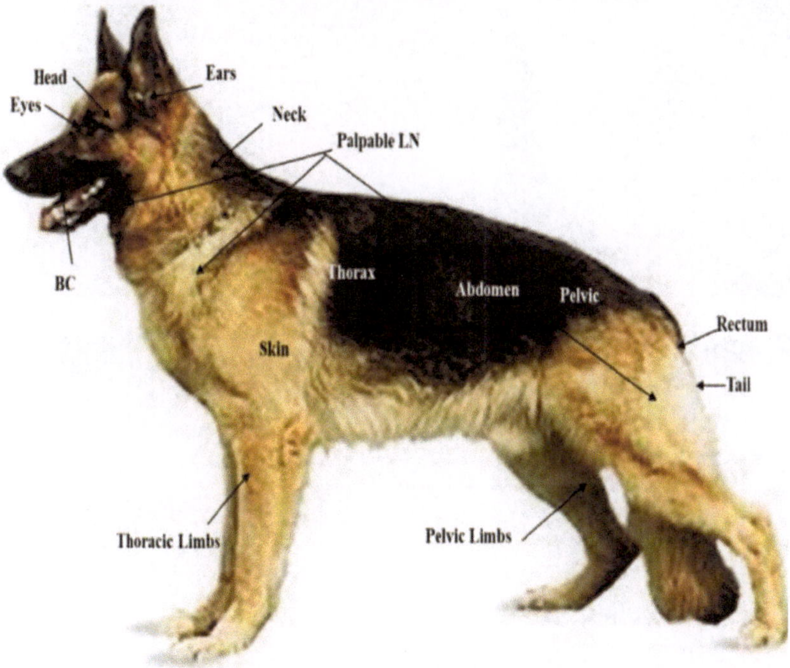

Fig.44. Diagram showing points of examination from head to toe in order of sequence (courtesy Varshney, 2025).

Examination of head and neck begins with assessing the shape and symmetry of the skull followed by examination of eyes (position, movement, vision, eye lids, conjunctiva, papillary response to pen torch light, palpebral and menace response, lens position and its clarity, eye discharge or blepharospasm, color of the eye), ear (response of pinna to sound, skin of the pinna,swelling at ear pinna (Fig. 45A), ear canal for discharge (Fig. 45 B), nose (shape of external nose, symmetry, nares for the discharge and patency), jaws (fracture of mandible -Fig.45C), buccal cavity (mucus membrane for color, moistness, petechiae/ecchymosis; gums for capillary refill time, color and integrity of palate, teeth, tonsils; salivary mucocele etc (Fig. 45 D), neck (mandibular lymph nodes (Fig. 45E) and mandibular salivary glands, flexion and extension of neck, thyroid glands, jugular vein for jugular pulse).

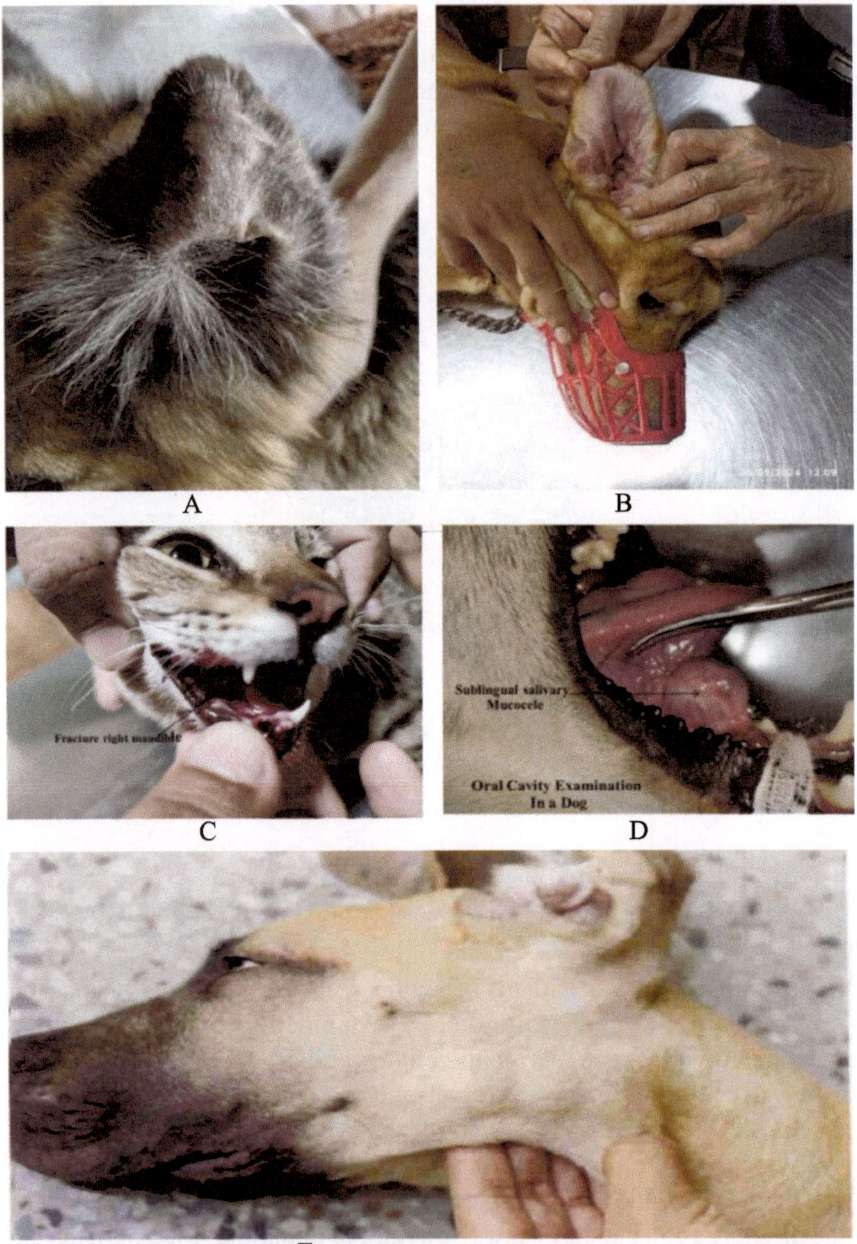

Fig.45. Examination of head and neck. Soft swelling on ear pinna (A) later diagnosed as aural hematoma. Ear pinna erythema, scaling and discharge from ear canal (B) diagnosed as otitis media and pinna dermatitis. Hanging lower jaw (C) due to fracture of right mandible. Sublingual salivary mucocele (D)in a dog detected on buccal cavity examination. Swollen lymph nodes (E) in a dog. Figure A,D, and E are with the courtesy of Varshney, 2025.

The color of mucus membrane (Fig.46) provides important information about perfusion. Pink mucus membrane (Fig.46 A) reflects adequate perfusion of peripheral tissue and a healthy state. Poor perfusion, loss of blood (anemia) or vasoconstriction is suggested by pale colored mucuc membrane (Fig.46 B). In hypoxemic pets (inadequate oxygenation) mucus membrane becomes cyanotic or congested (Fig.46 C). Brick red color of the conjunctiva is indicative of increased perfusion or vasodilation as seen in cases of fever, hyperthermia, early shock, or systemic inflammation. Yellowish (icteric) mucus membrane (Fig.46D) is suggestive of increased level of bilirubin owing to hemolysis, hepatic diseases or biliary diseases. Petechiae on mucus membrane are suggestive of coagulation disorder such as coagulation factor deficiency, platelet disorder, or disseminated intravascular coagulation.

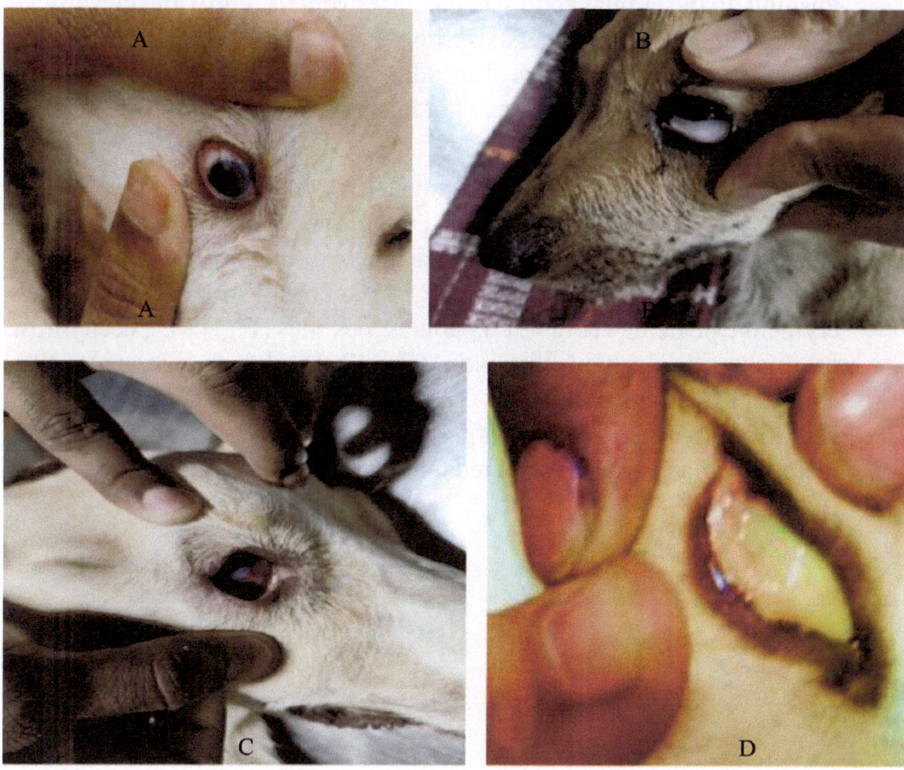

Fig.46. Examination of the color of the mucus membrane showing different colors of mucus membrane of the eye. Normal pink colored mucus membrane **(A)**. Pale or whitish mucus membrane **(B)**. Congested mucus membrane **(C)**. Icteric eye mucus membrane **(D)**. (courtesy Varshney, 2025).

Jugular vein is examined for its pulsation and pressure as it is a direct reflection of right ventricular filling pressure during systole and diastole and right atrial pressure during systole. Jugular vein collapses normally when head is tilted up at an angle of 45^{0}, during inspiration or immediately after 2nd heart sound. There is no distension of jugular veins in healthy dogs. Distended jugular vein, unable to collapse,is indicative of right heart failure, chronic obstructive pulmonary disease, pulmonary fibrosis or ventricular filling defect due to pericardial disease. Hepatojugular reflex is a valuable adjunct in the diagnosis of congestive heart failure. Though, jugular veins can be examined in standing, sitting or sternal recumbency, standing position is preferred for examination of jugular vein. Hepatojugular reflex test is conducted on standing dogs and cats by compressing cranial abdomen and observing the jugular veins. Pushing up the cranial abdomen increases venous return and elevation in jugular venous pressure persists throughout the period positive pressure is applied to the abdomen. The added jugular distension or positive hepato-jugular reflex indicates right heart failure, tricuspid valve disease or pericardial disease.

Examination of integument is conducted by running both hands on the entire body surface.It gives an idea about the condition of the skin and hair coat. The presence of alopecia, masses (cutaneous/subcutaneous- Fig. 47 A), ectoparasites (ticks - Fig. 47B, fleas- Fig. 47C, lice), scales, erythema, petechiae (Fig.47D),papules, pustules (Fig.47 E), ecchymosis or thickened elephant like skin (Fig.47 F), feline mammary fibro epithelial hyperplasia (Fig.47 G) and canine mammary tumor (Fig.47 H) can be ascertained. Skin turgor can also be assessed. Entire chain of mammary glands can be palpated for detecting any mass or inflammation if any.

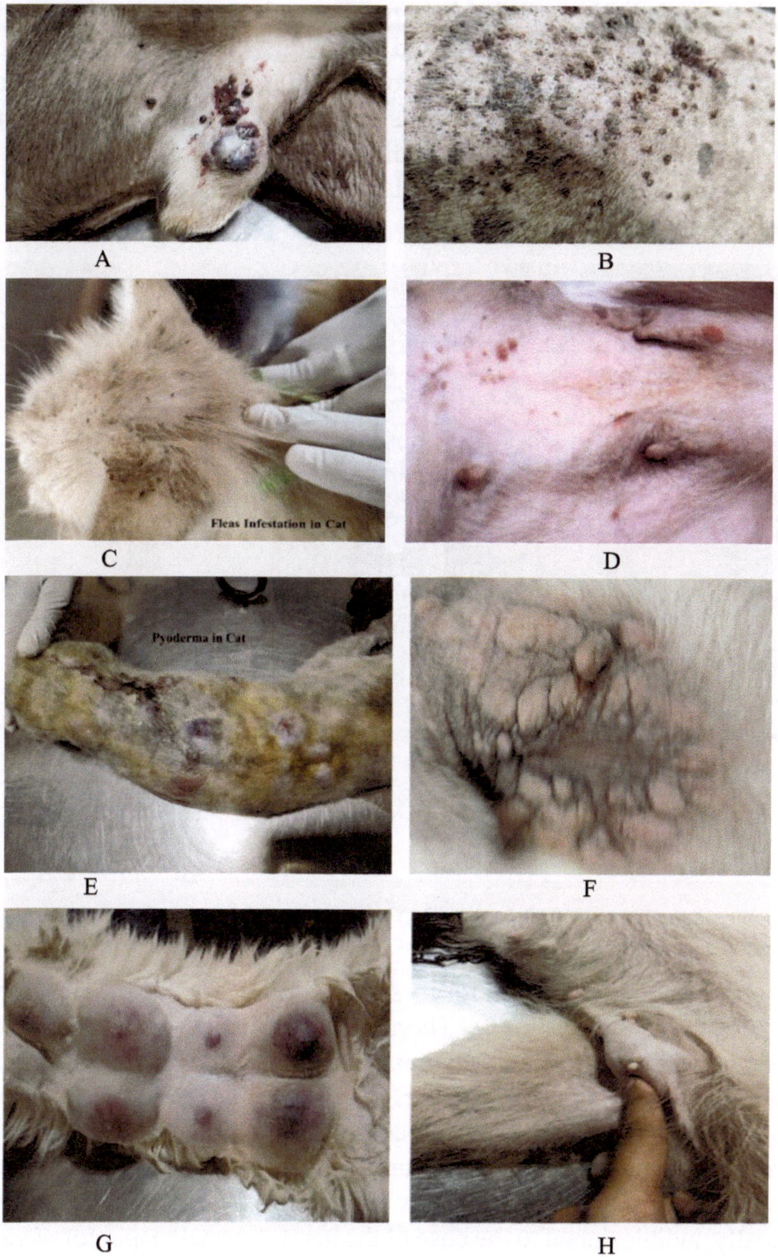

Fig. 47. Examination of integument showing cutaneous mass diagnosed as wart **(A)**, tick infestation in dog **(B)**, fleas infestation in cat **(C)**, erythema and petechiae **(D)**, pustules **(E)**, elephantine skin (F), feline mammary fibro epithelial hyperplasia **(G)** and canine mammary tumor **(H)**. Figures (**C, D, E, F and G**) are with the courtesy of Varshney, (2025).

Both thoracic limbs are examined from scapula to toe. Any abnormality such as pain,no weight bearing and knuckling (Fig.48A) is noted,if any. Joint and muscles are palpated for detecting any swelling, pain and heat. Foot pad, nails and nail beds are examined for color, symmetry, erosion, foreign body, mass, interdigital cyst/ hematoma/ mass (Fig. 48 B) /wound or dermatitis. Prescapular lymph nodes are palpated for their size and shape.

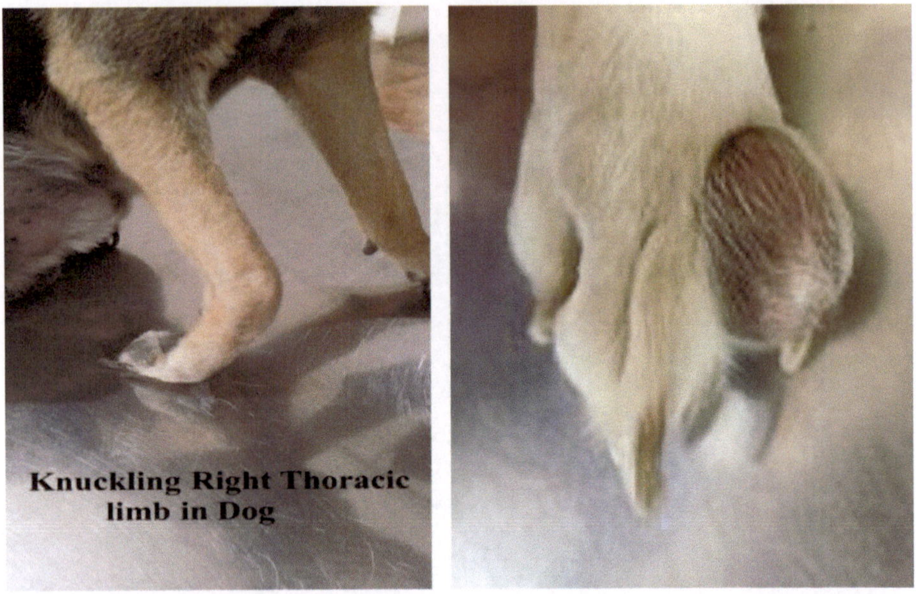

A B

Fig.48. Examination of thoracic limb in dogs. **(A)** Knuckling of right thoracic limb. **(B)** Mass in interdigital space (courtesy Varshney, 2025).

Thoracic cage is examined for its symmetry, integrity of the ribs, spinal hyperesthesia or deformity. Thorax is auscultated for assessing lung sounds ;and heart rate and rhythm. It can assist in detecting inspiratory or expiratory dyspnea, vesicular (normal) or adventitious (abnormal sounds such as wheezes, crackles) sounds. Normal lung sounds are bronchial sounds arising from turbulent airflow generated within the normal trachea and larger airways in the lungs. Normal lung sounds are usually audible dorsal to the base of the heart. Normal inspiratory sounds are soft and low pitched whereas expiratory sounds are even softer and lower in pitch. Lung sounds may be increased or decreased. Panting increases loudness and intensity of bronchial sounds. Adventitious sounds are abnormal or pathological sounds. Adventitious lung sounds are classified a continuous or discontinuous.Wheezes are the example of continuous adventitious sound and crackle are the example of discontinuous adventitious sounds. Crackles are associated with pulmonary edema. Wheezes are associated with bronchitis and feline asthma. Cardiac auscultation should be done on both sides of the thorax in standing position at all valve areas, base and apex as described earlier. It can also be done in recumbent pets (Fig. 49).Normal heart sound is LUBB-DUBB – PAUSE. While murmurs are abnormal extra sounds commonly occurring between the lub and the dub having shooshing or wooshing quality. Murmur are caused by an increase in turbulence or change in the way the blood flows through heart. Murmurs are audible in cases of structural defect in the valves, anemia, fever, infection, dirofilariasis,hyperthyroidism or hypertension. Most heart murmurs are audible parasternally in cats. Auscultation of the heart and evaluation of the femoral artery pulse should be done simultaneously to detect pulse deficit.

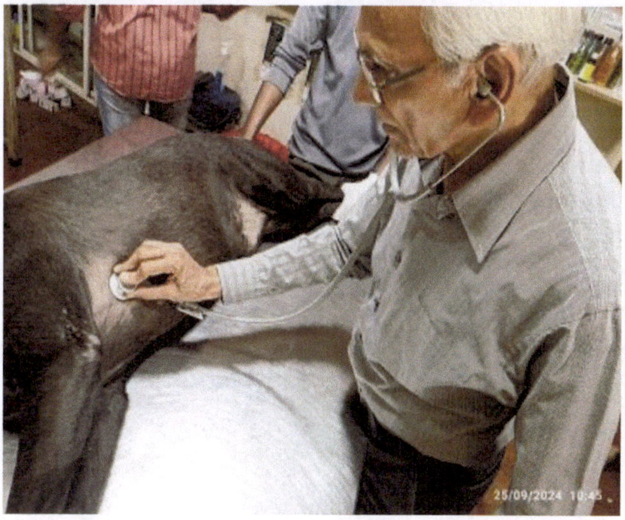

Fig. 49. Cardiac auscultation in a recumbent dog.

Abdomen is palpated using light but forceful touch moving the fingers in cranial to caudal and dorsal to ventral directions. For abdominal palpation tips of the fingers are used to detect mass, if any or to evaluate the size and shape of the organs. Abdominal distension (Fig.50) or pain is also noted. In healthy dogs and cats liver is not palpable as it does not extend beyond costal margin. Kidneys are located in dorso-cranial abdomen. In dogs kidneys are fixed in retroperitoneal space and right kidney is more cranial than the left kidney. In dogs, only caudal end of the kidneys can be felt. While in cats, each kidney can be felt and palpated as kidney are more mobile. In many healthy dogs tail of the spleen can be felt on the left ventral abdominal floor. Intestines are palpated throughout the abdominal cavity. In dorsal abdomen colon can be palpated. Fecal matter in the colon may be mistaken as mass. Fully distended urinary bladder is palpable in the caudal abdomen. The inguinal lymph nodes are generally not palpable in healthy subjects.

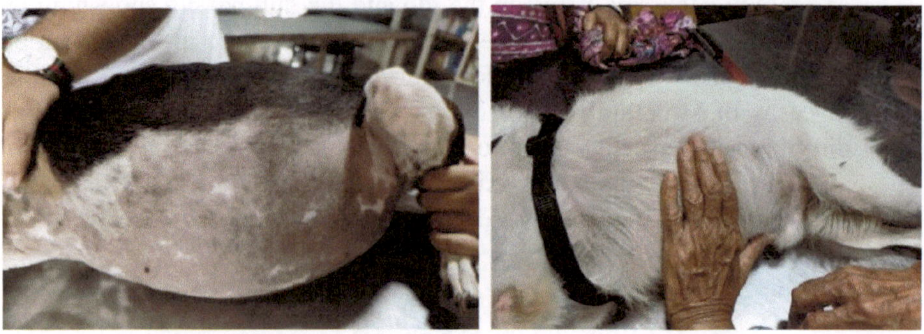

Fig. 50. Showing distended abdomen due to ascites in dogs.

Pelvic limbs are palpated in the manner similar to thoracic limbs. Any abnormality such as pain, limping, edema, knuckling, if any, is noted. Joint and muscles are palpated for detecting any swelling, pain, heat. Foot pad, nails and nail beds are examined for color, symmetry, erosion, hyperkeratosis (Fig.51) foreign body, mass, interdigital cyst/ hematoma/ mass/wound or dermatitis.

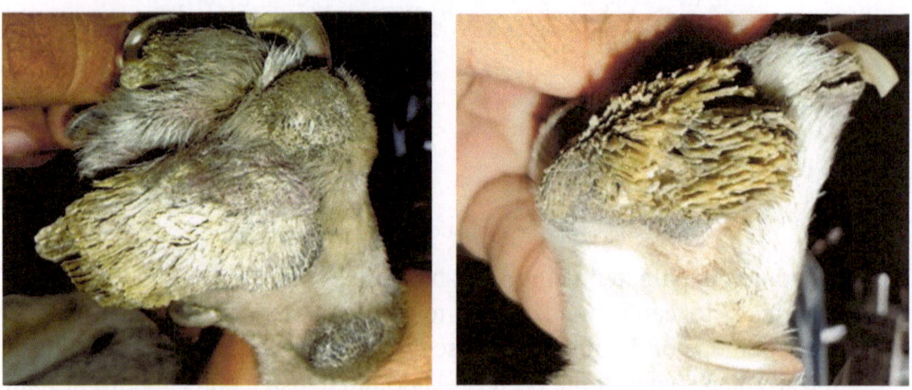

Fig.51. Pug showing thickened hard skin on the foot pads, hair like projections growing off the pads of the paws (frond like skin growth) characteristic of hyperkeratosis

Tail is examined for its movement, hair loss and skin lesions. Palpation of the tail can detect the presence of any tumor or swelling,if any. Lumbosacral pain can be detected by applying gentle pressure to the base of the tail.

Popliteal lymph nodes (Fig. 52) can be palpated behind the stifles and other lymph nodes can also be palpated at their location shown in Fig.52.

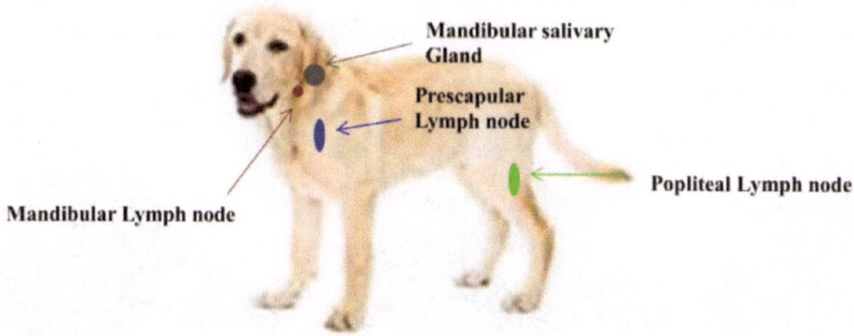

Fig.52. Showing location of the lymph nodes in the dog (Courtesy Varshney, 2025).

External genitalia and perineal area should also invariably be examined. Vulva, vagina, prepuce and penis should be examined for discharge, swelling, redness or any growth (Fig.53 A, 53 B). Examine the symmetry of the testicles and whether both testicles have completely descended or not (Fig.53C). Any lesion (swelling, wound etc.) on testicles should also be noted. Penis is exteriorized and examined for venereal granuloma (Fig.53 B). Perineal area is examined for swelling, mass, or fistula.

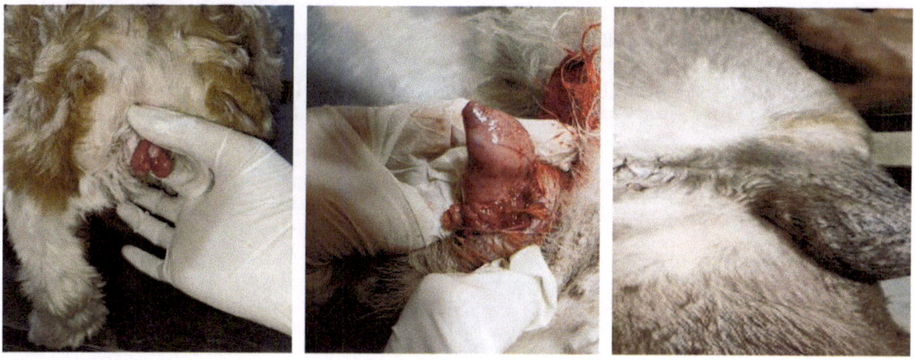

A B C

Fig.53. Examinations of genitals. A mass is visible in the vagina **(A)** and at the base of the penis **(B)**. Scrotal examination is showing scrotal sac dermatits **(C)**.

Integrity of caudal nerves of the tail, the prudendal nerve, spinal cord segment and associated nerve roots can be ascertained by stimulating the perineum with hemostat using a gentle prod which will cause flexion of the tail and contraction of anal sphincter.

Per rectal examination (Fig.54) is conducted to evaluate rectal wall thickness, anal glands, pelvic urethra, prostates in male, and vaginal tract in females. The position of anal glands is normally at 4 and 8 –clock positions. During per rectal examination, feces can also be examined for color and consistency.

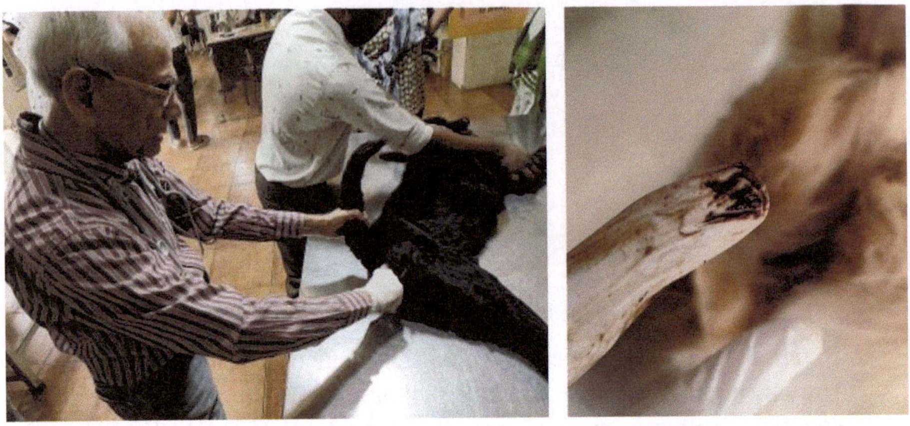

A B

Fig. 54.Per rectal examination in a dog. A lubricated finger of the gloved hand is gently inserted into the rectum through anus to palpate the organs below and anal sacs just at the entry at 4 and 8 O' clock position (A). After taking out the finger the feces on the finger (B) can also be examined for color, consistency, mucus or blood. The color of the feces on the finger (B) is tarry black suggesting melena. It may be due to bleeding in the stomach or anterior part of the intestine.

References and Literature Reviewed

Varshney, J.P. (2025). A Fully Illustrated Handbook on Clinical Diagnostic Processes in Canine Medicine.NIPA GENX Electronic Resources and Solutions P. LTD., New Delhi-110034.

7

Medication in Farm and Pet Animals

Medication refers to administration of drugs. The route, dose and frequency of the drugs administration is most important in providing health care and welfare of the animals. Wide range of body size and their weight in animals and rearing practice makes the treatment and drug administration strategies a bit complex. Because of this complexity wide range of drug formulary and drug delivery strategies has been developed in animals for health care. Route of drug administration varies from oral administration, parenteral administration to local application on skin and mucus membrane. Similarly forms of the drugs varies from liquids, semisolids to solids for oral administration; from clear solution, suspension, emulsion to powder for parenteral administration; and from powder, cream, paste, suspension, ointment, solutions to emulsifiable concentrates for local application.

Oral Administration of Drugs

All three forms of drugs viz. liquids (solution, emulsion, suspension), semisolids (paste), and solids (tablets, powder, capsule, granule, premixe medicated block) are used for oral administration. Drugs in solution can be administered orally in cattle using drenching bottle (Fig. 55).

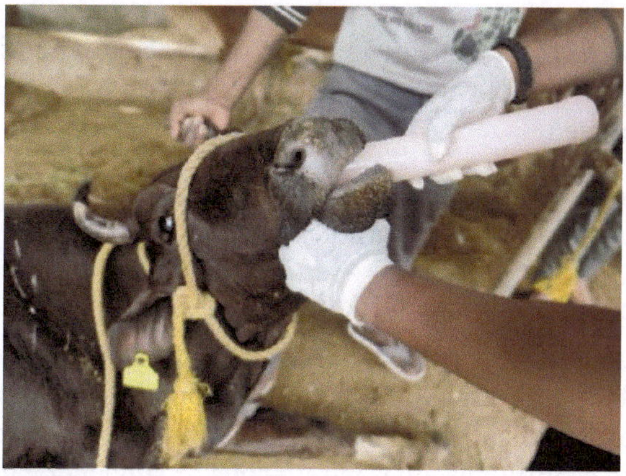

Fig.55 . Administration of liquid medicine orally in a cattle using drenching bottle.

A solution is defined as a mixture of two or more components that mixes well forming a single phase. Solutions are less irritable to gastric and intestinal linings and absorbs quickly. However, some drugs in solution have unpleasant taste.

In suspension, insoluble coarse particles of the drugs are coarsely dispersed in a liquid. Poorly soluble or insoluble drugs are administered as suspension. Taste of the drugs in suspension is less noticeable and suspensions are well shaken before use.

Emulsion contains two non-mixing liquids, one of that is dispersed throughout the other in the form of droplets as oil in water.

In paste, the powdered drug is dispersed in a liquid or fatty base. Pastes may be tasty or tasteless. Paste are easy to administer by owners and are popular in feline and equine practice.

Active ingredients in the tablets are mixed with the filler. Tablets are of two types viz. swallowed as whole or chewable. Chewable tablets are preferred in dogs and cats. As compared to liquid forms of the drugs, tablets are more stable (physically as well as chemically). Poorly water-soluble drugs in the tablet are associated with low absorption rate. Some drugs are associated with local irritation of mucus lining of stomach and intestines.

Capsules are made of gelatin and have filling of the drug and filler. Capsules may be of hard gelatin filled with solid formula or soft gelatin filled with drug in liquid form or semisolid form. Bad flavor or bitter drugs are used in the form of capsule.

Drug in powder form has a mixture of drug powder mixed with powdered filler. These are added to the food. Powders are more stable and dissolve quickly. Unpleasant taste of the powder may lead to rejection of feed when powdered drug is mixed in feed.

Granules are powdered particles of larger size consisting of more than one form of medication of different particle size.

Administration of Drugs as Injection

Drugs used parenterally are prepared as solution, emulsion, suspension or in powder form. Unstable drugs in solution are prepared in dry powder form. Their solution or suspension is made by adding water just before use. Injections are administered as intramuscular, intravenous or subcutaneous (Fig. 56).

Intramuscular injection in cattle and buffaloes is generally preferred in neck muscles or gluteal muscle (rump). Subcutaneous injection should be given in front of the shoulder. Intravenous injection is given in the jugular vein.

In goats intramuscular injection is preferred in neck muscles and intravenous injection is given in the jugular vein. The common site for subcutaneous injection in goats are the extra (loose) skin under the fore leg area or behind the elbow, over the shoulder blade, in the flank area and on the side of the neck.

The sites for intramuscular injection in dogs and cats are the quadriceps (muscles on the front of the thigh),lumbo-dorsal muscles (muscles either side of the lumbar spine) or the triceps muscles (behind the humerus).The muscles at the back of the thigh should generally be avoided due to possibility of the damage to the sciatic nerve running in this area. Subcutaneous injections are given in dogs and cats in the skin that can be comfortably tent and produce enough space for administration of the drug.The loose skin is pinched between thumb and fingers of the one hand and inserting the needle by the other hand in the skin fold as shown in the figure. The most frequent site for intravenous injection in dogs are the cephalic and lateral saphenous veins. Cats are most often given IV injections in the cephalic, medial saphenous and femoral veins. Jugular vein is also used for intravenous administration.

Cattle

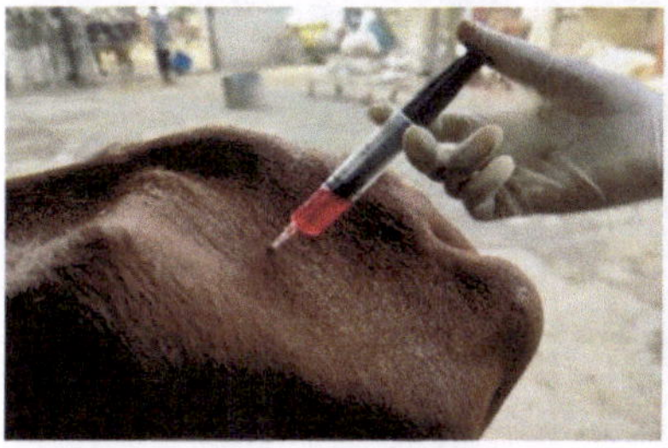

IM (rump)

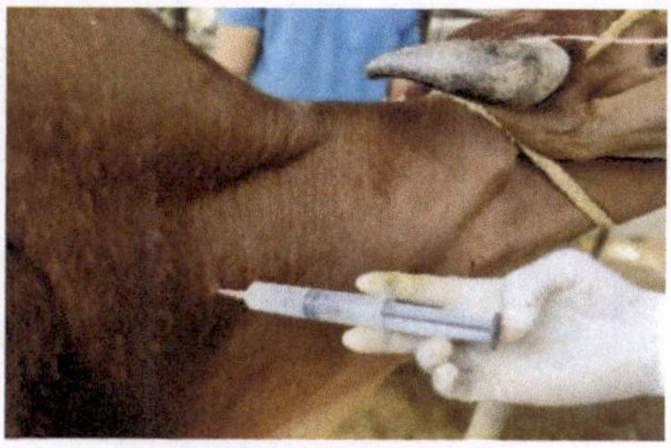

IM (neck muscles)

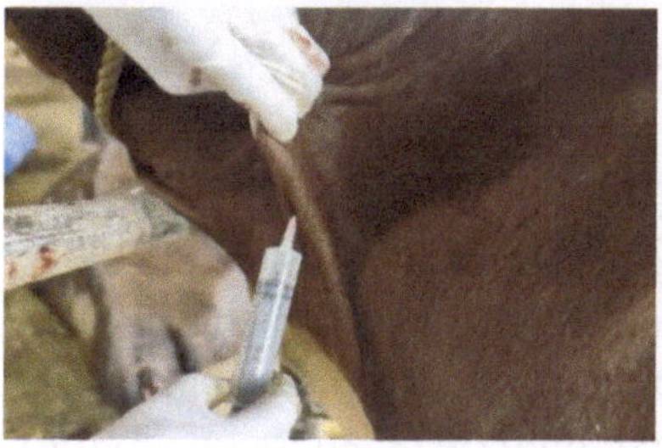

SC

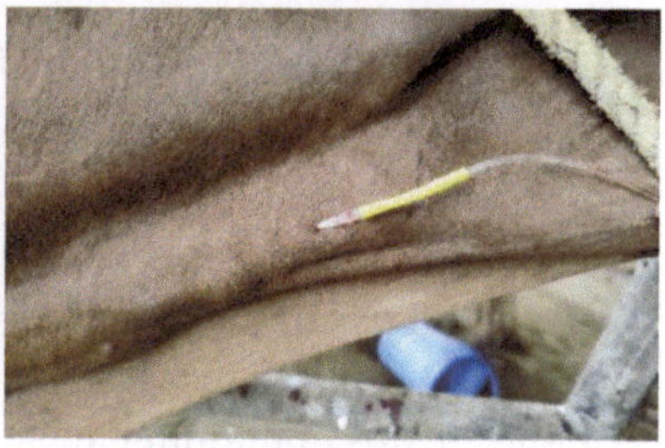

IV (jugular vein)

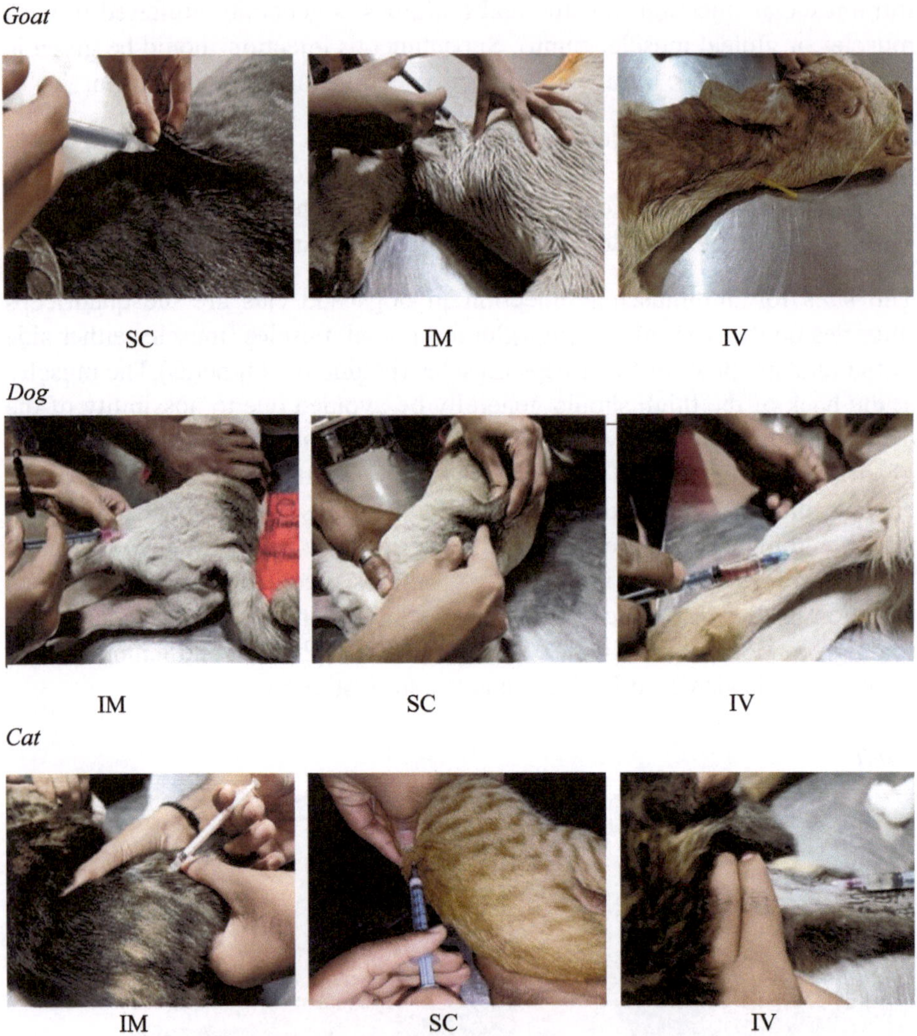

Fig. 56. Parenteral administration of drugs in cattle, goat, dogs and cats.

Drugs used locally on Skin and Mucus Membrane

These drugs are also known as topical drugs. Topical drugs vary in form from dusting powder, cream, ointment, paste, suspension concentrate to emulsifiable concentrates. Spot on or pour on formulations are becoming of special interest in veterinary practice. Transdermal patches and gels are the drugs that work by carrying drug across the skin barrier to the blood stream. Insecticidal collars, plasticized polymer resins containing an active ingredient, are being used for the control of ticks and fleas on dogs and cats. The active ingredient is released as vapor, dust or liquid depending on the chemical used.

Dusting powders are fine textured material containing drug(s) with base of talc, starch, or zinc oxide. These powders are good for general dusting or for skin folds. Powders should be avoided on wet surfaces as caking and clumping may occur.

Creams are semisolid emulsions (oil in water or water in oil). Oil in water creams are easily removable by washing or licking. Water in oil creams are cleansing and skin softening and spread easily.

Ointments are greasy preparations containing dissolved or dispersed drugs. They are good for chronic dry skin diseases.

Pastes are less greasy than ointments and do not seal wounds.

Solutions are liquid preparations containing drug(s) such as eye drops, ear drops.

Transdermal delivery gels deliver the active drug through the skin to the blood stream and are being used in the treatment of undesirable behavior, cardiac disease and hyperthyroidism in dogs and cats. The drug dose is applied to the inner surface of the ear pinna.

Transdermal delivery patch contains drug in the patch that is applied to the skin. The drug absorption continues for longer time. The patches are being used to relieve local pain in dogs, cats and horses.

Spot-on or pour-on solutions contain active ingredient with co-solvent and a spreading agent. The active ingredient is distributed to the entire body. These preparations are being commonly used in the management of ectoparasites in dogs and cats. The pipette containing the drug is held up right. The tip of the pipette is snapped off along the scored line. Hair coat of the pet between the shoulder blade is part off until skin is visible. The tip of the pipette is placed on the skin and the pipette is squeezed gently to empty the contents (Fig.57A and B).

Insecticidal collars (Fig.57 C) are being commonly used in dogs and cats in the management of ticks and fleas. The collars release active ingredients (from plasticized polymer resins) in the form of vapor, dust or liquid based on its chemical used. These collars are most often available in the form of plastic strips made of plasticized polyvinyl chloride containing an active substance (organophosphate pesticides, carbamates, synthetic pyrethroids and others).

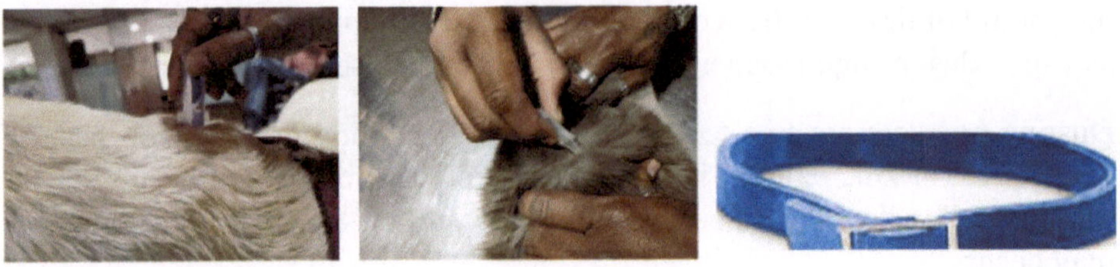

A B C

Fig.57. Application of pour on in dog **(A)** and cat **(B)** and Tick and Flea collar **(C).** The pour on is applied topically and directly on the skin usually along the back line. after parting the hairs between neck and shoulder blades in 2-3 different locations. Present day collars are adjustable and water proof and provides 4 to 8 months of continuous protection against ticks and fleas for dogs and cats.

8

Special Examination of Gastrointestinal System

Ruminant

Ruminant's gastrointestinal tract is unique and different from equines, canines and felines because of complex stomach (polygastric) compartmentalized as rumen, reticulum, omasum and abomasum leading to a different digestive process as well as their diseases. In gastrointestinal dysfunction of ruminant, a complete examination, starting from history, systemic states, visual inspection of abdomen, examination of buccal cavity and esophagus, examination of left and right abdomen, per rectal examination, radiographic examination, paracentesis abdominis to laparoscopy is recommended.

History

- A detail history, with respect to status of pregnancy, lactation, dietary nature (grazing, stall fed), diet (straw, forage/fodder, concentrate, supplements), period of illness, speed with which illness developed, abdominal pain, volume and consistency of feces, and any treatment given with its response, may provide some clues to diagnostic possibilities.

Systemic State

- Examination of vital signs should invariably be conducted as it may help in detecting severity of the disease.
- Acute grain/carbohydrate engorgement (rumen acidosis), acute diffuse peritonitis, acute intestinal obstruction and abomasal torsion are associated with dehydration and increased heart rate (tachycardia).
- Pallor mucus membranes are suggestive of gastrointestinal hemorrhage.
- Increased respiratory rate associated with gastrointestinal disorders/ diseases suggests pain, fluid accumulation or electrolyte imbalance.

- Grunting or moaning sounds, suggestive of pain, are heard in cases of acute diffuse peritonitis or viscus distension.
- Persisting complete anorexia is suggestive of chronic lesion with poor prognosis. Whereas return of appetite and cud chewing is a favorable sign.

Examination of Oral Cavity and Esophagus

- Mouth cavity can be examined visually and by finger palpation with or without mouth speculum.
- The patency of the esophagus can be ascertained by using stomach tube.

Inspection of Abdomen

- Abdomen is visually inspected for its contour or silhouette from back, left and right side (Fig.58).

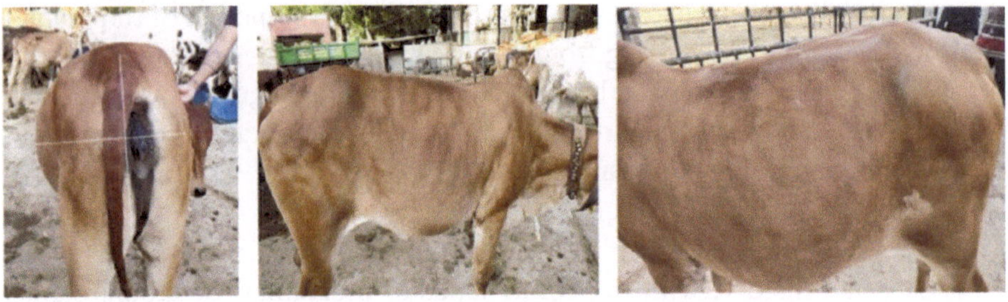

A B C

Fig.58. Visual inspection of the abdomen from the back (A), right (B)and left (C).

- The location of the abdominal contour may reflect cause of distension.
- Distension may be unilateral, bilateral (symmetrical or asymmetrical), or in dorsal/ ventral half.
- The degree of distension may vary from slight, moderate to marked.
- Marked distension of left abdomen can be seen in acute ruminal tympany, simple indigestion, severe grain overload and vagus indigestion.
- Normal to severe distension of right paralumbar fossa may be visualized in right abomasal dilatation and torsion.
- Normal to moderate distension of right lower flank is seen in abomasal impaction.
- Distension of left paralumbar fossa with ping in upper aspect of 9-12 ribs is characteristic of left abomasal displacement.

- Intestinal obstruction or paralytic ileus may be associated with slight to moderate distension of right abdomen.
- Normal to moderate distension of right flank with ping in right paralumbar fossa is seen in cecal dilatation and torsion.
- Bilateral lower abdominal distension with fluid waves may be due to ascites.
- Percussion and auscultation of the abdomen can assist in detecting and localizing gas filled viscera.
- Rectal examination is valuable in detecting any marked enlargement or abnormality in abdominal cavity.
- Abdomen is also examined using metal detector (Fig. 59) for metallic foreign bodies in reticulo-rumen

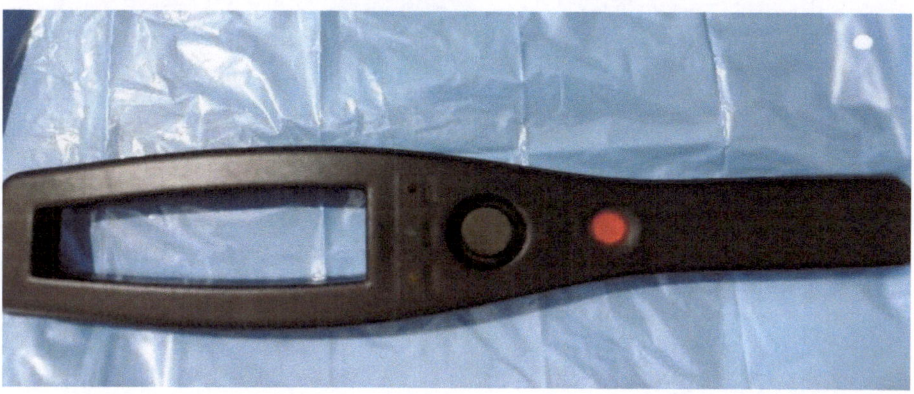

Fig.59. Metal detector for detecting metallic foreign bodies ih the reticulo-rumen.

- Silhouette of the animals is examined from the back and side. Silhouette of the contour changes in bloat (Fig.60 B), vagus indigestion (Fig.60 C), LDA (Fig.60 D and E), hydrops (F) and intestinal hernia (G).

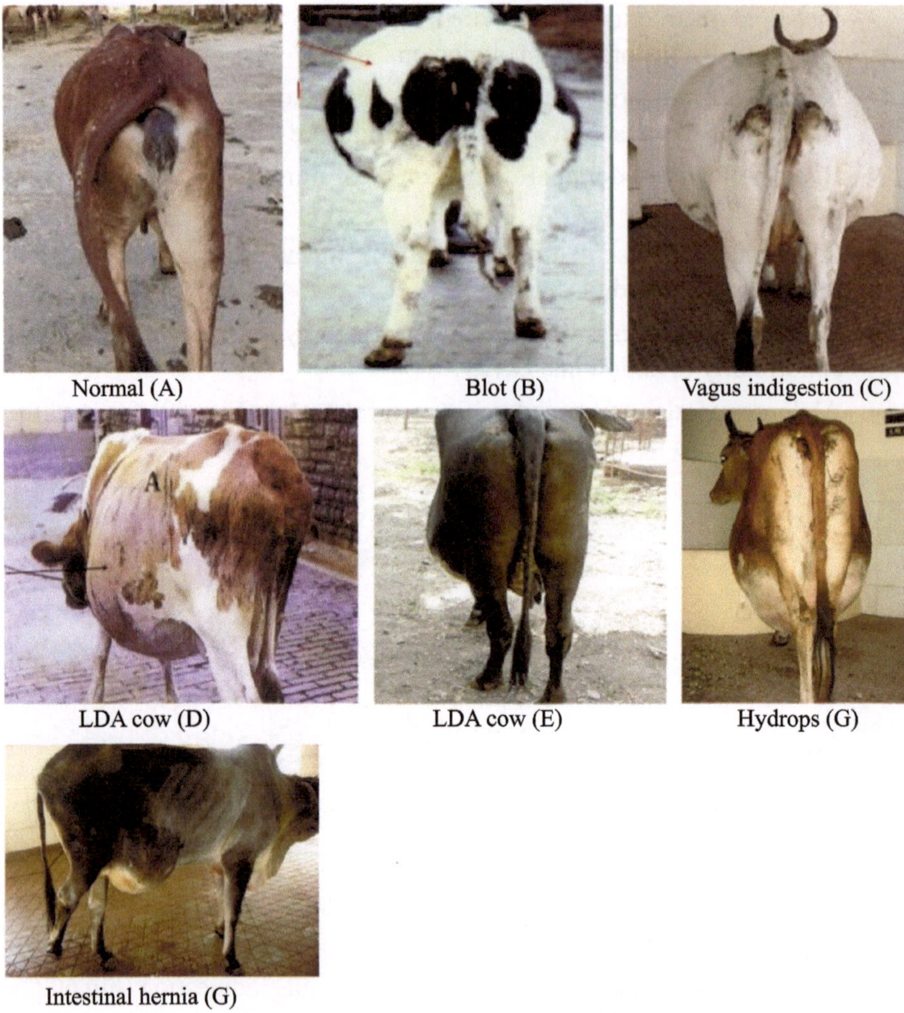

Normal (A) Blot (B) Vagus indigestion (C)

LDA cow (D) LDA cow (E) Hydrops (G)

Intestinal hernia (G)

Fig. 60. Silhouettes of the contour of the abdomen of cattle in health and diseases. Figures at **C,E,F** and **G** are with the courtesy of Dr. P.S. Chaudhary, Chief Surgeon, Nandini veterinary Hospital, Surat-395001 (Gujarat).

Examination of Rumen (Left flank)

- Rumen motility can be examined by simultaneous auscultation, palpation and observing left paralumbar fossa. Rumen contractions are associated with alternate rising and sinking of the left paralumbar fossa. The simplest way of counting rumen movements is putting a palm on left paralumbar fossa and counting alternate sinking and rising (Fig.61). In healthy bovines, rumen contractions are approximately 2-3 per 2 minutes.

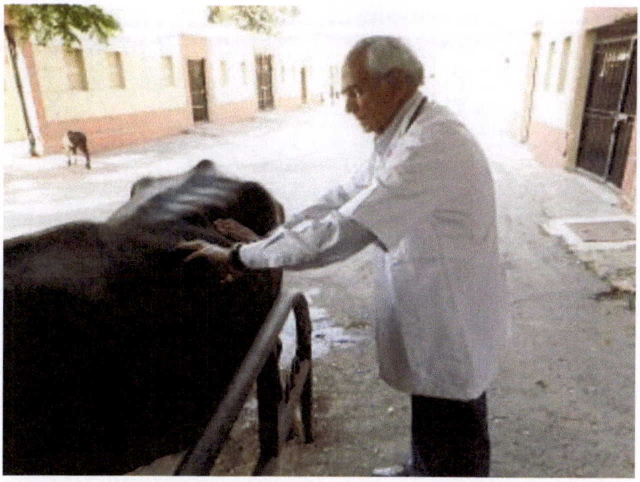

Fig.61. Examination of rumen movements in a buffalo by putting a palm on left paralumbar fossa and counting alternate sinking and rising.

- In vagus indigestion rumen contraction are incomplete and may vary from 3 to 7 per minute.
- Palpation of left rumen in left paralumbar fossa can assist in determining the nature of the rumen contents. In roughage fed animals, rumen gives the feeling of doughiness and also pits on pressure.
- Rumen contents of straw fed (large quantity) or dehydrated animals gives the feeling of firmness on palpation of rumen.
- In grain engorgement or rumen acidosis, rumen contents are soft and porridge like. Deep palpation of the rumen may reveal fluid fluctuation.
- Gas filled atonic rumen (tympany) is tense, resilient and tympanic on percussion.
- Auscultation of rumen of roughage based normal animals reveals booming -crackling sounds. These sounds become less distinct in animals fed with grains and less roughage. Fluid splashing indicates grain engorgement or an atonic rumen with prolonged anorexia(in cases of chronic diffuse peritonitis, omasal or abomasal impaction).
- In vagus indigestion the rumen is hypermotile (rumen contraction 3-6 per minutes). In some other cases of vagus indigestion, the rumen may be atonic.
- Pings (Fig. 62) detected on the auscultation and percussion of left paralumbar fossa (from mid-point of 9th rib to 12th rib) are suggestive of left sided abomasal displacement.

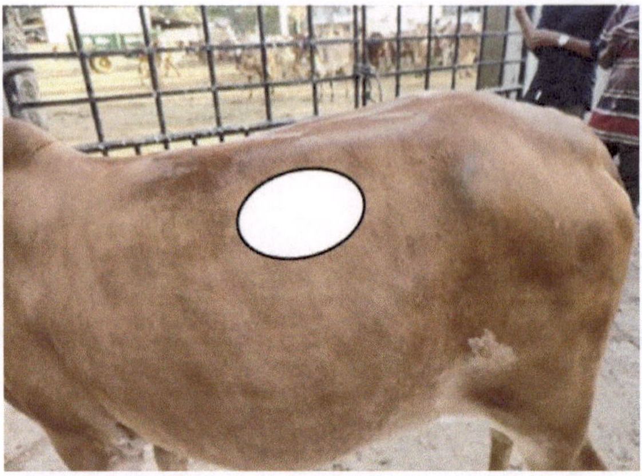

Fig.62. Site for detection of pings in LDA in cows. Area marked.

Examination of Right Abdomen

- Right abdomen is visually inspected for any distension (may be due to a viscus filled with gas, fluid or ingesta; or gravid uterus).
- The right abdomen should also be examined employing a combination of deep palpation, ballottment and simultaneous percussion and auscultation.
- Pings on percussion of right abdomen (Fig. 63 A) are suggestive of right sided abomasal dilatation and torsion, cecal dilatation and torsion (Fig. 63B), coiled colon torsion, intestinal tympany, intussusception, pneumoperitoneum, intestinal tympany in post parturient cows or torsion of the root of mesentery in calves.

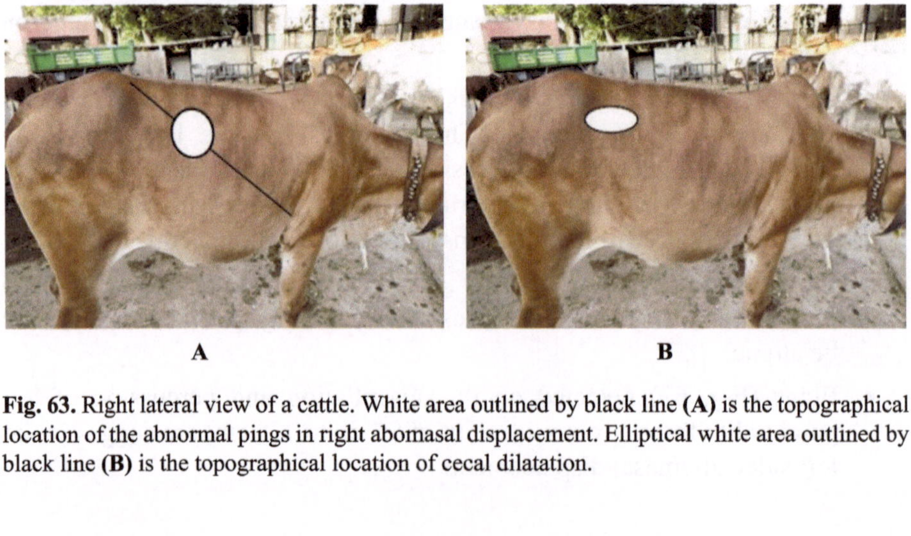

A B

Fig. 63. Right lateral view of a cattle. White area outlined by black line **(A)** is the topographical location of the abnormal pings in right abomasal displacement. Elliptical white area outlined by black line **(B)** is the topographical location of cecal dilatation.

- Fluid splashing sounds on ballottement and auscultation of right flank may be due to fluid filled intestines(enteritis, acute intestinal obstruction), fluid filled abomasum (right side dilatation).
- Palpation of a firm viscus in right flank (caudal or ventral to right costal arch) is indicative of omasal impaction, abomasal impaction, enlarged ventral sac of rumen or enlarged liver.

Radiographic Examination of Abdomen

- Abdomen of the ruminants can also be examined using radiography. For detecting foreign material, metallic objects and plastic material, radiography can play an important role (Fig.64).

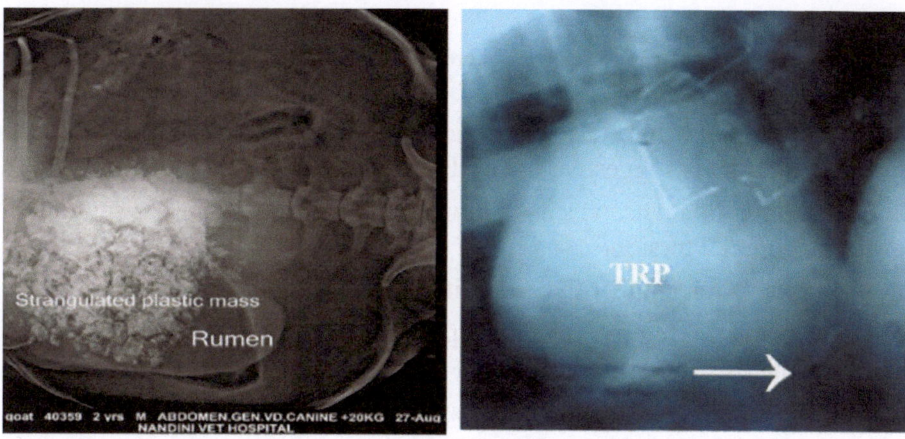

A B

Fig. 64. Abdomen radiographs of a goat and a buffalo showing radio-opaque material in the rumen of the goat (A) and metallic object in the rumen of the buffalo.

Rectal Palpation

- Rectal examination is important in ruminants in detecting some of the specific abnormalities of digestive tract.
- On rectal palpation, L-shaped rumen (more distension of ventral sac of rumen as compare to dorsal sac- Fig. 65) is suggestive of vagus indigestion.

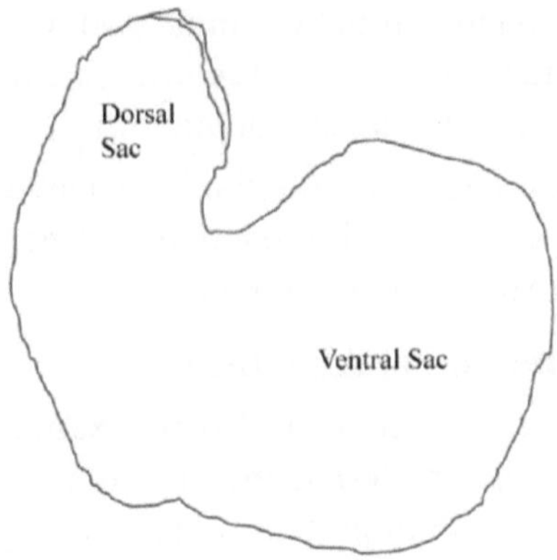

Fig.65. Rough sketch of rumen showing more distension of ventral sac in cases of vagus indigestion on rectal palpation.

- Palpation of a long distended movable organ with blind end on rectal exploration is suggestive of cecal torsion.
- Palpation of a tense viscus in lower right half of abdomen is suggestive of abomasal torsion.
- Rumen is felt smaller in cases with left abomasal displacement.
- Palpation of lumps in abdomen and pelvic cavity is suggestive of lipomatosis.
- Intussusception is usually not always palpable.

Gross Fecal Examination

- Gross examination of the feces is valuable in providing clues for differential diagnosis of gastrointestinal diseases.
- Feces should be examined for color, amount, odor, consistency, presence of undigested or poorly digested feed material (degree of comminution), and presence of other substances (mucus, blood, fibrin etc.).
- A reduction in the amount of feces can be due to a decrease in feed and water intake. Frequently passing of an increased amount of feces with higher water contents is suggestive of diarrhea. Failure of passing feces for > 24 hrs may indicate physical or functional intestinal obstruction or paralytic ileus.
- The color of the feces is influenced by the nature of the feed, bile concentration and passage rate. Feces of milk fed calves is yellowish and turned to pale brown on hay or straw feeding. Cattle fed on green forage, hay or large amount of grain pass feces of dark olive green, brown olive

or gray olive color respectively. Diarrheic feces are pale than normal. In hemolytic anemia, the color of feces is dark olive green or black-green because of increased amount of bile products. Abomasal and small intestinal bleeding (hemorrhages) leads to black tarry feces. Red brown color distributed throughout the feces or frank blood adhering to feed material (dysentery) is indicative of bleeding in the rectal or large intestine.

- Freshly voided feces of cows and buffaloes is not malodorous. In case of putrefaction/ fermentation of the ingesta, the odor of the feces becomes malodorous. Smell of the feces becomes fetid in cattle suffering from salmonellosis.
- Consistency of the feces is influenced by the water content, type of the feed and length of transit time in the tract. Feces of the calves on milk are medium to firm porridge-like. Feces of normal cows and buffaloes is medium porridge –like. Dehydration leads to more solid consistency to firm balls appearance.

  Pasty feces are seen in cases of left side abomasal displacement. In vagus indigestion and chronic peritonitis, feces become sticky and tenacious.
- Appearance of poorly digested plant material in the feces is suggestive of rumination failure or accelerated passage of ingesta through reticulo-rumen. Small walnut size chunks of undigested plant material in feces are seen in many cases of acute traumatic reticulo-peritonitis. Kernels of grains may be detected in feces of cows and buffaloes ingested large quantities of unprocessed wheat or barley.
- Mucus on the surface of fecal balls indicates increased transit time of ingesta in the large intestine. Mucus plugs in rectum are suggestive of paralytic ileus (functional obstruction). Clear watery mucus in feces may be seen in cases of enteritis.
- Long strands of fibrin may be seen in fibrinous enteritis.
- Blood in feces may have its origin from swallowing of blood coughed up from lung hemorrhage (occult blood), abomasal hemorrhage (black tarry feces), small intestinal hemorrhage (uniformly dark red) or large intestinal(cecum or rectum) hemorrhage (frank blood).

Abdominal Pain Detection

- Grunting (an expression of pain) is observed in cases of acute local or diffuse peritonitis, severe pneumonia, pleurisy, and lung emphysema.
- Grunts (elicited on deep palpation of the abdomen using closed hand) may be heard on tracheal auscultation. Detection of grunts, on application

of pressure to abdomen, is suggestive of peritoneal lesion. Pressure on abdomen is applied, just behind the xiphoid sternum by a wooden bar held horizontally by two persons and lifting it quickly, and auscultating the trachea simultaneously for grunts (Fig. 66). Pinching of wither is also used to elicit a grunt (Fig.67) to detect anterior abdominal pain. It causes the animal to depress its back. In cases of peritonitis, depreesion of back is associated with grunt. The test is done by grasping a fold of skin over the wither. In normal circumstances, the grasping of the skin fold over the wither causes the animal to dip the point. The animal becomes reluctant to make a movement and grunt on wither pinching when there is peritoneal inflammation or irritation due to penetrating foreign body.

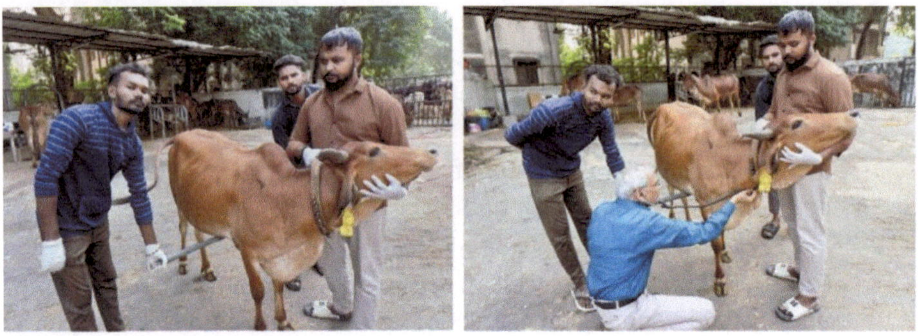

A B

Fig.66. Pressure application just behind xiphoid sternum using a pipe and lifting it quickly (A) and simultaneous auscultation of trachea (B) for detection of grunts.

Fig.67. Wither pinching test in Buffalo (A) and cattle (B) for detecting grunts.

Equines

Gastrointestinal tract of horses is examined using the techniques of visual inspection for abdominal contours (changes occurs in horses with distension or gauntness), intestinal auscultation for the presence or absence of borborygmi, abdominal palpation and percussion, rectal palpation, passing of stomach tube and paracentesis abdominis.

Abdominal Auscultation

Abdominal auscultation is valuable in horses but of limited value in pigs. Auscultation of stomach and intestines can provide information about the frequency and amplitude of gastrointestinal movement (peristalsis) as well as nature of intraluminal contents. After feeding or excitement there is an increase in amplitude and frequency of gastrointestinal movement. Intestinal sounds heard over right ventral abdomen are loud and booming in character (borborygmi). Small intestinal sounds over left abdomen are faint rushing sounds. Increased frequency and intensity with fluid quality of intestinal sounds is heard in horses with enteritis. Borborygmi are of loud and crackling nature in spasmodic colic; of decreased intensity and frequency in large intestine impaction; completely absent in thromboembolic colic (verminous aneurysm and infarction of colon); and decreased to no sound and infrequent fluid tinkling in intestinal obstruction. Fluid tinkling sounds in right flank may be heard in cases of intestinal stasis or intestinal atony.

Palpation

Deep palpation and percussion through abdominal wall has limited value in horses as only fetus can be palpated.

Passage of Nasogastric Tube

Passage of nasogastric tube is an important technique in horses to relieve gastric distension with fluid as gastric reflux is common in horses with colic.

Rectal Palpation

Gastrointestinal, urinary and genital tracts, vessels, peritoneum and pelvic structures, amount and nature of feces can be examined on rectal palpation. It is advisable to use lubricant and avoiding force while conducting rectal palpation to avoid straining and injury. Paralysis and ballooning of the rectum, intestinal loop distension, hard masses (cecal or colonic impaction), intestinal obstruction, or tight bands of mesentery (leading to segmental displacement) can be detected on rectal examination. Anterior mesenteric artery and some of its branches can be felt. In case of verminous aneurysm, thickened, pulsatile vessel with uneven surface and pain may be detected on rectal palpation.

Per Rectal Sonography

Per rectal ultrasonography along with fecal examination is a valuable diagnostic aid in the diagnosis of iliac thrombosis due to migrating larvae of *Strongylus vulgaris*. Detection of a pedunculated echogenic intraluminal mass originating from aorta is suggestive of iliac thrombosis (Fig.68). Caudal end of the spleen is palpable in left abdomen in horses. Inguinal ring may also be palpable. Vagina, cervix and uterus can be examined through rectal palpation for gross abnormalities (metritis, retained placenta or ruptured uterus).

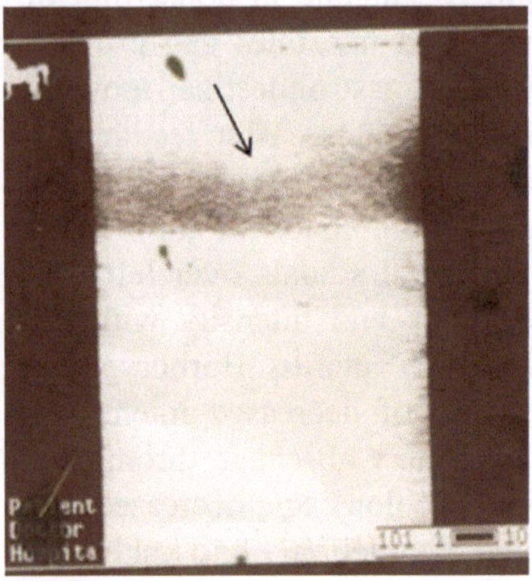

Fig.68. A trans rectal sonogram of an eight year old mare with mild recurrent colic, posterior ataxia, hind limb incoordination, relative coolness of hind limbs(distal to tarsus), and per rectal detection of a firm area of aorta just cranial to aorto-iliac quadrification showing a pedunculated echogenic intraluminal mass originating from dorsal aortic wall just in front of aortic-iliac quadrification suggesting iliac thrombosis.Faecal examination showed preponderance of strongylid eggs. This iliac thrombosis was due to migrating larvae of Strongylus vulgaris (Hoque et al.,2002).

Paracentesis Abdominis

Peritoneal fluid can be collected through paracentesis abdominis.

Canine

Diseases of gastro-intestinal tract are very common in dogs and cats. Gastro-intestinal tract starts from mouth to anus. The first step in making diagnosis of gastro-intestinal disease is to localize the problem. Whether it is buccal cavity, esophagus, stomach, or intestines. Cursory history for seemingly simple complaint like vomiting, anorexia or loose feces may be unrewarding because gastrointestinal diseases may be primary (concerning with only gastro-intestinal tract and its functions) or secondary (due to underlying disease/ disorder of organs other than gastro-intestinal tract). Therefore, for ascertaining the cause of gastrointestinal signs (anorexia, nausea, vomiting, regurgitation, constipation, loose feces, bloody feces), a thorough examination of all body systems is more logical. As with other body systems, a comprehensive systematic approach is needed for diagnosing diseases of gastro-intestinal tract also. The techniques for the diagnosis of gastro-intestinal diseases include examination of history, physical examination, urine examination, fecal examination, pancreatic function tests, intestinal function tests, application of imaging tools (radiography, ultrasonography, and endoscopy) and cytology or histopathology.

History

Detail history regarding day to day eating and elimination habits, type of food being fed, change in food if any, water intake, urination, deworming status, vaccination status, weight loss/gain, vomiting/regurgitation, constipation/ diarrhea and any change in behavior should be obtained. Signalment details should invariably be recorded. History with respect to diet (treats, table food, supplementation of vitamins and or calcium, flavored food, food with corn or wheat), frequency (how many times a day) and amount of feed being offered is of diagnostic concern.

Physical Examination

Physical examination is conducted to localize the problem and distinguishing between vomiting and regurgitation; and between small and large intestine diarrhea. The dog should be examined with respect to posture, behavior, attitude, vital signs (temperature, respiration, pulse), hydration status (skin turgidity, capillary refill time, color of mucus membrane), body condition score and abdominal status. Distended abdomen (Fig.69) may indicate ascites, tumor, hepato or splenomegaly, gas accumulation, gastric dilatation/ volvulus. Salivation and lip smacking may hint nausea, gastric or esophageal reflex or hepatic encephalopathy. Regurgitation (megaesophagus, esophageal stenosis, gastric motility disorder), excessive salivation (foreign body in dogs, hepatic encephalopathy in cats), oral ulceration (CRF), and halitosis (dental disease, foreign body) are suggestive of abnormalities related to upper gastro-intestinal tract. Increased borborygmi are suggestive of enteritis or inflammatory bowel disease (lower gastrointestinal tract). Large volume diarrhea indicates involvement of small intestine's while small volume diarrhea suggests large intestine's involvement. Similarly, melena suggests hemorrhage/ bleeding in upper gastrointestinal tract and fresh frank blood in feces suggests its origin from lower gastrointestinal tract. Palpation of abdomen and rectal palpation should be done as a part of clinical examination.

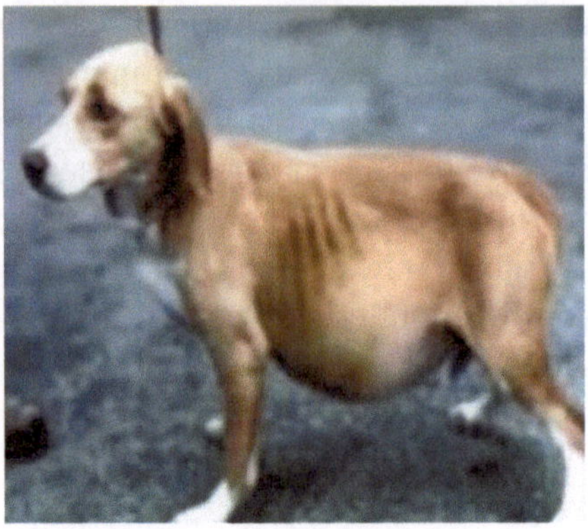

Fig. 69. A dog is with distended abdomen.

Urine Analysis

It may provide an overall assessment of health and also inclusion or exclusion of some diseases (kidney disease, urinary tract infection, diabetes). Low volume of urine may be due to fever or dehydration.

Fecal Examination

Feces are routinely subjected to direct examination (color, consistency, stickiness, mucus, coccidian oocysts or Giardia trophozoites), parasitological examination (presence of parasitic ova, cysts) and cytological examination (for organisms). Colorless or pale (acholic) feces are seen in bile duct obstruction or in exocrine pancreatic insufficiency. Melena (tarry or black feces) indicates hemorrhage/ ulcer in upper gastro-intestinal tract. Fresh blood in feces may be due to hemorrhage in lower gastrointestinal tract or parasites. Feces should also be looked for the presence of foreign material if any. Greasy feces suggests mal absorption syndrome (protein -losing enteropathy) or exocrine pancreatic insufficiency. *Giardia* trophozoites and coccidian oocysts can be detected on direct microscopic examination of feces. Fecal examination using flotation and zinc sulfate centrifugation technique should be done for identification of parasitic ova to rule out or rule in parasitism as a cause of diarrhea. Fecal cytology is recommended for identifying cells (neutrophils, eosinophils), spores of *Clostridium* (safety pin shaped), *Campylobacter, Cryptosporidium*, Mycobacteria or fungal infection (Histoplasma).

Pancreatic Function tests

In exocrine pancreatic insufficiency feces are voluminous and rancid; undigested striated muscle fibers (Fig.70) are detected microscopically in new methylene blue stained fecal slides; X-ray film strip incubated with a mixture of fresh feces and sodium bicarbonate for 60 minutes at 37^{0}C remains undigested (Fig.71) and feces stained with Lugol's iodine (2%) reveals predominance of blue-black granules (Fig. 72). Detection of undigested fat in feces, decreased proteases activity (presence of striated muscle fibers and no digestion of x-ray film) and decreased amylase activity in feces (no digestion of x-ray film) lend support to the presumptive diagnosis of exocrine pancreatic insufficiency. Serum lipase, amylase and serum trypsin like immunoreactivity (TLI) tests are conducted for evaluating pancreas function. Canine pancreatic elastase has also shown promise in the diagnosis of acute pancreatitis.

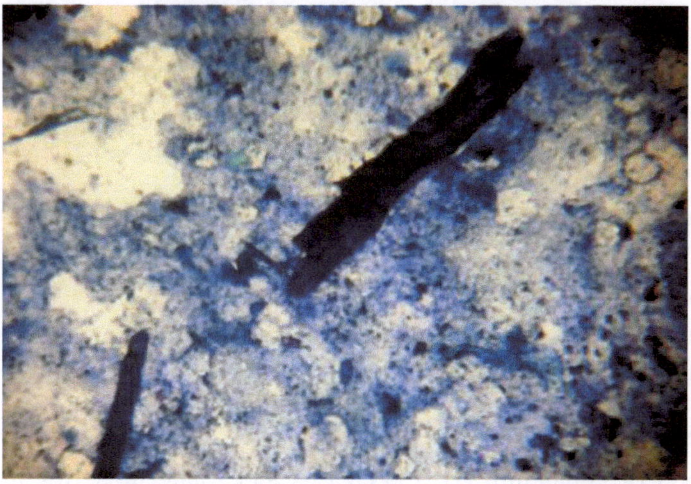

Fig.70. Feces stained with New Methylene Blue showing striated muscle fibers (Courtesy Varshney, 2025).

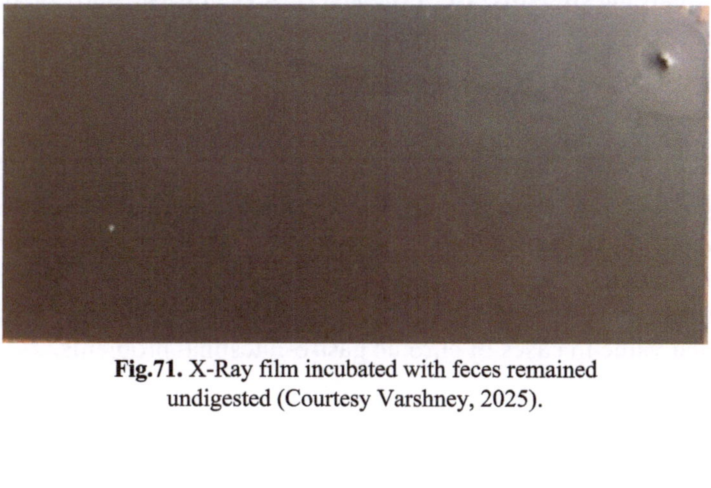

Fig.71. X-Ray film incubated with feces remained undigested (Courtesy Varshney, 2025).

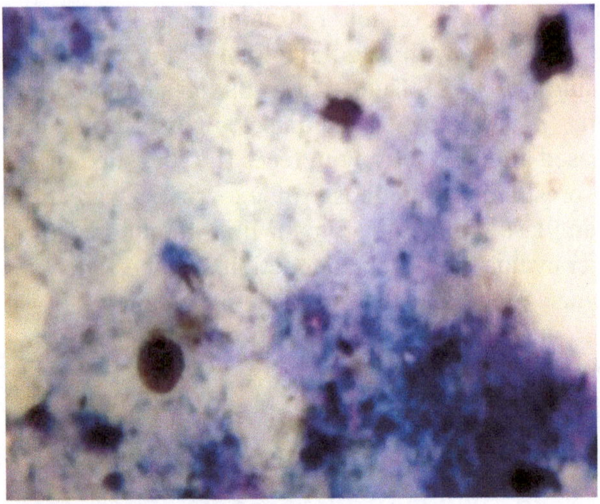

Fig. 72. Feces stained with Lugol's showing preponderance of blue black granules (Courtesy Varshney, 2025).

Cytology

Pancreatic cytology or histopathology provides definite diagnosis of pancreatitis or pancreatic tumor.

Intestinal Function Tests

Serum level of cobalamin (B_{12}) decreases and folate increases in bacterial overgrowth or exocrine pancreatic insufficiency. Low level of folate is indicative of upper small intestine diseases and low level of cobalamin is indicative of lower small intestine disease. Low levels of both folate and cobalamin are seen when entire small intestine is affected.

Abdominal Radiography

In cases of acute and severe vomiting, radiography can be of immense value in ruling out or ruling in foreign bodies, gastric distension, gastric dilatation /volvulus (Fig.73 A), intussusception, or tumors. Contrast radiography is very valuable in detecting foreign bodied (Fig.73 B), large hemorrhaging/ perforating ulcers, esophageal stenosis (Fig.73C), mega-esophagus (Fig.73 D), diaphragmatic hernia (Fig.73 E), paralytic ileus. Abdominal radiography is of limited value in cases of chronic gastro-intestinal problems.

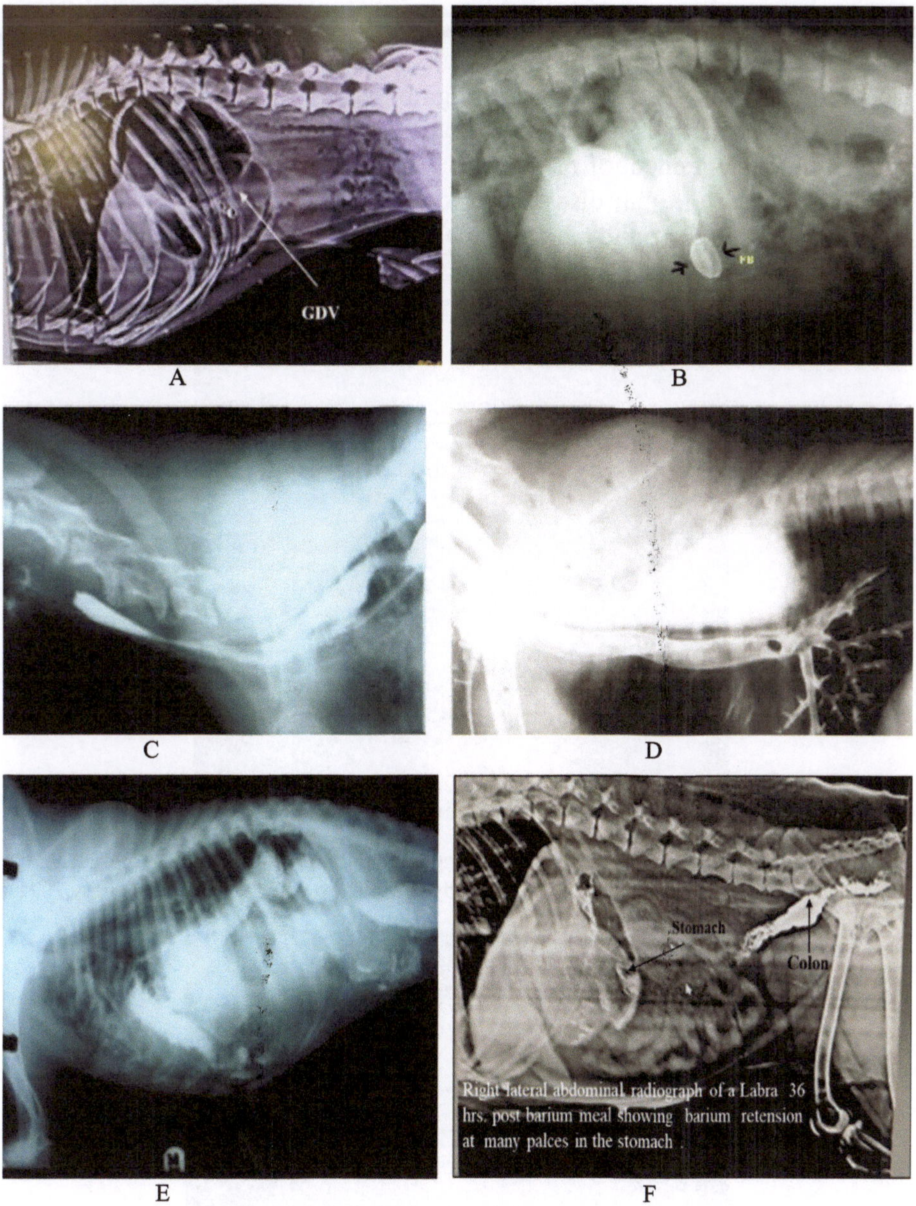

Fig.73. Abdominal plain (A) and contrast radiographs (B,C,D,E,F) of dogs showing gas distended gastric shadows and tissue density separating it into two separate compartments suggesting gastric dilatation with volvulus (A), foreign body in stomach (B- courtesy Varshney, 2025), esophageal stenosis (C), mega esophagus (D), diaphragmatic hernia (E- Courtesy Chaudhary et al. 2010)., gastric ulceration or gastric mucosal aberrations (F).

Abdominal Ultrasonography

Abdominal ultrasonography is a valuable diagnostic test in detecting gastric ulceration, gastric masses, gastric foreign bodies (Fig.74), gastritis, intussusception (Fig. 75), intestinal foreign bodies, intestinal perforation, intestinal masses, and enlargement of mesenteric lymph nodes. Normal pancreas is not visible in ultrasound, but diseased pancreas can be visualized. Pancreas enlargement and fluid around pancreas with change in pancreatic echogenicity is suggestive of pancreatitis and necrosis.

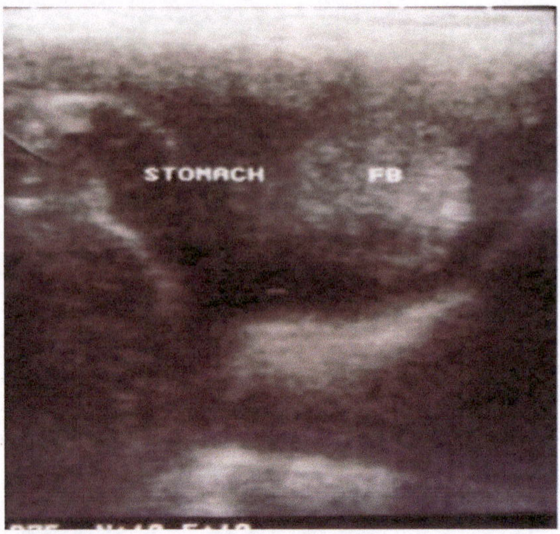

Fig.74. Abdominal sonogram of a dog showing irregular shaped material in stomach with acoustic shadowing suggesting some foreign body (Courtesy : Varshney and Chaudhary,2022).

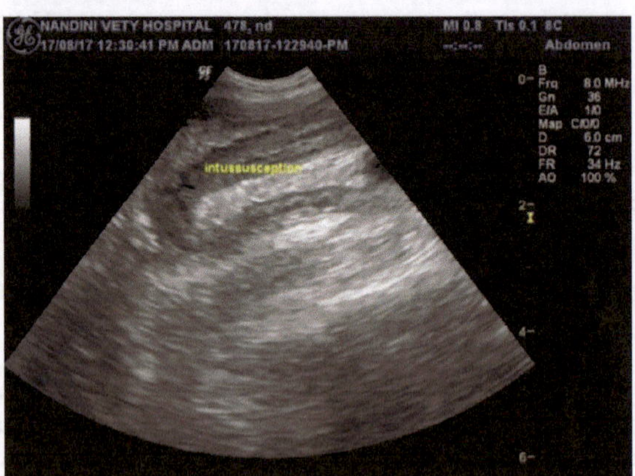

Fig.75. Trans abdominal sonogram of a dog showing multilayered appearance in longitudinal axis suggesting intestinal intussusception (Courtesy : Varshney, and Chaudhary, 20222).

Gastroscopy and Endoscopy

Gastroscopy (Fig.76 A) and endoscopy/ laparoscopy (Fig.76 B) techniques are valuable in localizing disease in suspected cases of gastrointestinal disease (generalized gastro-intestinal disease, disease of stomach, disease of duodenum or disease of colon) to localize the lesion. The endoscopic techniques can also assist in taking biopsy.

A B

Fig.76. Showing Gastroscope **(A)** and Laparoscope **(B)**. Courtesy Varshney, 2025

References and Literature Reviewed

Chaudhary, P.S., Varshney,J.P. and Deshmukh, V. V. (2010). Diaphragmatic hernia in dogs and its surgical repair. Intas Polivet 11: 311-315.

Hoque, M. Varshney, J.P., Singh, G.R. and Maiti, S.K. (2002). Sonographic detection of aorto-iliac thrombosis in horse and its management.J.Rem.Vet.Corps.41:17-21.

Varshney,J.P. and Chaudhary, P.S. (2022). Ultrasound of the Stomach. In: Ultrasound in Veterinary Medicine. Fundamentals and Applications. NIPA GENX Electronic Resources and Solutions P.Ltd. New Delhi- 110 034.

Varshney,J.P. (2025). A Fully Illustrated Handbook on Clinical Diagnostic Processes in Canine Medicine. NIPA GENX Electronic Resources and Solutions P.Ltd. New Delhi-110 034.

9

Collection and Examination of Rumen Liquor

Collection of Rumen Liquor

Rumen fluid/liquor can be collected through various methods such as stomach tube, or rumenocentesis. Collection of rumen fluid in clinical practice is indicated for the diagnosis of reticulo-ruminal diseases as well as for therapeutic transfaunation.

Stomach Tube

Passing of stomach tube (Fig.77) is a traditional method of collection of rumen fluid. Passing of stomach tube in bovines is a routine procedure. The animal is restrained in the cattle crush using head bail with or without halter and a rope.

Equipment: Flexible stomach tube, water based lubricant, appropriate size mouth gag, vacuum pump, fluid collection vessel. The length of stomach tube is to be estimated from the nares to rumen and is marked. The tube is well lubricated.

Procedure: Animal is restrained in a cattle crush with neck extended and head in elevated position. Mouth is opened inserting hand into the gap between incisors and molars and gag is inserted into the mouth using other hand. Lubricated tube is inserted just past the gag to the back of the throat. As soon as animal swallows gently advance the tube into esophagus. Continue advancing the tube gently till it enters the rumen. At cardiac sphincter some resistance may be experienced. With continuous gentle pressure tube can be advanced further. As soon as tube reaches rumen, smell of the rumen liquor may be experienced at the outer end. Desired amount of rumen fluid may be collected in the vessel with the help of vacuum pump. Stomach tube is kinked and removed smoothly. Gag is then removed carefully. Animal is watched for any adverse sign, if any and attended accordingly.

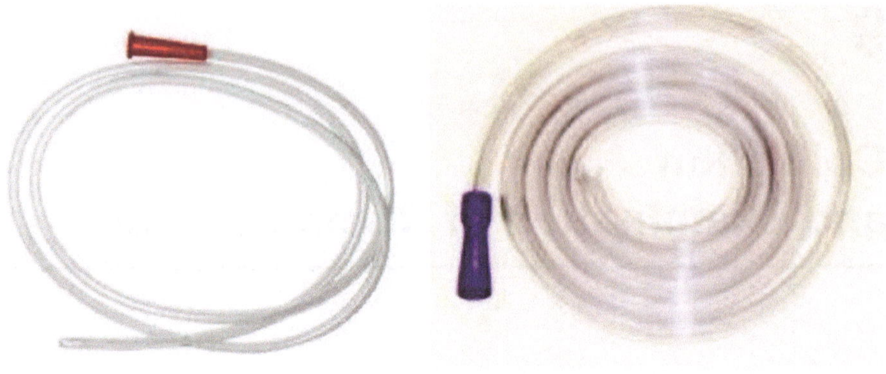

Fig. 77. Showing stomach tubes

Rumenocentesis

It is a simple minimal invasive technique of rumen fluid collection. It involves insertion of a rumen needle (16 gauge,5-6 " long stainless needle Fig. 78) into ventral abdomen for aspirating the fluid.

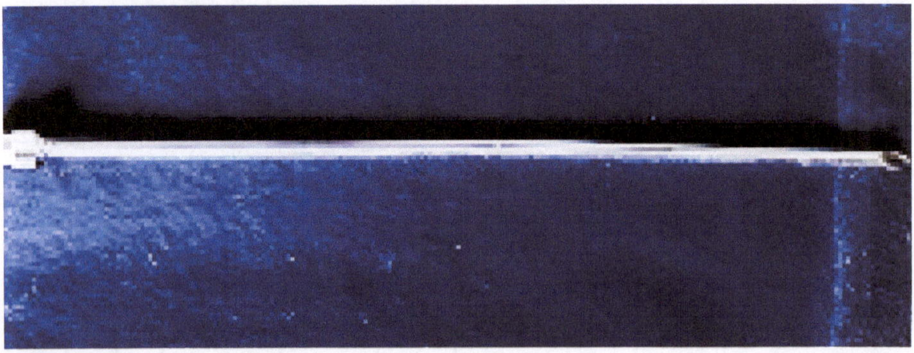

Fig. 78. Showing rumen needle. It is 5 to 6 inches long 16 g needle.

Site of Rumen fluid Collection : The site for collection of rumen liquor is 15-20 cm caudo-ventral to costochondral junction of the last rib (Fig.79). The site for collection of the fluid is prepared by clipping and scrubbing (using alcohol or spirit).

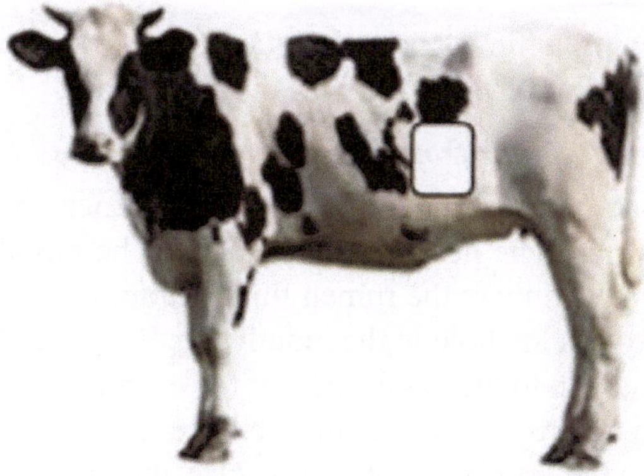

Fig. 79. Showing the site (marked by square) of rumen fluid collection.

Time of Collection: In herds where concentrate feeding is done, collection of the rumen fluid should be done 2-5 hours after the feeding of the concentrate.

Sedation or Anesthesia: Generally no sedation or anesthesia is required. If required light sedation with 20-25 mg Xylazine intravenously can be used for about 500-600 kg animal.

Procedure: The collection of fluid can be done in standing position on cattle in a cattle crush. The needle is inserted in to the rumen with a smooth thrust at 15-20 cm caudo-ventral to costochondral junction of the last rib (Fig.80). The fluid can be aspirated using the syringe. Occasionally needle is obstructed by the ingesta that can be cleared by pushing air through the needle. When needle is obstructed do not create negative pressure with in the syringe as it can increase the pH.

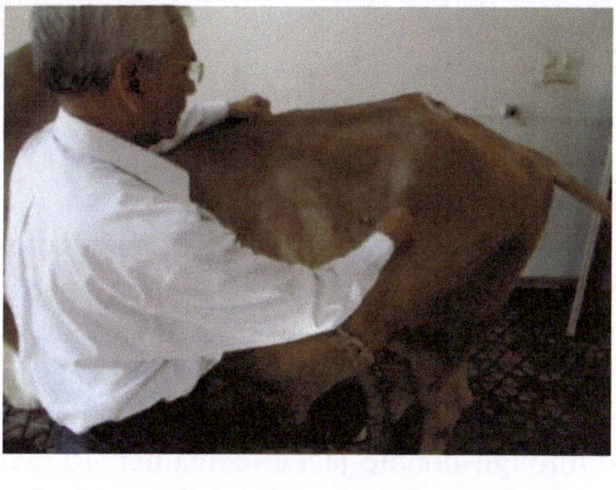

Fig. 80. Insertion of rumen needle into the rumen of a cow for rumen fluid collection.

Procedural Risks

Rumenocentesis being an invasive technique, it is associated with the risk of peritonitis, abscess in body wall, or injury due to restrain.

Collection of Rumen Liquor from Sheep and Goats

The sheep and goats are restrained in between the knees by a person at the level of the fore limbs. The head is raised above. The mouth gag is inserted across the mouth and intubate the rumen fluid extractor pump by introducing suction strainer through the hole in the mouth gag. It is passed over the tongue and then the epiglottis in to the cranial part of the esophagus by exercising care not to enter it in to the trachea. Once the suction strainers is in the esophagus, pass it further down the esophagus until it reaches the rumen. The tube passing through the esophagus can be visualized on the left side of the neck of the animal. As the suction strainer is dipped into the rumen contents, the suction pump is operated to collect the rumen liquor.

Rumen liquor from sheep and goats for clinical purpose can also be collected by rumenoentesis (Fig.81) as described above for cows.

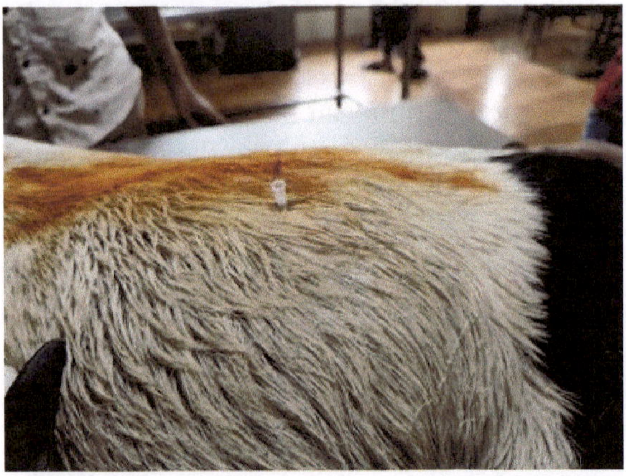

Fig. 81. Insertion of rumen needle in to the rumen of a goat for collection of rumen fluid.

Examination of Rumen Liquor

Rumen fluid contains flora and fauna. Bacteria and protozoa are about 40-60% of total microbial matter in the rumen. Rumen fungi are also present in the rumen. The rumen fluid should be examined as soon as possible after its collection to minimize the effect of cooling and air exposure on the protozoal activity and pH of the fluid. If rumen fluid is to be transported to distant lab, it should be done through double jacket container. Rumen samples kept at room temperature for 8-9 hrs and in refrigirator for 24 hrs can be utilized for biochemical estimations.

Physical Properties of the Ruminal Fluid

(a) *Color* -Normally the color of rumen liquor is influenced by the diet. On green forage/ fodder, rumen fluid is of greenish color. It is straw colored on dry roughage or straw. Milky gray (Fig.82) and dark brownish are abnormal colors. Rumen liquor of milky gray color is seen in cases of rumen overload or rumen acidosis on high grain diet. Dark greenish colored rumen fluid is seen in cases of prolonged rumen stasis or protein putrefaction.

Fig. 82. Showing watery consistancy and milky gray color of rumen liquor of cow diagnosed with rumen acidosis.

(b) *Consistency*- Normally rumen fluid is slightly viscus in consistency. It turns to watery when ruminal bacteria and protozoa are inactive or dead. Froathy consistency of the rumen fluid is suggestive of froathy bloat or vagus indigestion. Watery with foamy consistency indicates rumen decomposition.

(c) *Smell or odor*- Normal rumen liqour is of acceptable aromatic odor. Ammonia like smell of rumen liqour is suggestive of urea poisoning. Putrefied smell is indicative of protein putrefaction. Intensely sour smell of rumen liquor is characteristic of grain engorgment/carbohydrate overfeeding due excessive lactic acid production.

Chemical Properties of the Ruminal Fluid

(a) *pH*- It is a chemical property determined by H + ions. It can be measured with pH meter or in clinical settings it can be measured by pH paper strips (Fig.83). pH should always be measured immediately after collection. pH is influenced by type of feed ingested. On forage /fodder based diets pH in ruminants varies between 6.0 to 7.0. On grain feeding ruminal pH decreases. pH <5.5 is regarded as indication of ruminal acidosis. Lower pH can be seen in vagus indigestion, intestinal obstruction or due to abomasal reflex. Elevated pH > 8.0 suggestive of protein putrefaction or urea poisoning.

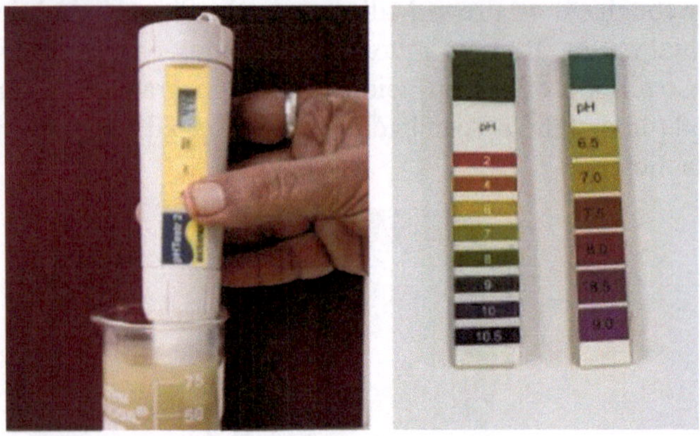

A B

Fig. 83. Measuring pH of rumen fluid with digital pH meter (A). pH paper strips can also be used for measuring pH of the rumen fluid in clinical settings.

(b) *Sediment Activity Test* – This test should also be conducted immediately after collection of rumen liquor. It is an indication of the activity of rumen protozoa. The rumen liquor is taken in to a test tube. Allow it to stand in a test tube. Time taken for completion of sedimentation of the fine particles and floatation of coarse particles in noted. Sedimentation time of 3-9 minutes is regarded as normal time. Very rapid sedimentation without floatation is seen in cases of ruminal acidosis or prolonged anorexia and indicates inactivity of ruminal microflora. Absence of floatation and sedimentation can be seen in cases of frothy bloat and also in few cases of vagus indigestion.

(c) *Methylene Blue Reduction Test or Redox Potential*- It is indicative of anaerobic microbial fermentation. The test is performed by mixing 20 ml ruminal fluid with 1.0 ml of 0.03% methylene blue, keeping it at room temperature for noting time taken in changing the color of the mixture. Rumen liquor of a healthy cattle on hay and grain diet usually takes approximately 3 minutes to decolorizes leaving a narrow ring of blue color at the top. Prolongation in methylene blue reduction time indicates delay in anaerobic microbial fermentation as seen in cases of rumen acidosis and indigestion on roughage feed.

(d) *Cellulose Digestion Test*- The test is the reflection of the activity of cellulitic bacteria. The test is performed by mixing 10 ml of rumen liquor with 0.3 ml of a solution of 16% glucose and hanging a thread of pure cellulose tied with a button or glass bead at its free end and incubating at 37⁰C. Time is noted when the bead is dropped free at the bottom. Fully active rumen microflora takes around 24-30 hrs. to digest pure cellulose. The test is not very accurate. Delay in digestion of the cellulose indicates the reduction or inactivity of cellulitic organisms.

Microscopic Examination of the Ruminal Fluid

(a) *Assessment of Motility of Rumen protozoa*- The motility of the rumen protozoa is examined by keeping a drop of rumen liquor on the microscopic slide, covering it with cover slip and examining it under low power (10 x 10). The motility of the protozoa is graded as 0 (no activity or very weak or sporadic), + (sluggish movement and very few number of protozoa), ++ (good motility with little crowding) and +++ (highly motile protozoa with very good crowding). Normally fauna or protozoa are motile. Sluggish or no motility of protozoa is abnormal.

(b) *Classification of Protozoa*- Freshly prepared and fixed film of rumen liquor is stained with Lugol's Iodine. The protozoa are categorized for clinical purpose on the basis of their size (Fig.84), morphology, number and size of flagellae, presence of vacuoles macro or micro nucleus, skeletal plates and various pines and projections. Iodine staining of rumen protozoa for idophilic polysaccharide, though reflects on the state of carbohydrate digestion in animals, is not of much clinical significance. Iodophilic activity of protozoa is identified by black coloration of starch contents in the protozoa (Fig.85). It can be graded as 0, +, ++, +++ depending on quantity of starch contained. The quantum of starch content of protozoa reflect the degree of digestive disease.

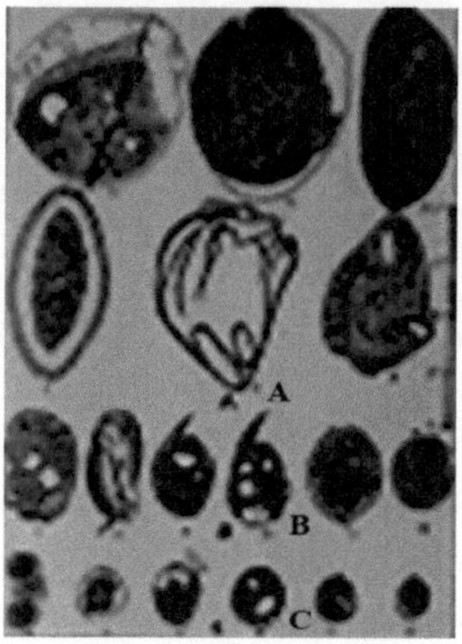

Fig. 84. Drawing of rumen protozoa and their sizes. A. Large B. Medium C. Small.

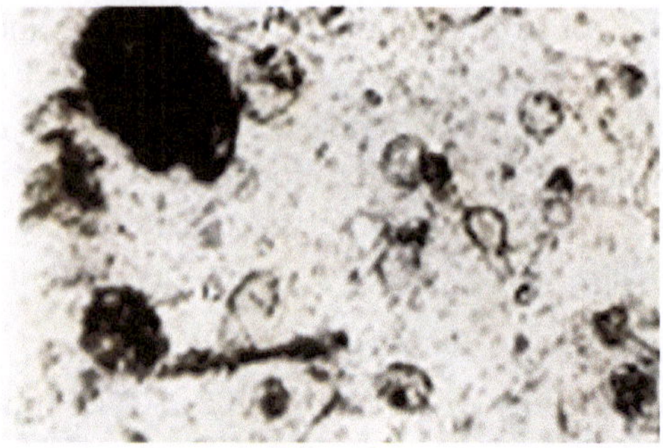

Fig. 85. Microphotograph of rumen protozoa stained with Lugol's iodine (x 250).

(c) *Examination for Bacteria*- There are numerous type of bacteria in the rumen such as those break down cellulose, starch, sugar, protein (proteolytic) or those producing methane. Slides prepared from rumen fluid sediment (filtered, centrifuged, discard supernatant and use sediment for slide) are air dried and stained with Grams stain (Fig.86), Iodine, PAS, nigrosine, Congo red for bacterial identification. Normally there is preponderance of Gram–ve microflora in rumen liquor of healthy cows (85B). In rumen acidosis there is preponderance of Gram +ve bacteria (Fig. 86 A).

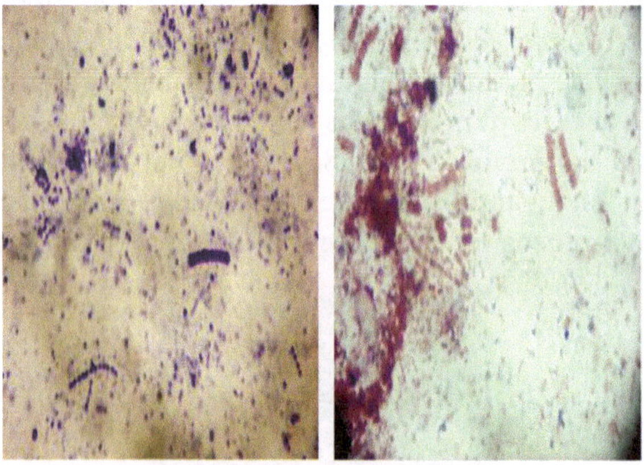

A **B**

Fig. 86. Showing preponderance of Gram +ve microflora in rumen liquor of a cow suffering from rumen acidosis (A). Normally there is preponderance of Gram –ve microflora in rumen liquor of healthy cows (B).

(d) *Examination for Fungi*- Fungi are present as yeast. Take a drop of supernatant fluid on the slide, mix it with a drop of Iodine and apply cover slip over it and examined under microscope

10

Clinico-pathological Evaluation of the Liver

Liver is one of the largest and most sensitive organ in the body performing important role in digestion and metabolism; regulation of nutrient; and production of proteins, enzymes, hormones, clotting and immunological factors. Liver also plays significant role in detoxification. Liver is affected in both primary and secondary liver diseases. Clinical signs of the hepatic dysfunction appears only after a large portion of the liver (about 75%) is inactive because liver has a large reserve of function. Therefore, a systematic clinico-pathological examination of the liver becomes mandatory to identify liver abnormalities.

Clinical Signs of The Liver and Biliary Disease

Jaundice (pre hepatic, hepatic, post hepatic or extra hepatic), nervous signs (hyperexcitability, convulsions, coma, weakness, muscle tremors, dullness, compulsive walking head pressing, mania), edema and emaciation, diarrhea or constipation, photosensitization, hemorrhagic diathesis, abdominal pain, alteration in liver size (enlargement or reduction), displacement of liver (diaphragmatic hernia) or rupture of liver (accidental) are associated with diseases of liver and biliary system. The clinical manifestations are vague and many of them are nonspecific making the definite diagnosis a difficult task. Clinical manifestations generally create suspicion of liver disease, its confirmation calls for special examination of the liver.

Special Examination of the Liver

Special examination of the liver is conducted employing the technique of liver palpation, biochemical tests of liver function, radiography, sonography and biopsy.

Palpation and Percussion

Normally liver is not palpable in large ruminants. The rounded and thickened edge of the enlarged liver can be felt behind the costal arch on right side.

Percussion of the area of liver dullness gives just a speculation of the liver size. Deep percussion or palpation over the area of liver dullness in posterior thoracic region on the right side can be performed to detect the pain, if any. Palpable enlarged liver in cattle may be due to right sided congestive heart failure, multiple abscesses in the liver or diffuse hepatitis.

Liver Biopsy

Liver biopsy can be obtained by blindly inserting a True -Cut type biopsy instrument (14G, 4-6 inches for bovines) in 10th intercostal space at a point at which imaginary line from tuber coxae to elbow intersect the 10th intercostal space on the right side (black line in Fig.87). An alternative way to find the site is in 10th intercostal space at a point at which imaginary line from the greater trochanter (hip) and parallel to the ground (yellow line Fig. 87) intersects the 10th intercostal space (Fig. 87). Usually, there is a space of about two inches within the 10th intercostal space and between these two imaginary lines (red triangle). Anywhere within that space is suitable for the biopsy site. The site is prepared as for surgical intervention and 2-4 ml of lidocaine (xylocaine) is infiltrated under the skin and into the intercostal muscles at the site. A stab incision (using Bard Parker blade) is made to accommodate biopsy needle. Needle is inserted parallel to the ground slightly angled forward towards shoulder. The needle penetrates the skin, intercostal muscles, and diaphragm before entering the liver. There is a popping sensation as the needle punctures the diaphragm. After taking biopsy, needle is withdrawn. The procedure is generally safe. Nevertheless there is a small risk of inducing necrotic hepatitis.

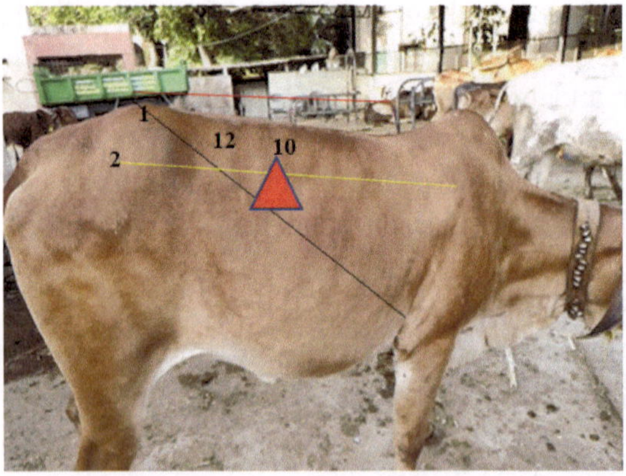

Fig. 87. Showing the site for insertion of the biopsy needle (red triangle). 1. Tuber coxae (hook).2.Greater Trochanter. 12 and 10 intercostal space. Black line is joining the point of tuber coxae to elbow. Yellow line is imaginary line from greater trochanter drawn parallel to the ground.

Radiographic Examination

Radiographic examination of liver in large animals is generally limited to research investigation.

Ultrasonographic Examination

Ultrasonographic examination (Fig.88) is being increasingly used to assess the liver with respect to its size, position, and parenchymal pattern.

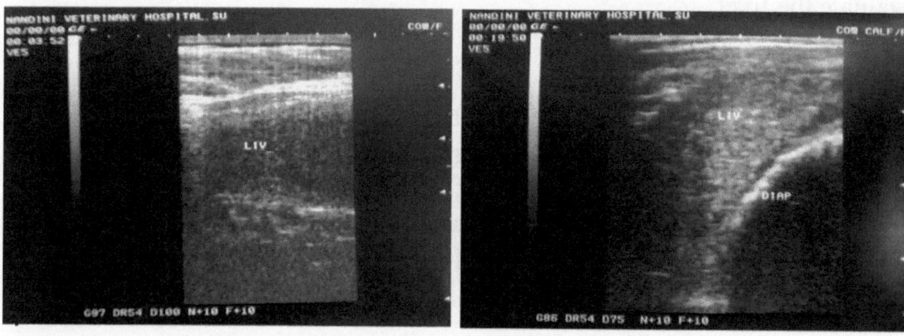

A B

Fig. 88. Ultrasonogram of the liver of cow calves showing normal liver echo texture (A) and hperechoic changes in the echo texture (B).

Biochemical Tests

Biochemical test are of diagnostic value in hepatic disorders as clinical manifestations are of very limited value. There is no specific test that can specify exact nature of hepatic lesion. Therefore combination of tests are conducted to arrive at the diagnosis. Serum bile acid, arginase and gamma -glutamyl transferase are sensitive indicators of cholestasis and/or hepatocellular necrosis in horses. In cattle, SDH (sorbitol dehydrogenase), GGT (gamma-glutamyl transferase), SGOT (serum glutamic-oxaloacetic transaminase) or AST (aspartate amino transferase) and BSP (bromsulphalein) clearance are sensitive indicators of hepatocellular insult. Not a single test provides complete information. Therefore a battery of test is needed to reach a definite conclusion.

Laboratory tests, used in the diagnosis of liver diseases, fall in four categories viz. 1. Those measuring excretory rate of parenterally administered substance (such as bromosulfalein), 2. Those measuring liver ability to detoxify substances, 3. Those measuring changes in the level of liver enzymes, and 4. Nonspecific tests providing indirect assessment of hepatic function (such as blood glucose, serum proteins, clotting factors and urine analysis).

BSP Clearance test

This test was developed originally by Rosenthal and White (1925). It has been used in cattle, sheep and horses. BSP solution (40 mg/ml of 0.9% sodium chloride sol.) is injected intravenously (@ 2mg/kg B.Wt.) Time needed to reduce original BSP level to half is taken as standard BSP half time. It varies from 2.5-5.5 minutes in healthy cattle, around 2 minutes in healthy sheep and horses. In goat BSP test along with OCT and prothrombin time can be used to evaluate the liver.

Icteric Index

Though it is not considered as liver function test, it is used to measure the degree of jaundice. The color of the plasma is compared with standard solution of potassium dichromate. The color of plasma of different species of animal varies, depending on carotene concentration, from pale in sheep, deeply colored in horse to less deeply colored in cattle. Despite hyperbilirunemia in cattle, marked jaundice in infrequent.

Serum Enzymes

Sorbitol dehydrogenase (SDH) is most selective and preferred test for detecting liver insult. Its level varies from 0-8.0 U/L in horses, 32 ±26 U/L in cows. The enzyme level is low in healthy horses and cattle. Its increase is an indication of hepatocellular injury or necrosis (reversible or irreversible, diffuse or focal). Half life of the enzyme is short (about 4 hrs.). Continuous increase in SDH level indicates on going hepatocellular injury or necrosis. It is a good indicator of liver damage in early stage in cattle and sheep. It is also a preferable test in detecting liver insult in horses.

Aspartate aminotransferase (AST) previously termed as SGOT or L- alanine aminotransferase (ALT) previously termed as SGPT do not have much specific diagnostic value of liver damage in cows.

Arginase is a specific indicator of liver damage and is better option in cattle as its level in other organs is not appreciable. Because of short half-life, it is a good indicator of an acute liver disease.

Though, variations occur in the level of serum alkaline phosphatase (SAP), wide variations in normal cattle make the interpretation difficult. It is used to evaluate excretory function of the liver in horses. For biliary obstruction, serum alkaline phosphatase test is preferable.

GGT is another enzyme that is widely distributed in equine kidney, pancreas and liver. It is used for the diagnosis of hepatobiliary diseases in cattle, sheep and horses. GGT level can fluctuate widely in horses.

GLDH (glutamate dehydrogenase) is present in ruminants liver in high concentration and increases with liver damage. Therefore it can also be used to evaluate liver in ruminants.

OCT (ornithine carbamoyl-transferase) is another liver specific enzyme that increases even in chronic liver disease with active necrosis (not in healing phase).

Serum bile acid

Bile acid is the end product of cholesterol metabolism by the liver and has been used to detect liver damage in humans. It has shown promise in detecting liver damage in experimental chronic copper poisoning in sheep. It is also a sensitive and specific indicator of wide variety of liver diseases in cattle and horses. Diurnal variations occur in the level of serum bile acid in healthy cattle.

References

Rosenthal, S.M. and White, E.C. (1925). Clinical application of bromsulphalein for hepatic function. J.Am. Med.Assoc. 84:1112.

11

Special Examination of Urinary System in Farm Animals, Dogs and Cats

Farm Animals and Horses

Urinary system mainly comprises of kidneys, ureters, bladder and urethra. In farm animals, diseases of urinary bladder and urethra are diagnosed with more frequency than that of kidneys. Renal insufficiency is observed as a sequel in cases of nephritis (pyelo or embolic nephritis), nephrosis or amyloidosis. Urinary diseases are clinically manifested with pain, dysuria, and variation in urine flow (polyuria, oliguria to anuria) and changes in urine (proteinuria, hematuria, hemoglobinuria, myoglobinuria, glycosuria, ketonuria, crystalluria, pyuria, and cellular casts). Decreased body weight, loss of appetite, polyuria, polydipsia and ventral edema are seen in chronic renal failure. Like dogs and cats, urinary tract is not accessible in large farm animals for clinical examination. Trans abdominal palpation of kidneys is feasible in calves and small ruminants. Rectal palpation, urine examination and renal function tests provide diagnostic assistance in the diseases of urinary tract in large animals.

Clinical Manifestations

Dysuria and abdominal pain are generally seen in urinary tract disease. Acute pain from urinary tract disease is not common. Pain associated with urethral obstruction and bladder distension is manifested by tail switching, kicking at belly, and straining. Pain in cystitis, vesicle calculi and urethritis is accompanied by frequent small volumes of urine with grunting.

Rectal Palpation

Rectal examination in cattle, buffaloes (Fig. 89) and horses can provide valuable information about the urinary tract (size and pain). Left kidney is palpable in cows and buffaloes. In horses, caudal pole of the left kidney can be palpated. Empty urinary bladder and ureters are not palpable. However, distended urinary bladder with thickened wall can be felt in the midline at the anterior end of the pelvic cavity.

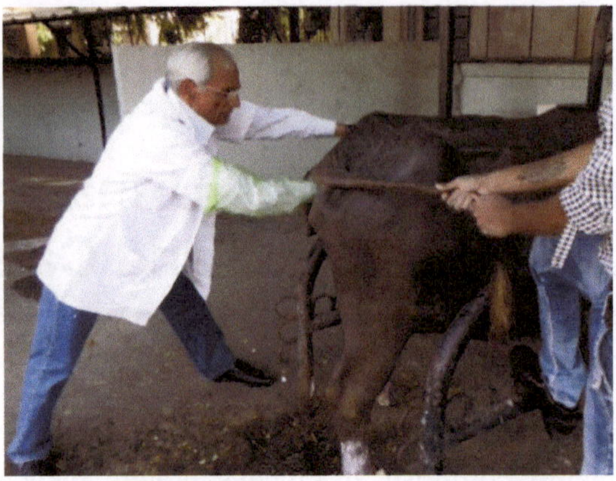

Fig. 89. Rectal examination in a buffalo. Distended urinary bladder, thickened wall urinary bladder and left kidney may be palpated.

Catheterization

Catheterization in young calves, rams, boars is not feasible. Cows and mares can be catheterized with rigid small diameter catheters. Male horses can be catheterized when penis is relaxed. Female goats can be catheterized (Fig.90).

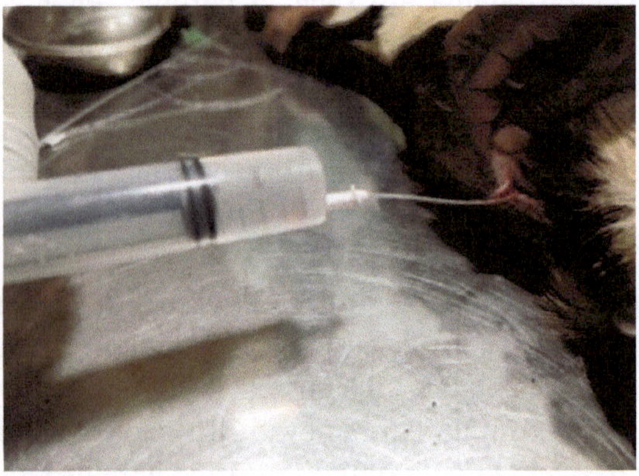

Fig. 90. Catheterization of a female goat.

Analysis of Urine

Urine analysis is an important part of special examination of the urinary system. Collection of urine is a bit difficult in large animals. For microbial examination, catheterized urine sample is preferable.

Proteinuria is also called as albuminuria and is seen in cases of urinary tract infection, glomerulonephritis, renal infarction, tubular nephrosis and amyloidosis. In cases of fever and toxemia, a small amount of protein in the urine can be detected. Highly alkaline urine in herbivores can yield false positive reaction to protein on urine dipstick test.

The presence of casts (tubular structures) in the urine indicates inflammatory or degenerative diseases of the kidneys and may not be seen in all cases of renal disease. These are better detected in fresh urine samples as these may be dissolved in highly alkaline urine.

Hematuria can occur due to pre-renal (trauma to kidneys, septicemia or purpura hemorrhgica), renal (acute glomerulonephritis, pyelonephritis, renal infarction, renal artery embolism) and post renal (cystitis or urolithiasis) causes. Enzootic hematuria in cattle is due to urinary bladder tumors owing to bracken fern ingestion. Microscopic hematuria is detected on the microscopic examination of the urine. Blood in urine gives positive reaction to biochemical tests for hemoglobin and myoglobin. False hemoglobinuria may occur in cases of hematuria because of lysis of erythrocytes but microscopic examination may detect RBCs in the urine. Urine is of deep brown color with positive reaction to biochemical test for hemoglobin and urine sediment is without RBCs debris in cases of true hemoglobinuria. Urine with hemoglobinuria is generally negative to Dipstick test for protein unless concentration of hemoglobin is very high in the urine. Hemoglobinuria is due to intravascular hemolysis and it occurs only when plasma levels exceed the renal threshold.

Myoglobinuria is suggestive of severe muscle damage as seen in horses with azoturia. Myoglobinuria is associated with clinical and biochemical changes of acute myopathy.

The presence of pus or leukocytes in the urine is termed as pyuria and is indicative of inflammation or infection of kidneys or urinary bladder. It is detected on microscopic examination of urine sediment revealing higher number of WBC per high power field. Culture examination of the urine with pyuria usually reveals the growth of the organism.

Crystalluria without clinical signs of irritation and large amount of crystals is of no special significance in herbivores as crystals of calcium carbonate and triple phosphate are usually seen in normal urine. Large number of crystals are suggestive of concentrated urine (dehydration) and development of urolithiasis in future.

Glycosuria with ketonuria in dogs is generally seen in diabetes mellitus that is uncommon in large animals. Glycosuria is detected in ruminants in cases of enterotoxaemia due to *Clostridium perfringens* type D or in cases of acute tubular necrosis. Parenteral administration of glucocorticoids, adrenocorticotrophic hormone or dextrose solution is also associated with glycosuria

Ketonuria is more common than glycosuria in ruminants and is seen in diseases such as starvation, acetonemia in cows, and pregnancy toxemia in ewes and does. Lactating cows may show a small amount of ketonuria.

Urine Flow

Determination of exact amount of urine flow per day is practically not feasible in large animal practice. Normal daily production of urine is variable and depends on many factors. Frequency of the urine may be ascertained.

Renal Function Tests

Renal function tests fall in four categories viz. test conducted on urine, tests conducted on blood, test conducted on urine and blood and specialized techniques (radiography, ultrasonography, and endoscopy)

Tests conducted on Urine

i. Specific gravity of urine is the simplest test to monitor functioning of renal tubules (fluid conservation and solute excretion). In normal animals specific gravity of the urine varies between 1.028 and 1.032. It decreases in chronic renal diseases.
ii. Urine osmolality, an alternate to urine specific gravity, measures solute concentration in the fluid and is a direct assessment of renal tubular function of conservation and excretion.
iii. Changes in urine enzymes, not functional test, are an early indicator of tubular degeneration.

Tests conducted on Blood

- Blood urea (BU) or blood urea nitrogen (BUN) and serum creatinine are the important tests of renal function especially glomerular filtration. Increase in the concentration of both indicators occurs when 60 to 75 % nephrons are dead or inactive. Both indicators may increase in prerenal uremia. In cows, prerenal lesions may cause higher increase in blood urea than in cases of renal disease. Glomerular filtration rate is more valuable test of renal function.

Tests conducted on Urine and Blood

- Water deprivation and Fractional excretion of filtered sodium are conducted on urine and blood.

Imaging Techniques

Radiography has a limited diagnostic value in urinary tract diseases in large animals. Endoscopy and ultrasonography has been described in the evaluation of urinary tract diseases but are not very common.

Biopsy Technique

Percutaneous renal biopsy can be undertaken in cows and horses. In horses ultrasound guided biopsy can be used for collecting kidney sample.

Dogs

History

History of dogs with chronic renal failure is nonspecific. Anorexia, polydipsia, polyuria, weakness and loss in body weight are generally narrated by the owners. The disease is more common in geriatric dogs.

Clinical Manifestations

Generally clinical symptoms of chronic renal failure are inapparent or very mild. Isothenuria (urine with an osmolality equal to that of plasma) and azotemia (increased concentration of non-protein nitrogenous substances such as creatinine and urea in the blood) do not develop until 65 to 70 % nephrons are dysfunctional or damaged. Polyuria, polydipsia, loss of weight, lethargy, vomiting and diarrhea are commonly seen in cases of chronic renal failure. Physical examination may show change in body condition score, dehydration, heart murmurs, hypertension, and anemia. Fundus examination may reveal vessel tortuosity and retinal detachment owing to systemic hypertension. Hematochezia or melena on rectal examination may be evident owing to gastrointestinal uremic ulcers.

Biochemical Indicators

Though serum creatinine and blood urea or blood urea nitrogen are important biomarkers of renal damage, their levels are also influenced by some external factors. Muscle wasting decreases creatinine level. BUN level increases in patients with gastric bleeding and on high protein diet; and decreases in dogs with mal nutrition or on protein restricted diet. Pre renal (decreased renal perfusion in dehydration, hypovolemia or hypotension) or post renal (ureteral obstruction) factors do increases azotemia (nitrogenous waste products blood urea, creatinine).Glomerular filtration rate is the gold standard test for renal function but not routinely used in clinical practice. Creatinine (mainly) and to lesser extent blood urea nitrogen are correlated with glomerular filtration.

Recently SDMA (serum symmetric dimethyl arginine) and CysC (cystatin C) has shown promise in detecting renal insult at an early stage. Of these CysC is the biomarker that can detect the earliest stage of CKD i.e. Stage 1 (Kim *et al.*, 2020).

International Renal Interest Society (IRIS) has recommended staging of renal failure based on serum creatinine and or SDMA as stage 1 (Dogs creatinine < 1.4 mg/dl, SDMA < 18 µg/dl; Cats creatinine <1.6 mg/dl, SDMA <18.0 -µg/dl), stage 2 (Dogs creatinine < 1.4-2.8 mg/dl, SDMA < 18.0-35.0 µg/dl; Cats creatinine <1.6 – 2.8 mg/dl, SDMA <18.0- 25.0 µg/dl), stage 3 (Dogs creatinine < 2.9-5.0 mg/dl, SDMA < 36.0-54.0 µg/dl; Cats creatinine <2.9 – 5.0 mg/dl, SDMA <26.0-38.0 µg/dl), and stage 4 (Dogs creatinine >5.0 mg/dl, SDMA >54.0 µg/dl; Cats creatinine >5.0 mg/dl, SDMA >38.0 µg/dl).

Ultrasonography

It is a valuable tool in diagnosing and staging renal failure in dogs and cats. The most important ultrasonography feature of chronic renal failure/ chronic renal disease is diffuse increase in renal cortical hyper-echogenicity (Fig.91), abnormal ratio of cortico-medullary junction and pyelectasia .These abnormalities increase significantly with the progression of IRIS stage (Perondi *et al.,* 2020). Detection of renal and urinary calculi is facilitated by ultrasonography.

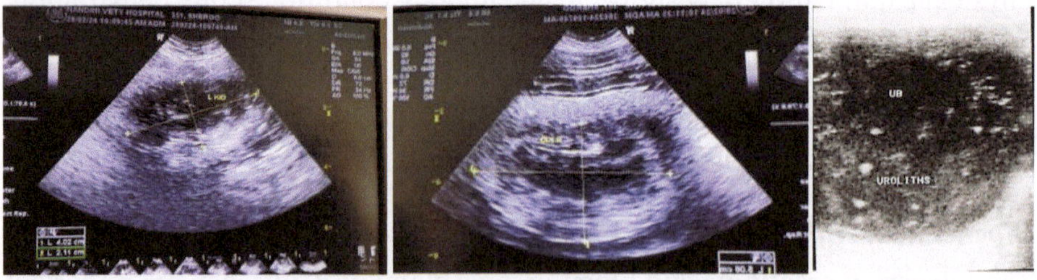

A B C

Fig. 91. Ultrasonograms of dogs showing diffuse increase in renal cortical hyper echogenicity of both kidneys (A and B) suggesting chronic renal disease; and calculi in the urinary bladder ©.

Ultrasonography can assist in differentiating acute and chronic renal disease. Chronic renal disease is characterized by small kidneys with irregular margins, hyper echoic cortices and poor corticomedullary differentiation. Whereas acute kidney disease is characterized by kidneys of normal or large size with normal architecture, hydronephrotic, and hyperechoic sonographic changes. Acute kidney disease is generally associated with perinephric fluid.

Radiography

It may be helpful in detecting stones or tumors.

Urine Analysis

Urine analysis is an important diagnostic technique that helps in diagnosing kidney and urinary tract diseases. Urine is subjected to routine examination including microscopy; and culture examination including antibiotic sensitivity. Routine urine analysis includes tests for specific gravity, pH, color, turbidity, glucose, ketone, bilirubin, blood and protein. Urine sediments are examined microscopically for pus cells, erythrocytes, crystals, other cells, casts and organisms. Urine in sterile container is used for culture examination and antibiotic sensitivity. For urine analysis fresh urine sample is preferable. If delay is expected, urine samples should be kept in refrigerator.

Normal urine is transparent and of yellowish or amber color. Its intensity depends on the volume of the urine produced and concentration of various products. Abnormal urine color may be caused by presence of endogenous or exogenous pigments. Red color of the urine (Fig. 92) is due to the presence of RBCs in the urine. Coffee colored urine may be due to the presence of hemoglobin.

Fig. 92 .Showing red colored urine. Microscopic examination of the sample revealed the preponderance of RBCs (Courtesy Varshney, 2025).

Normal urine smells like ammonia. Bacterial infection may cause strong smell due to pyuria.

The urine specific gravity (SG) is highly variable depending on the fluid and electrolyte balance of the body. In healthy dogs it is 1.015. It increases in dogs with dehydration. Urine with low specific gravity in dehydrated or azotemic dogs may be due to renal failure, hypo- or hyperadrenocorticism, hyperkalemia, diabetes mellitus, hyperthyroidism. Specific gravity of the urine is also low in dog being treated with diuretic therapy. It also remains low (1.010) in dogs with diabetes insipidus. In diabetic dogs with glycosuria, specific gravity of the urine increases despite increased urine volume.

pH of the urine of healthy dogs is acidic and is influenced by diet, drugs and disease. Urinary infection with urease producing organism leads to alkaline urine pH.

No protein is detected in the urine of healthy dogs and cats. Proteinuria (presence of protein in the urine) can be due to pre-renal (fever, strenuous exercise, hyperthermia/heat stroke, hyperproteinemia or seizures), renal (nephritis) or post renal (inflammation, hemorrhage or infection) etiology. Whenever proteinuria is detected, it is to be interpreted along with the findings of specific gravity, urine pH and sediment microscopy. Alkaline urine may give false positive reaction and the presence of Bence –Jones protein may give false negative reaction.

Normally glucose is not present in the urine of healthy dogs and cats because renal threshold for glucose is around 180 mg/dl in dogs and 240 mg/dl in cats. With euglycemia, the amount of filtered glucose is less than the renal threshold, and all of the filtered glucose is reabsorbed in the proximal renal tubules. The presence of glucose in the urine (glycosuria) can be ascribed to hyperglycemia (owing to diabetes mellitus, excessive endogenous or exogenous glucocorticoids, or stress) or from a proximal renal tubular defect (primary renal glucosuria or Fanconi syndrome). Whenever glucosuria is detected, blood glucose level should also be monitored. High concentration of ascorbic acid or formaldehyde in urine may give false negative results. The contamination of urine sample with hydrogen peroxide, chlorine, or hypochlorite (bleach) may give false positive results.

Ketone bodies are not detected in the urine of healthy dogs. Routine test for ketone detects acetone and acetoacetic acid but beta- hydroxybutyrate is not detected. The presence of ketone bodies in the urine (ketonuria) is suggestive of ketosis (secondary to diabetes mellitus in dogs), low carbohydrate diet in cats or prolonged fasting/starvation.

The presence of bilirubin in the urine (bilirubinuria) indicates that the amount of conjugated bilirubin has exceeded the renal threshold. It may occur in liver disease or hemolysis. A small amount of bilirubin in the concentrated urine may be normal. False positive result may be associated with pigmenturia. Urine containing large amount of ascorbic acid may yield false negative result.

Healthy dogs may normally have a small amount of urobilinogen (a product formed from bilirubin by intestinal microflora). Its level is increased in dogs with hyperbilirubinemia and it may absent in cases of biliary obstruction. Occult blood test detects intact erythrocytes, hemoglobin and myoglobin. A positive occult blood test indicates hematuria, hemoglobinuria or myoglobinuria.

For microscopic examination, the urine is centrifuged at 1,000–1,500 rpm for 3–5 min. The supernatant is decanted leaving a very small volume of urine (0.5 ml) and the sediment. The sediment is suspended in the small volume of the urine by tapping the tip of the conical tube against the table several times. A few drops of the sediment are transferred on to a glass slide and covered with a cover slip and examined under low (for crystals, casts and cells) and high power (for cells and bacteria) of the microscope. New methylene blue or modified Wright stain can be used to identify cells in air dried slides. Erythrocytes in unstained slides appear as small and round bodies having a slight orange tint and smooth appearance (Fig.93). Number of erythrocytes in urine of healthy dogs remain <5 per high power field. Increased number of erythrocytes in the urine is suggestive of hematuria. It may be due to urinary infection and or crystalluria. Urine collection by catheterization or cystocentesis may also cause hemorrhage. White cells in the urine are also termed as pus cell (Fig.93). Their number in the urine of healthy dogs remains limited to 4-5 per high power field. Increase in white blood cells (leukocytes) per high power field in the urine is suggestive of urinary tract inflammation, infection, trauma or tumor. Transitional epithelial cells may be detected in urine as a contaminant or due to transitional call carcinoma. Squamous cells may be seen in cases with squamous cell carcinoma. Casts (Fig.93) are elongated cylindrical structures. Hyaline casts, epithelial cellular casts, granular casts and waxy casts may be seen. Hyaline casts are seen in dogs with fever, exercise and renal disease. Epithelial cellular casts indicate renal tubular disease. Fatty casts are not common but may be seen with disorder of lipid metabolism. Few hyaline or granular casts are considered as normal but their large number is suggestive of renal disease. The presence of organisms in aseptically collected urine indicates infection. Small number of the organism may be due to contamination from lower urogenital tract. Organisms can be confirmed by staining as well as culture examination. Sometimes fungal hyphae, yeast and parasitic ova may be seen in the urine sediment. Crystals (Fig.93) in the urine suggests crystalluria. Crystalluria and urolithiasis are not the same. Uroliths may be without crystalluria. Crystals in the urine are influenced by the pH of the urine, urine temperature, concentration of the crystallogenic material and the time lapse between urine collection and examination. In dogs, urine crystals of struvite (Fig.93), calcium oxalate (Fig.93), urate, cysteine (Fig.93), tyrosine (Fig.93), biurate (Fig.93) and bilirubin are commonly seen. Fat droplets (Fig.93) in the dog's urine can be mistaken as erythrocytes. But these vary in size and tend to float on a different plane of focus. These droplets are not pathological. In the urine of male dogs detection of spermatozoa is a common finding.

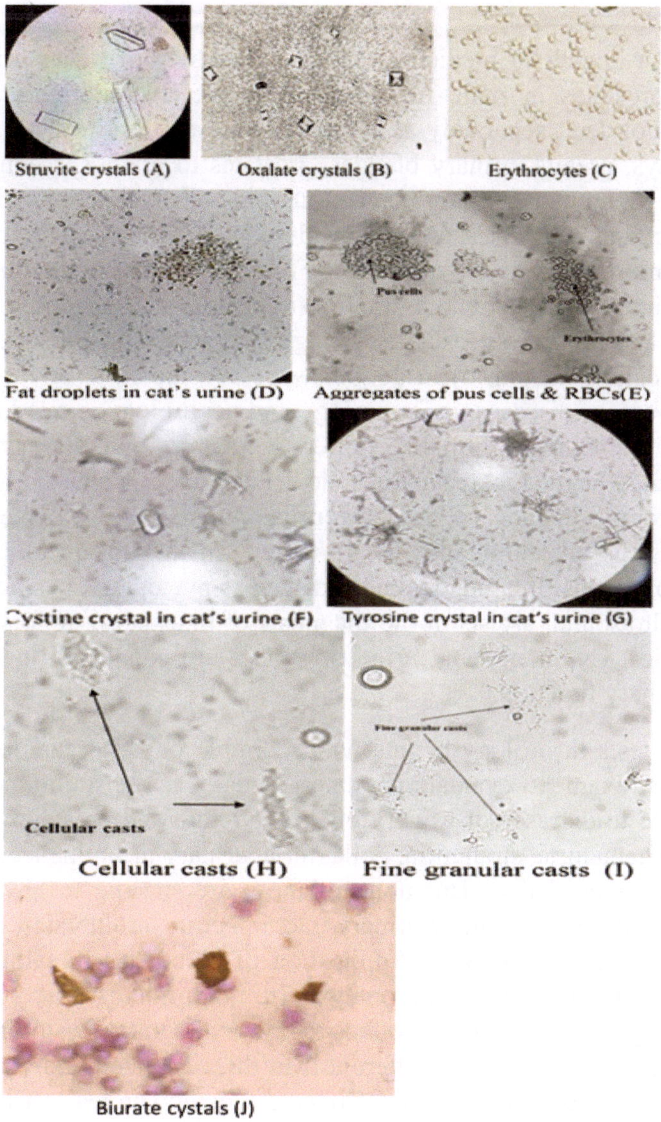

Fig. 93. Microscopic examination of urine sediments showing struvite crystals (colorless three dimension prism –like crystals), oxalate dehydrate crystals (colorless square whose corners are connected by intersecting line- looking like an envelop), erythrocytes (small biconcave without internal structure), fat droplets (round, variable size and refractive), aggregates of pus cells (round approximately 1.5 to 2 times of the size of RBC, have refractive internal granulity) and erythrocytes, cystine crystal (colorless hexagonal shape with equal or unequal sides) tyrosine crystal (yellowish needle like massed together), cellular casts, fine granular casts and biurate crystals (yellow to brown spherules with irregular projections- thorn apple appearance). Figures A, B, C,D, E and J are with the courtesy of Varshney,(2025).

Urology

Urology is a branch of medicine related to diseases of lower urinary tract including prostates in male in human medicine. In veterinary medicine, urology is not dealt separately as a specialty but entire urinary system right from kidneys, ureters, urinary bladder, prostates to urethra is considered as one system. In humans, urinary bladder, urethra and prostate falls under lower urinary tract and their diseases are considered under the specialty of urology. With the same analogy, diseases of urinary bladder (cystitis, crystalluria, calculi in U.B., urinary bladder tumor), urethra (urethritis, stenosis due to calculi or neoplasm) and prostate (prostatitis, benign hyperplasia of prostate, prostatic abscess, prostatic neoplasm) in dogs may be considered under urology. But symptomology of diseases under urology and nephrology is overlapping and almost similar making it difficult to differentiate upper and lower urinary tract involvement. Diagnostic techniques described under nephrology are also employed for the diagnosis of the lower urinary tract diseases. Urine collection and blood tests are commonly the first steps in diagnosing and screening urologic conditions. Imaging techniques,such as pyelogram, cystography, CT scan and ultrasound, assists in the diagnosis of blockages, tumors and other abnormalities. Cytometry and urine flow tests are conducted in humans to monitor urinary function.

Diagnostic tests in urology begins with a simple physical examination and digital rectal exam (to evaluate prostate glands). The most common clinical signs related to diseases of urinary bladder, urethra and prostates (in males) in dogs are hematuria, stranguria, pollakiuria, urinary incontinence, or urinary scalding (Bartges, 2004). But none of these are specific to any particular disease and fail to differentiate urinary tract infection, urolithiasis, micturition disorders or prostate diseases. Application of blood test (creatinine, BUN, prostate specific antigen), urine analysis (routine, microscopic and culture) and imaging techniques (ultrasound, x-ray, pyelogram, cysto-urethrogram and cystoscopy) is in no way different than that described under nephrology for the diagnosis of renal diseases. Prostate specific antigen test is primarily used to screen for prostate cancer. Radiography can detect well-formed calculi (Fig.94). Ultrasonography is of great advantage in the diagnosis of cystitis, bladder tumor, bladder calculi, crystalluria, hematuria, and prostate diseases (Fig.95). Urine examination is imperative for detecting urinary tract infection, nature of crystalluria and occult hematuria.

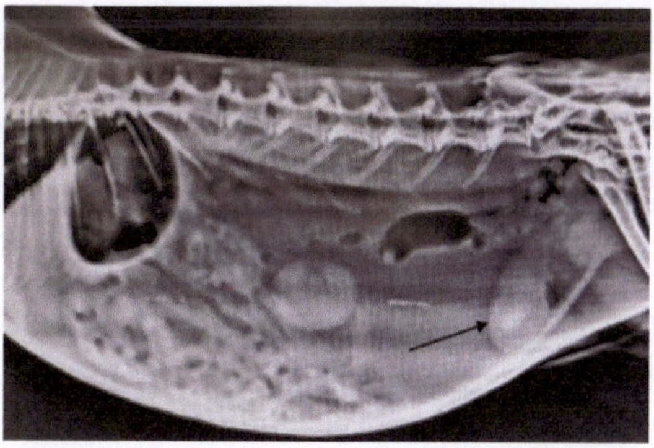

Fig.94. Radiograph showing calculi in the urinary bladder marked with black arrow (Courtesy Varshney, 2025).

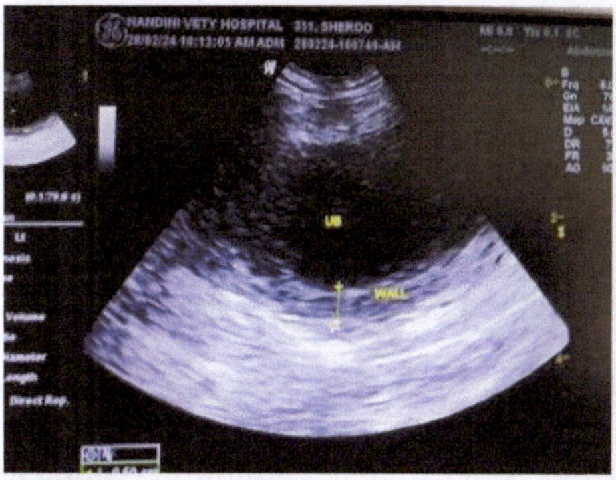

Cystitis

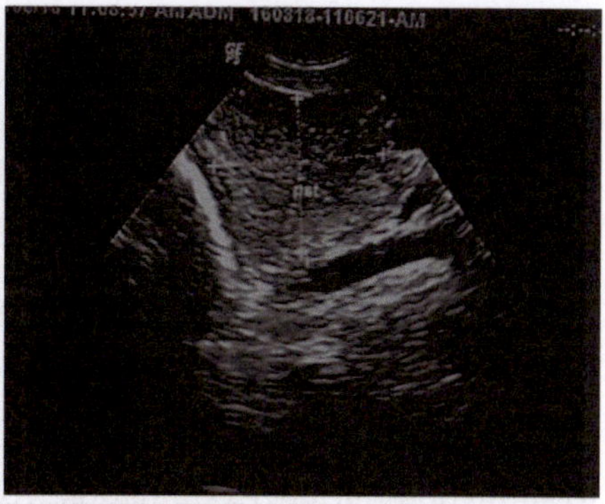

Benign Hyperplasia of Prostate

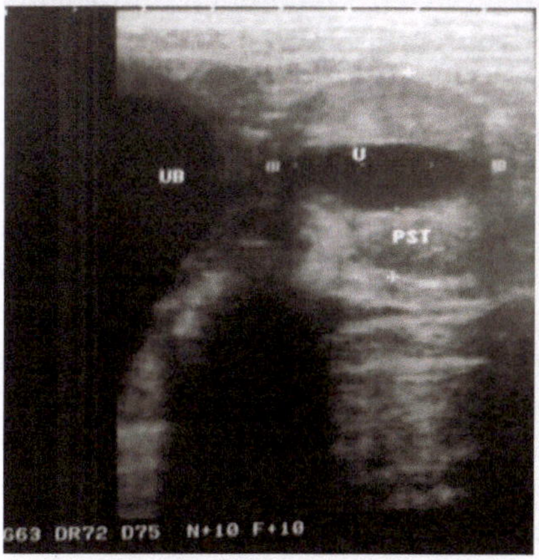

BHP pressing the urethra

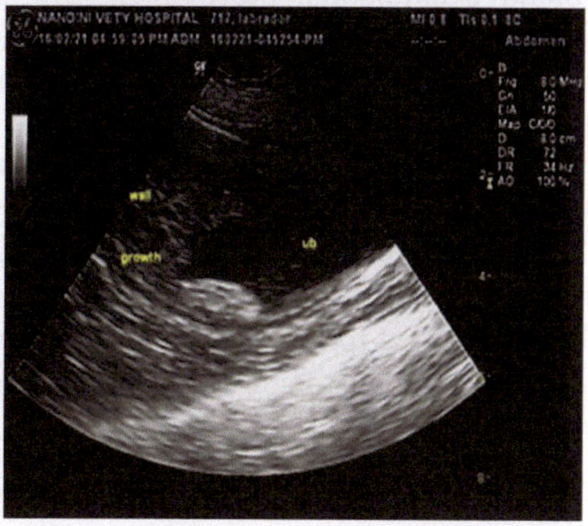

Tumor in Urinary Bladder

Fig. 95. Ultrasonograms of dogs showing different diseases (cystitis, benign hyperplasia of prostate and urinary bladder tumor). Figures are with the Courtesy of Varshney, 2025).

Cats

Feline kidneys are more moveable as compared to that of canine. Right kidney is more cranial but its cranial pole is not buried in the liver. In feline lower urinary tract diseases (FLUTD), affecting bladder and urethra, are very common and their incidence has been reported varying from 4.5 to 8.0% (Lekcharoensuk *et al.*, 2001; Longstaf *et al.*,2017).

Feline Lower Urinary Tract Diseases

Etiology of FLUTD varies from feline idiopathic cystitis (FIC),urethral plugs, crystalluria, urolithiasis, urinary tract infection (UTI), tumor, congenital defects (urethral stricture, urachal remnant), trauma, neurological disorders (reflec dyssynergia) to behavioural disorders. Because of similar clinical signs, a systematic and special examination of urinary tract is mandatory for etiological diagnosis. Himalayan cats are at an increased risk of urolithiasis. Male cats are prone to urethral obstruction because of narrow urethral diameter. Tumours are less common in cats. Cats with certain metabolic disorders, urolithiasis, and those undergone urinary tract procedures, such as catheterization, or perineal urethrostomy have an increased risk of urinary tract infection. In cases of feline idiopathic cystits, information related to cat's environment, diet, litterbox management, access to outdoor and stress is important in the management.

Examination approach for the diagnosis of urinary tract diseases in cats is almost similar to that of dogs described above starting from history, clinical examination, urine examination (routine, microscopic, culture), complete blood examination, biochemical evaluation (BUN, creatinine), ultrasound examination, radiographical examination (plain, urocystography, contrast urethrography) to cytology. In the absence of ultrasonography, excretory urography is of the routine diagnostic value. Instead of iodinated contrast being injected intravenously, it can also be injected directly into the renal pelvis (antegrade ureteropyelography) or into the ureter distal to an obstruction (retrograde pyelography). Urethrography is less commonly done in cats due to the short length of the penile urethra. Ultrasonography has indications similar to those listed for excretory urography but has the advantage of being able to display internal structure, allowing differentiation of solid and cystic masses.

A thorough physical examination is essential. A distended, painful bladder that cannot be expressed is an important clinical finding in cats with urethral obstruction. Cats with nonobstructive FLUTD generally have a small or minimally distended bladder.

Abdominal survey radiography is a valuable diagnostic aid in all cats with lower urinary tract signs as it can detect distended urinary bladder (Fig.96) and uroliths. Struvite and calcium oxalate liths are radiopaque and therefore assessment of their presence, location, number, and size is feasible on radiography.

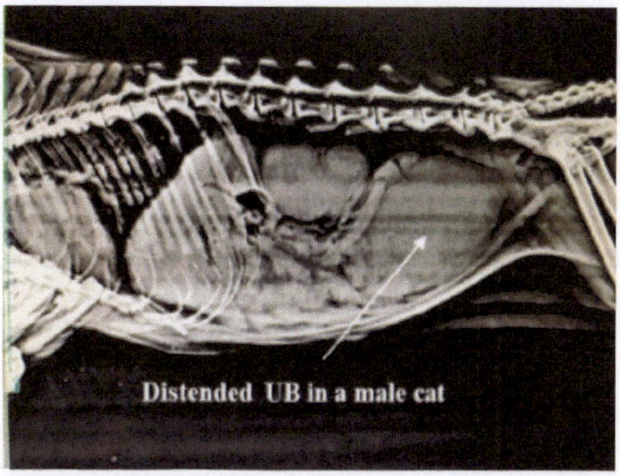

Fig.96. Abdominal radiograph of a male cat (right lateral recumbency position) revealing distended urinary bladder.

Contrast urethrography is the best imaging method to evaluate for urethroliths. It is important in cat with chronic, persistent lower urinary tract signs to rule out small stones before a diagnosis of feline idiopathic cystitis is arrived at.

Urine culture and antibiotic sensitivity testing is indicated in cases of pyuria and bacteriuria detected on urine analysis to confirm the etiological organism. In cases of FLUTD, urinary tract infection is relatively uncommon in cats.

Abdominal ultrasonography, concurrently with radiography, is of additional value to assess the size and number of uroliths. It can also detect thickened bladder wall (Fig. 97), urachal remnant, or bladder mass if any.

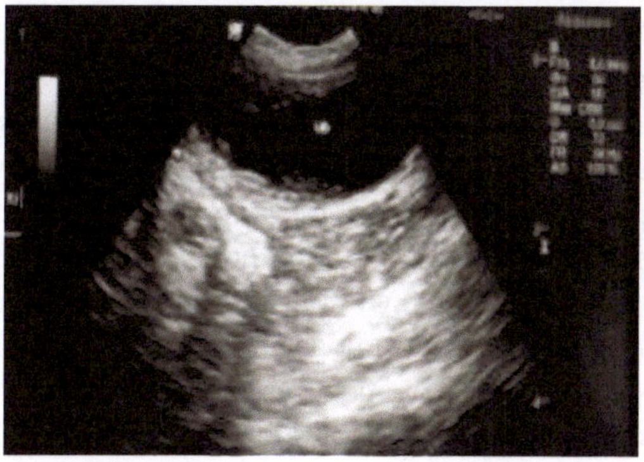

Fig.97. Transabdominal ultrasonograph of a male cat showing thickened urinary bladder wall (> 3 mm) suggestive of cystitis.

Cystoscopy is an advanced technique and can be performed at refferal centres or veterinary college clinics. It has limitations in male cats because of narrow urethra. It provides visualization only.

Struvite and calcium oxalate crystals (Fig. 98) are commonly seen in cats. Struvite crystals are more common in young cats and calcium oxalate crystals in middle-aged to older cats. Uroliths may occur with or without crystalluria. Struvite liths are associated with neutral to alkaline urine. In cats, struvite crystalluria or uroliths are not associated with urinary tract infection whereas in dog it is associated with urinary tract infection. Calcium oxalate crystals are associated with neutral to acidic urine.

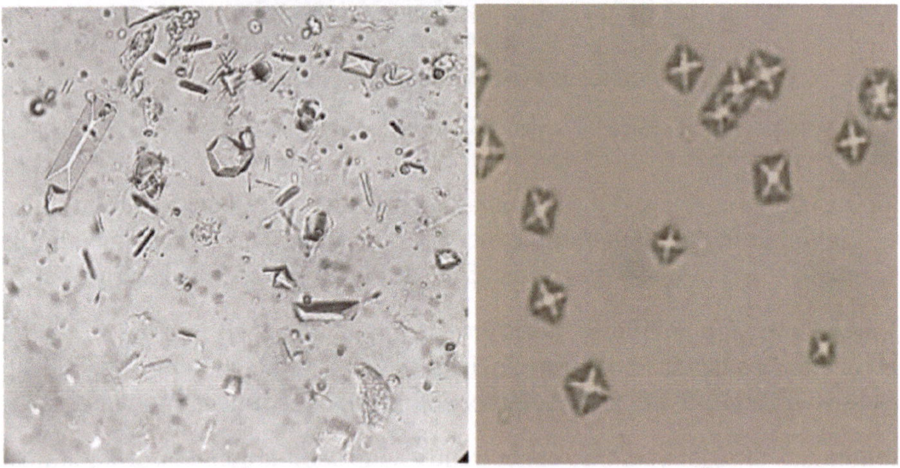

A B

Fig.98. Showing crystals of Stuvite **(A)** and calcium oxalate **(B)** in cats' urine.

Feline Idiopathic Cystitis

Feline idiopathic cystitis (FIC), idiopathic FLUTD or feline urologic syndrome occurs initially at 2 to 7 years with an acute signs of lower urinary tract inflammation that resolve spontaneously after 4 to 7 days in 80% to 90% of cases. No single test can lead to confirmatory diagnosis of FIC. Therefore diagnosis of FIC is based on exclusion of other etiologies of FLUTD. Negative culture, bladder wall thickening without any evidence of crystals / calculi/ or any mass on sonographic and radiographic examinations are suggestive of feline idiopathic cystitis. Anxiety and stress have been described as important factor in the genesis of feline idiopathic cystitis. FIC may have variable presentations, including urethral obstruction, frequently recurring episodes, or chronic persistent signs. Spontaneous resolution of clinical signs may be mistaken for response to empirical therapy with antibiotics leading to misdiagnosis.

Urethral obstruction is often due to urethral plugs consisting of a matrix of mucoprotein and inflammatory debris and aggregates of crystals (predominantly struvite). A large, firm bladder on abdominal palpation is a common finding in cats with urethral obstruction. Urethral catheterization (Fig.99) is necessary for collection of urine for urinalysis, culture, and sensitivity testing. Complete blood count and serum chemistry profile (including serum potassium level) and an electrocardiogram is also advisable to detect the associated hyperkalemia that can cause cardiotoxicity.

Fig. 99. Urine collection through catheterization in a male cat.

References

Bartges, J.W. (2004) . Diagnosis of urinary infections. Vet. Clin.-Small. Anim. 3:4923-4933.

Kim, J., Lee, Cang-Men. and Kim, Ha-Jung (2020). Biomarkers for chronic renal disease in dogs: comparison study. J.Vet.med.Sci. 82:1130-1137.

Lekcharoensuk, C., Osborne, C.A. and Lulich, J.P. (2001). Epidemiologic study of risk factors for lower urinary tract diseases in cats. J. Am. Vet.Med.Assoc. 218:1429-1435.

Longstaff,L., Gruffydd-Jones, T.J., Buffington, C.T. Casey,R.A. and Murray, J.K.(2017). Owner-reported lower urinary tract signs in a cohort of young cats. J.Feline Med. Surg. 19:609-618.

Perondi, F., Lipi, I., Marchetti,V.,Bruno,B., Borrelli, A., and Citi,S. (2020). How ultrasound can be useful for staging chronic renal disease in dogs: Ultrasound findings in 855 cases. Vet. Sci. 7(4):147. doi: 10.3390/vetsci7040147. PMID: 33019496; PMCID:PMC7712280.

Varshney, J. P. (2025). A Fully Illustrated Handbook on Clinical Diagnostic Processes in Canine Medicine. NIPA GENX Electronic Resources and Solutions P.Ltd. New Delhi-110 034.

Part II
Diseases of Cardio-Pulmonary, Nervous, Musculo-skeletal System, Eye and Ear

12

Special Examination of the Respiratory System

Besides physical examination, there are many diagnostic techniques for making a specific diagnosis of respiratory diseases. The specialized techniques are used by the specialists. These specialized examination techniques include collection of nasal and brochoalveolar samples either using nasopharyngeal swab or flexible fiber endoscopy and their laboratory examination; pleuroscopy; thoracocentesis; thoracic radiography; and thoracic sonography.

Collection of Respiratory Tract Discharges/Fluid

The respiratory tract discharges (nasal, bronchoalveolar fluid) can be collected using nasopharyngeal swabs and flexible fiber optic endoscope or flexible tube.

Nasopharyngeal swab collection

Nasopharyngeal swabs have been used to identify PI3 and IBR infections. *Pasteurella* isolates may be commensals. Bronchoalveolar samples are more exact than nasopharyngeal swabs in identifying the etiological agents of pneumonia in cattle. Long handled sterile swabs are used. Swabbing is to be done vigorously to obtain samples of mucosal cells required for virus isolation. Early clinical cases are preferable. Material for virus analysis from the swabs must be placed in viral transport medium.

Trachio-bronchial/ broncho-alveolar lavage (BAL) collection

This is a simple technique to obtain a broncho-alveolar sample for various tests such as rapid identification of viral antigens by indirect fluorescent antibody tests (IFAT), bacteriological culture and cytology. Sampling of severely affected animals should be avoided because of the added stress caused by the procedure.

Procedure - Broncho-alveolar samples can be obtained through a fibreoptic endoscope or a 50 ml catheter tipped syringe and a sterile 90 cm long flexible

6 mm diameter tube. Sterile gloves, sterile saline, local anesthetic gel and viral transport medium are the other pre-requisites. Local anaesthetic gel is applied to the inner surface of a nostril. The distances to the larynx and the base of the neck are measured and marked on the tube. The tube is gently passed intranasal via the ventral meatus to the larynx. Once the larynx is reached, the tube is pushed quickly forwards on inspiration to gain entry to the trachea through the glottis. When the tube enters the trachea, breathing is felt and heard at the end of the tube accompanied by some mild coughing. Successful entry of the tube into trachea (not into the esophagus) may require repeated attempts. The tube in trachea is advanced to the base of the neck until there is resistance. At this point the tube is at the bifurcation of the major bronchi (the carina) of the trachea. The 50 ml catheter syringe containing 20ml of warm sterile saline is attached to the proximal end of the tube and the saline is injected. Suction is immediately applied to the syringe to withdraw as much of the injected saline as possible. About 10 ml is usually retrieved. Gross examination may reveal cellular debris, suggesting inflammatory changes in the trachea and bronchi, and occasionally lungworm larvae may be observed. Some of the refluxed saline should then be decanted into the viral transport medium and some placed in a sterile tube for bacterial culture. Some of the aliquot should be placed in EDTA for cytological examination. Analyses may include IFAT for PI3, RSV, IBR and BVD, and bacterial culture for *Pasteurella* and *Haemophilus*. To avoid sample contamination by nasopharyngeal commensals the technique can be modified by placing a short wider bore 8 mm tube through the ventral meatus to the entrance of the glottis and then feed the 6 mm collection tube through this wider bore tube.

The flexible fiber optic endoscope not only assist in collection of tracheal lavage but also visualizes the condition of the airways. Contamination of the endoscope during passage through upper airways may occur. The flexible fiber optic endoscope has facilitated the examination of upper respiratory tract (nasal cavities, nasopharynx, guttural pouches, palatal arch, epiglottis, larynx, and trachea and major bronchi) of horses. The technique can be used to examine upper respiratory tract of cattle also.

Pleuroscopy

It is the technique used for visual inspection of the pleural cavity. Pleuroscopy can be done using rigid or flexible fiber optic endoscope.

Collection of tracheal mucus

Tracheal mucus can be collected from cattle using a sputum cup for bacteriological examination. Though the technique is simple, good restrain and sedation is needed to draw tongue and view laryngeal orifice.

Laboratory Examination of Collected Respiratory Secretions

The nasal swab and /or tracheo-bronchial lavage collection are examined in the laboratory for microbial and cytological analysis using various techniques as per requirement.

Radiography

Radiography is of limited value especially in older animals. In cattle the presence of the heart just anterior to the elbow and the rumen caudal to the diaphragm prevents reliable interpretation of the area of lung overlying or adjacent to these structures. The interstitium of the lungs is normally prominent and is not a sign of interstitial pneumonia. In the adult animal the increased thickness and size of the thorax result in magnification of the normal and abnormal making interpretation difficult and unreliable. Heavy musculature of the shoulder overlying the thorax restricts diagnostic imaging to the caudal lobe. In addition, powerful radiographic equipment is required and is usually only available in referral centers. Radiography in calves can be quite rewarding, although the forelimbs have to be pulled well forward to facilitate the examination of the anterior lobe. Chemical restraint of compromised patients may lead to an unacceptable risk. Radiography has been used to assist the diagnosis of reticular wires which have penetrated the thorax, fractures of the ribs, lungworm, enzootic calf pneumonia, diffuse fibrosing alveolitis, pulmonary abscessation, tuberculosis, emphysematous bullae, pneumothorax, pleural effusions and pleural adhesions. Radiography has a potential to differentiate atelectasis from consolidation, interstitial pneumonia from exudative pneumonias. It can also help in identifying neoplasms, pleural effusions hydropericardium and space occupying lesions in the thorax.

Ultrasonography

This is a non-invasive technique that can provide useful diagnostic information, although experience is required to interpret the images correctly. The pleura and pulmonary surfaces of both lung fields can be examined through the intercostal spaces between the 7th and the 11th ribs. Sector transducers are the most useful, but linear transducers are more readily available. The highest resolution is obtained with a high frequency transducer (usually 7.5 MHz) and the greatest depth of penetration obtained with a low frequency transducer (usually 3.5MHz). To obtain diagnostic images, the hairs over the lung field has to be clipped and contact gel is to be applied. Structural changes and changes in lung density (such as those found in pneumonia) can be detected. Free fluid, although rare in cattle, can be detected. Focal abscesses can be identified.

Thoracocentesis

It is rarely performed in cattle as accumulation of large amounts of fluid in the pleural cavity is rare. Samples of fluid from the pleural cavity can be obtained by inserting a spinal needle or a catheter into the pleural cavity under local anesthesia via an intercostal space using strict asepsis at ventral site caudal to the heart. Equipment required for thoracocentesis are hair clippers, an 18 BWG (1.20 mm) 9 cm long lumbar spinal needle, skin disinfectant, alcohol, swabs, EDTA tube and plain tube. The side of the thorax to be sampled is selected by auscultation and percussion. The distal 5th intercostal space within the lung field is the site of choice as it is the most ventral accessible point of the pleural cavity. However, care is needed to avoid cardiac puncture. Ultrasound guided pleural fluid aspiration is more safe and is recommended. Whichever site is used, the area is clipped and surgically prepared. Sedation and local anesthesia is required. Local anesthesia can be given by using 1–2 ml of 2.0 % lignocaine at the site. An 18 BWG (1.20 mm) 9cm spinal needle with stylet is used. The needle is pushed carefully through the skin and intercostal muscles into the pleural cavity at the prepared site. The stylet is then removed and a syringe attached and negative pressure applied to assist the withdrawal of any pleural fluid. Care is required to avoid cardiac puncture if the 5th intercostal space is used. Samples are collected in to a plain and EDTA tubes for bacteriology and for cytology respectively. Collected samples should also be examined grossly for abnormalities.

Lung Biopsy

Lung biopsy is a useful technique to identify pathological changes within the lung. The technique is reserved for cases with respiratory problems remained undiagnosed with other techniques. Lung biopsy does have limitation that the biopsied area may not have pathology. This technique is valuable in conditions having generalized lung pathology. Chances of fatal hemorrhage in animals with pulmonary hypertension and the formation of pneumothorax should always be carefully considered while using the technique. However, percutaneous lung biopsy has been used safely in cattle using a Tru-cut needle. Equipment required include 10% buffered formalin, surgical scrub, alcohol, lignocaine 2%, needles and syringes, a Tru-cut biopsy needle (Fig. 100), sterile disposable gloves, antibiotics and tetanus antitoxin. The lung biopsy can be obtained in the standing position or in lateral recumbency in the sedated animal. Common sites for the biopsy are the 6th, 7th, 8th or 9th intercostal space in the mid-lung field. The hair is shaved or clipped over the proposed site and the site is prepared aseptically. 1–2 ml of 2.0% lignocaine is injected subcutaneously and then a further 1–2 ml of lignocaine 2% is injected intramuscularly slightly more caudally into the intercostal muscles. A cutaneous stab incision is made with a scalpel blade into the anaesthetized skin. The skin is then moved caudally and the Tru-cut needle thrust through the cutaneous stab incision and through the intercostal muscles and pleura into the stroma of the lung lobe ensuring that the skin incision will not directly overlie the deeper wound and helps to minimize the possibility of pneumothorax. The cutting stylet of the Tru-cut needle is then advanced to cut the sample and the cover is advanced over the cutting stylet to retain the biopsy sample. The Tru-cut needle is then removed. The biopsy sample is placed in 10% formal saline and submitted for histopathology analysis. Prophylactic antibiotic and tetanus antitoxin should be administered.

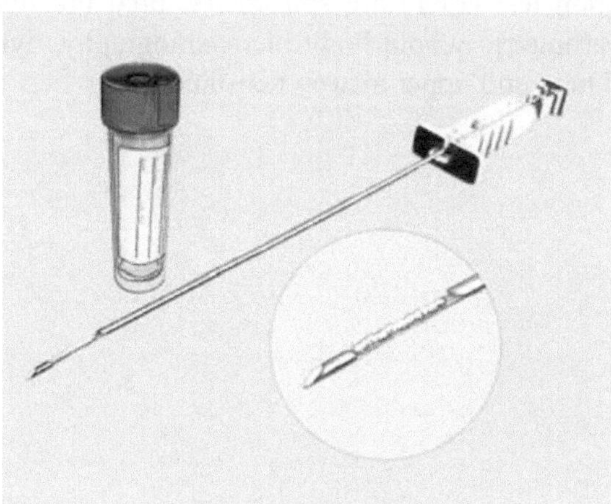

Fig. 100: Tru-cut biopsy needle.

Blood Gas Analysis and Packed Cell Volume (PCV)

Blood gas analysis needs to be performed rapidly and is usually available at referral centers or veterinary colleges. Hand-held cow side machines are also available but are expensive. Blood gas analysis can be helpful in defining the severity of hypoxia or characterizing acid/base disorders. Arterial samples are the most useful, and the auricular artery or coccygeal artery are usually used. The percentage oxygen saturation of arterial hemoglobin can be measured using a pulse oximeter. Pulse oximeters measure the absorption of red and infrared light in non-pigmented skin. The degree of light absorption is a function of the saturation of the arterial hemoglobin. Pulse oximeters with protective sheaths to block out extraneous light sources have been used in calves. The sensors were attached to shaved non-pigmented skin of the tail, the non-pigmented areas of the scrotum or the vulva. Pulse oximeters are portable, non-invasive, easy to use, and inexpensive and can be used with success in field conditions. Evidence of polycythemia from an elevated PCV may indicate chronic hypoxia.

Pulmonary Function Tests

Pulmonary function tests are being routinely used in human respiratory medicine. Its use in animals is not common. However, pulmonary function test is valuable in horses to know the efficiency of the lung before it is impaired. These tests are also of interest in dogs and cats. Pulmonary function tests evaluate respiratory mechanics, lung parenchymal function / gas exchange and cardiopulmonary interaction. In small animals, available techniques for pulmonary function test (PFT) are spirometry, tidal breathing flow-volume loop analysis, barometric whole body plethysmography, dynamic and static compliance, and lung and upper airway resistance.

13

Special Examination of Cardiovascular System

Heart of bovine, caprine, equine, canine and feline is a four chambered organ having left and right atrium, left and right ventricles, mitral valve, tricuspid valve, aortic valve, pulmonary valve, arteries, arterioles, veins and venules. Heart starts its functioning even before birth and continues till death. Its pumping action helps in providing nutrients as well as oxygen to all organs and parts of the body and to remove the waste products of the metabolism from the different organs of the body. Malfunctioning of the heart (pumping action and /or current generation and conduction) adversely affects nutrient and oxygen availability to organs and removal of waste products from the body lowering the quality of life and making the survival reasonably difficult. Heart functioning can be examined through physical examination of the heart such as general inspection (looking for abducted elbow, jugular distension or pulse, edema of ventral abdomen or limbs, tachypnea /dyspnea), evaluations of vitals (temperature, pulse, respiration, mucus membrane color, capillary refill time), cardiac palpation and percussion (ascertaining position of apex beat, presence or absence of precordial thrill), femoral pulse, abdominal palpation and percussion (ascites, hepatomegaly), cardiac auscultation (rate, rhythm, intensity of heart sounds, and murmurs), pulmonary auscultation (abnormal respiratory sounds in caudal lung area), electrocardiography (arrhythmias, heart enlargement particularly in dogs), echocardiography (valvular disease, chamber size, septal defect, pericardial effusion etc) and cardiac biomarkers (cTn-I, NT pro –BNP etc.).

Cardiac Inspection

Inspection of left thorax (Fig.101) may give an indication of the contact of the heart with chest wall. The movement of the chest wall due to apex beat over 5th and 6th intercostal space in cattle is observed when the heart is in contact with the chest wall.

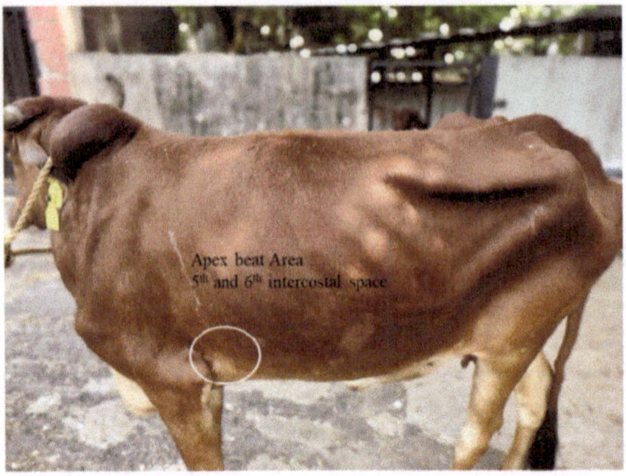

Fig. 101: Apex beat area on left thorax (5th and 6th intercostal space) in cattle.

Cardiac Palpation

Cardiac palpation (Fig.102) is conducted to detect thrills or palpable vibrations that may exist with cardiac cycle. It is usually performed by placing palm or fingers of one hand over cardiac area.

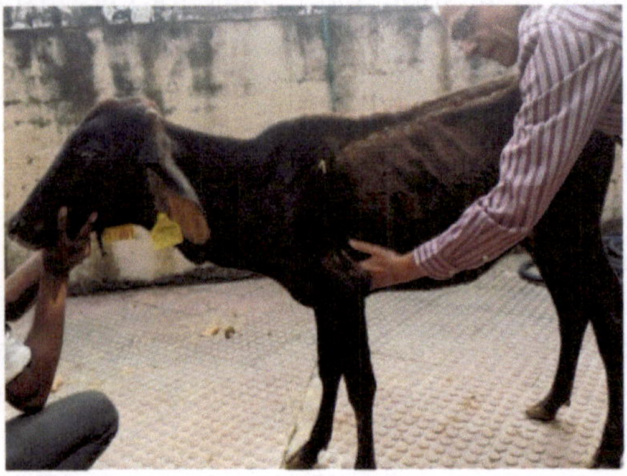

Fig. 102: Cardiac palpation in a cow calf by placing palm and fingers over cardiac area.

Cardiac Percussion

Percussion has limited diagnostic value in large animals as most of the heart is located under muscles of the fore limbs and thorax posing difficulty in resonance. Cardiac percussion (Fig.103) is started from top to bottom along the edge of triceps muscle, then in caudo-ventral, caudal and caudo-dorsal direction to ascertain the area of dullness if any. In cattle only apex of the heart lies caudal to the parts covered by shoulder muscles. Percussion in the ventral part of the thoracic lung percussion field provide only relative cardiac dullness. In fatty cattle heart percussion is difficult.

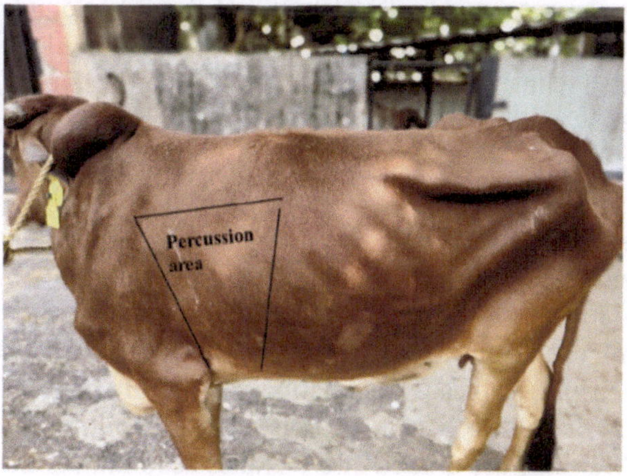

Fig. 103: Area of cardiac percussion in a cattle.

Cardiac Auscultation

Cardiac auscultation is an important part of clinical examination as it gives information about rate and rhythm of the heart; presence or absence of cardiac murmurs and muffling of heart sound; and intensity of heart sounds. In healthy cows heart rate varies from 60 to 80 beats per minute. In healthy sheep (60 to 120 bpm), goats (70 to 120 bpm) and swine (58 to 86 bpm) heart rate is also variable. The site for optimum cardiac auscultation in adult cattle is 4th and 5th intercostal space. For better results fore limbs should be manually extended forward and chest piece of the stethoscope is placed firmly with gentle pressure against the skin under the left elbow and triceps muscle (Fig.104). Cardiac examination should be conducted in quiet place.

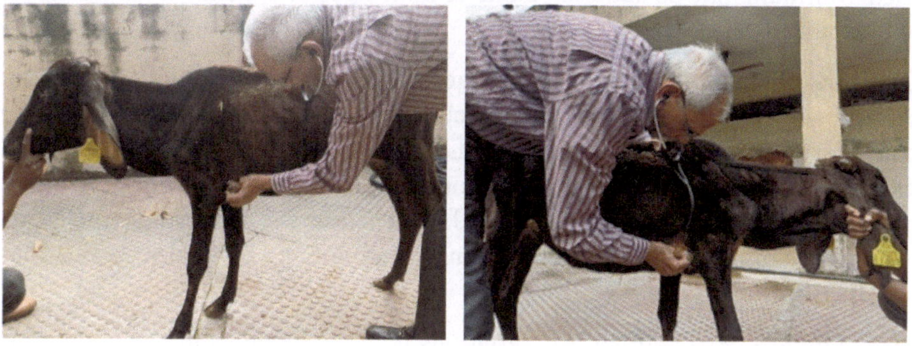

A B

Fig. 104. Cardiac auscultation in a cow calf on left thorax (A) and right thorax (B).

Heart sound is characterized by LUBB-DUBB-PAUSE. Sound of LUBB (S1) is associated with closure of mitral and tricuspid valves. While sound of DUBB (S2) is associated with closure of aortic and pulmonary valves. LUBB –DUBB sounds are clearly audible in normal healthy animals.

Murmurs

Abnormal heart sounds are called cardiac murmurs and are associated with endocardial lesions (valvular insufficiency). The murmurs are caused by turbulence in blood flow. Murmurs may be systolic, diastolic or continuous. Systolic murmur occurs during systole owing to insufficiency of atrioventricular (mitral and tricuspid) valves. Diastolic murmur occurs during diastole owing to insufficiency of valves. While continuous murmurs occurs during systole and diastole. Murmurs are graded into 6 grades according to their intensity.

Grade 1 murmurs are heard with difficulty and are very quiet.

Grade 2 murmurs are quiet but are heard easily.

Grade 3 murmurs are of equal intensity with the normal heart sounds.

Grade 4 murmurs are louder as compared to the normal heart sound.

Grade 5 murmurs are very loud with a thrill (a vibration) palpable through the chest wall.

Grade 6 murmurs are very loud and audible through a stethoscope held just off the chest wall.

Point of maximum intensity(PMI) of a murmur–PMI is located by moving the stethoscope carefully around the area of the chest wall between the 3rd and 6th ribs. Murmurs may be audible on one or both sides of the chest with a PMI on either side.

Pericardial frictional rubs– These are other abnormal audible vibrations heard over thorax owing to friction occurring with cardiac movement (systole and diastole) in an inflamed pericardial sac. Pericardial frictional rubs are heard in each cardiac cycle. Tinkling sounds in heart area are suggestive of the presence of gas on the surface of fluid enclosed in the pericardial sac. Heart sounds are muffled when there is an increased volume of fluid in pericardial sac (pericardial effusions).

Arrhythmias– Arrhythmias are variations in cardiac rate from that of normal range specified for the species. Marked increase in heart rate beyond the limit specified for the species is termed as tachycardia. A marked decrease in heart rate than the normal specified limit is called bradycardia.

Normal rhythm of the heart is LUBB-DUBB-PAUSE. Any variation in this rhythm is associated with rhythm disturbance and is abnormal. It also falls under the category of arrhythmia.

Heart intensity or loudness of heart sound – It depends on the contraction force (strength of heart beat), thoracic wall thickness, and factors altering the transmission of sound from the heart to the stethoscope.

Cardiac auscultation in horse

During cardiac auscultation in horse (Fig. 105), the entire precordium is auscultated. Valves should be auscultated at the point of maximum intensity (PMI) of valve sounds. Mitral valve is auscultated on the left side at 5th intercostal space dorsal to olecranon for apex beat. Aortic valve is also auscultated on the left thorax at 3rd intercostal space below the point of shoulder and medial to triceps muscle mass (heart base). Cranio-ventral to aortic valve at left heart base, pulmonic valve can be auscultated. On the right side of the thorax, tricuspid valve is auscultated at 3rd-4th intercostal space medial to triceps just above the olecranon. Heart rate in horses varies from 23 to 48 beats per minute.

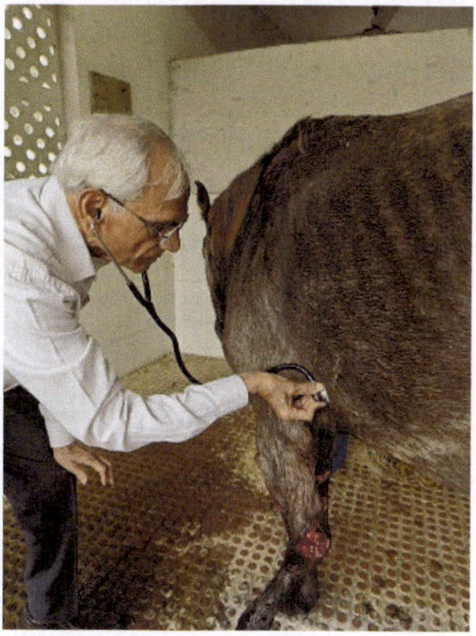

Fig. 105: Cardiac auscultation in a horse on left thorax.

Cardiac auscultation in dogs and cats

In dogs and cats, cardiac auscultation is done in a quiet room with relaxed mood in standing position at apex, at base and at the level of all valves' (mitral, tricuspid, aortic and pulmonary). The pet is gently restrained. Ventilation artifacts can be reduced by gently holding the mouth close. Mitral, aortic and pulmonary valves are auscultated on the left side of the thorax in standing position at 5th intercostal space around costo-chondral junction (opposite to elbow point in standing position), at 4th intercostal space dorsal to mitral valve (at the level of the point of shoulder), and at 3rd intercostal space at sternal border respectively (Fig.106). Tricuspid valve (Fig. 106) is auscultated on the right side of the thorax at 3rd to 4th intercostal space at costo-chondral junction.

In cats, mitral valve and pulmonic valves are auscultated on the left side at 5th/6th intercostal space above the sternum, and at 2nd/3rd intercostal space one third of the way up from sternum respectively. The tricuspid valve in the cat is auscultated on the right side of the thorax at 4th/5th intercostal space near sternum. Heart sounds are more intense in young dogs, thin/emaciated dogs as well as in disease conditions such as fever, anemia, or hyperthyroidism. Obesity, pleural effusion or pericardial effusion decreases the intensity of the normal heart sounds. Muffled heart sounds (decrease intensity of heart sounds) may be due to pericardial effusion, pericarditis, pleural effusion, diaphragmatic hernia, thoracic neoplasia, hypothyroidism or cardiac temponade. Obesity can also muffle the heart sounds because fat acts as an insulting layer dampening the heart sounds. In dogs, pericardial effusion results from neoplastic or idiopathic diseases. While pericardial effusion in cats is most often due to congestive heart failure secondary to cardiomyopathy.

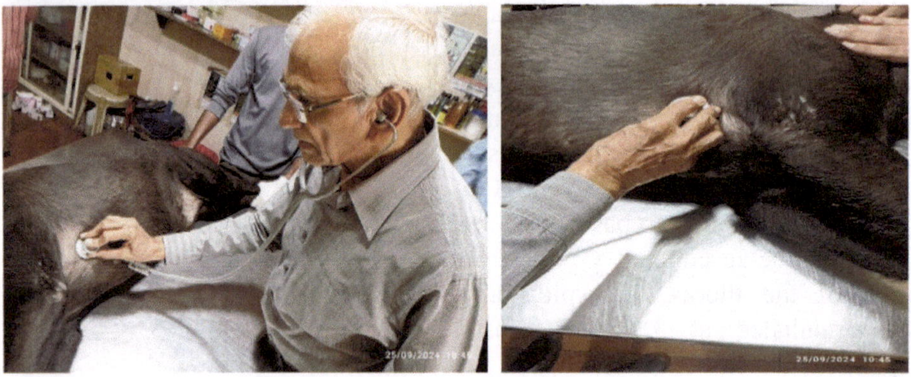

Fig. 106: Cardiac auscultation on both side of the thorax in a dog.

Jugular Vein Examination

Jugular vein is located in the jugular furrow on the left and right side of the neck. Examination of the jugular vein is very important in cattle and horses as it may provide valuable information about the functioning of the right heart.In healthy animals jugular veins are not visible (Fig.107 B) and palpable without manual compression at the level of the thoracic inlet. Visibly distended and palpable jugular veins without manual assistance (Fig.107 C) are pathological. Engorgement of both jugular veins is suggestive of compression or constriction of jugular veins, or congestive right heart failure. Sometimes, jugular pulse is seen. A small magnitude jugular pulse moving up about one-third of the way up the neck is normal in most of the animals. It should be differentiated from the transmitted carotid pulse that is not abolished by compression of the jugular vein at lower level. Jugular pulse associated with each cardiac movement needs to be differentiated whether it is physiological or pathological. Physiological jugular pulse is pre-systolic owing to atrial systole. While systolic jugular pulse occurring simultaneously with arterial pulse and first heart sound is pathological and is characteristic of tricuspid valve insufficiency. The most common etiology for jugular distension and jugular pulse in companion animals is right-sided heart failure.

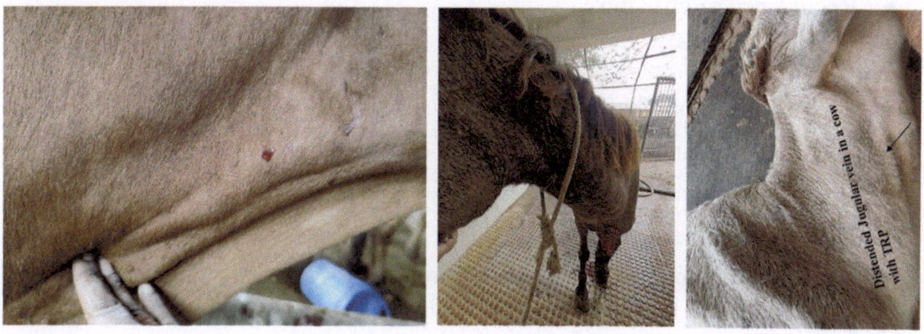

A B C

Fig. 107: Jugular vein examination. Distended jugular vein in the right jugular furrow on manual compression in a cow **(A).** It is normal as distension is with manual compression. Normal jugular vein is not visible in the left jugular furrow in the horse **(B)**. Distended jugular without any manual compression in a cow with traumatic pericarditis **(C)**.

Pericardiocentesis

Pericardiocentesis in dogs and cats

Pericardiocentesis is a sterile minimal invasive procedure to collect or remove excess pericardial fluid from the pericardial sac. It is a life saving procedure to relieve life threatening pericardial tamponade associated with severe pericardial effusion.

Indications of the technique

The technique has both diagnostic and therapeutic purpose.

- The technique is performed to collect pericardial fluid from cases of pericardial effusion for diagnostic purpose. The collected fluid is used for cytological and cultural examinations to detect different causes of pericardial effusions.
- The technique is indicated as an emergency procedure for therapeutic purpose in pets to relieve cardiac tamponade associated with severe pericardial effusion.

- In cats, pericardiocentesis is recommended in cases of feline infectious peritonitis, congenital or acquired myocardial disease, pericardial or heart base neoplasia, septic pericarditis or hemorrhages due to warfarin toxicity.

Potential risks of the technique

- Laceration of myocardium or coronary arteries.
- Incomplete drainage from the pericardial sac.
- Recurrence of effusions.

Equipment

- Clippers and disinfectant.
- Sterile gloves and drapes.
- Electrocardiograph.
- Ultrasound.
- Large intravenous catheter (10-14 gauge, 10-15 cm for large dogs; 16-18 gauge , 4-6 cm for small dogs and cats).
- 3- or 4-way stopcock
- IV extension tubing.
- 50 ml syringe.
- Sample collection containers (EDTA, plain, sterile for culture).
- Lidocaine or xylocaine (2%)

Procedure

Pericardiocentesis is a sterile procedure performed in dogs and cats in lateral or sternal recumbency from right side of the thorax to avoid injuries to lungs and carotid arteries.

- Sedation with butorphanol (0.2 mg/kg IV) with or without acepromazine 0.005 mg/kg IV or midazolam (0.2 mg/kg) can facilitate the procedure.
- Constant monitoring with continuous electrocardiography is needed during the procedure. If severe ventricular tachycardia develops, it should be treated with lidocaine (2.0 mg/kg IV).The technique of pericardiocentesis can be performed either as blind or under the guidance of an ultrasound.
- Blind pericardiocentesis is done between 3rd and 5th intercostal space on right side of the thorax. While ultrasound guided pericardiocentesis is done on the right side at a place of largest diameter of the effusion.
- The skin of the area over the 2nd to 8thintercostals space on the right side from the sternum to mid thorax is shaved, cleaned with alcohol or an antiseptic. Xylocaine (2%) is infiltrated in the skin, intercostals muscle and pleura of the site. The area around the site is draped with sterile drape.

- The intravenous catheter, extension set with three-way stopcock and syringe are prepared in a sterile manner. If catheter is large, a small stab incision can be made to facilitate the insertion of the cannula.
- Puncture should be made in the middle of the intercostal space so that trauma to intercostal vessel is avoided.
- The catheter is slowly inserted perpendicular to the skin through the skin, intercostals muscle, pleura and pericardial sac. Piercing of pericardial sac is indicated by the loss of resistance and the appearance of the fluid at the catheter hub. Pericardial fluid can be aspirated.
- In case of concurrent pleural effusion, straw colored fluid appears when needle/ catheter pierces pleural space. The catheter is advanced a little further and stylet is removed and connection to collecting system is made.
- If scratching or tapping sensation is felt, it indicates that the catheter is in contact with the heart warranting immediate slight retraction.
- Samples are collected in plain tube, EDTA tube and in sterile tube.
- When no more fluid is aspirated or needed, the catheter is retrieved.

Complications

- Usually pericardiocentesis is a relatively safe procedure. Nevertheless, ventricular premature beats/ ventricular tachycardia are the most common complication owing to myocardial injury by the catheter. Coronary artery laceration, ventricular puncture, pneumothorax or hemorrhages are the other less common complications that may occur during the process of pericardiocentesis. Pericardial effusion reoccurs commonly even after removal of excess pericardial fluid. Recurrence of pericardial effusion is within hours in case of ruptured hemangiosarcoma or may be after weeks in case of idiopathic pericardial effusion.

Limitations

- Expertise and procedural experience of the operator is of great value.

Pericardiocentesis in cattle

- The technique of pericardiocentesis in large animals (bovine) is employed for collecting and assessing pericardial fluid. A 12 cm spinal needle of 16 BWG (1.65 mm) is used for pericardiocentesis. The needle is inserted ,blindly or under the guidance of ultrasonography, through the thoracic wall into the pericardial sac on the ventral left thorax in the 5th intercostal space and fluid is allowed to flow or is aspirated using a syringe. Local anaesthetic (Xylocaine) is injected into the skin and muscle layers of the space between the 5th and 6th ribs (5th intercostal space). The area is prepared aseptically and the needle with syringe attached is advanced carefully towards the heart. Fluid, foul smelling in infection, is aspirated for cytology, culture and drainage purposes. When blind insertion of the needle is opted, care must be taken to avoid penetration of the myocardium with the needle.

Pericardiocentesis in horses

- Pericardial diseases are rare in horses. Nevertheless pericardiocentesis is recommended in suspected cases.
- Pericardiocentesis is indicated for cytological and bacteriological evaluation of pericardial fluid for ascertaining the etiological diagnosis.
- It is also indicated for therapeutic purpose in cases of pericardial effusions to provide immediate relief.
- Pericardiocentesis should not be attempted in horses when there is septic pleuritis or wounds overlying the site of the pericardiocentesis.
- It may be performed on either side, preferably under ultrasonographic guidance. It is usually performed at the left 5th or 6th intercostal space, in an area above the olecranon and below a horizontal line through the point of the shoulder.
- Other procedure is almost similar as described under other species.

Blood Pressure Monitoring

Blood pressure is the measure of force that heart uses to pump the blood in arteries around body. There are two types of blood pressure viz. systolic (when blood is pumped into arteries during cardiac systole) and diastolic (when the heart is in diastole and receiving the blood). Blood pressure can be monitored directly by arterial puncture pressure or indirectly. In large animals taking blood pressure directly by arterial puncture is not a practical approach in clinical practice and indirect determination of arterial blood pressure is also a difficult task because of paucity of locating a suitable artery where cuff can be applied and pulse return can be detected either by auscultation or palpation of pulse. Attempts have been made to take blood pressure in cattle from tail as in horses, but results do not correlate well due to anatomical difference. Systolic and diastolic blood pressures in cattle has been recorded as 100-140 mm Hg and 50-85 mm Hg respectively.

In horses, indirect measurement of arterial blood pressure has been feasible by applying rectangular self adhesive sphygmomanometer cuff containing an inflatable rubber bag to the base of the tail and detecting pulse sounds in the ventral coccygeal artery. On deflation of the cuff, first sound heard at emergence pressure is systolic blood pressure and second sound is diastolic blood pressure. In thoroughbreds systolic and diastolic blood pressures have been reported as 112 ± 16 mm Hg and 77± 14 mm Hg respectively. Varshney (1997) reported uncorrected coccygeal systolic and diastolic blood pressure in healthy horses as 133.81±3.68 and 84.24±5.05 mm Hg respectively. An increase in blood pressure (systolic 166.54 ± 8.3 ; diastolic 135.5± 1.89 mm Hg) has been observed in colicky horses (Varshney, 1997). While taking blood pressure in horses, animal posture is very important as lowering of head can significantly lowers both blood pressures as well as pulse pressure. In horses, hypertension has been associated with epistaxis, laminitis and painful fractures of distal bones of the limbs. Increase in systolic blood pressure has also been seen in horses with large intestine obstruction.

For non-invasive measurement of blood pressure in clinical practice in canines, Doppler ultrasonic flow detector, oscillometry technique and auscultative technique are being used. The major limitation of non-invasive measurement of blood pressure in canines is its inaccuracy particularly in hypotensive patients. Blood pressure monitoring is indicated in dogs with heart disease, renal disease, endocrinopathies, and syncope and in dehydration. Blood pressure monitoring is valuable indicator of ventricular wall stress and myocardial oxygen consumption.

Auscultative method

In auscultative method, the dog is restrained in right lateral recumbency and Velcro cuff is applied over hind limb on cranial tibial artery at the distal medial aspect of tibia (Fig.108). The width of the cuff should be around 30-40 % of the circumference of the fore limb, hind limb or tail from which arterial blood pressure is being measured. The cuff is inflated with air using hand bulb (as shown in the Fig. 108). The pressure varies in the cuff. The systolic and diastolic arterial pressures are indicated by the appearance of palpable beat with cuff deflation or appearance of throbbing of manometer or sound and ceasing of throbbing or sound respectively. In a study, Gaur and Varshney (2002) observed systolic and diastolic blood pressure in healthy dogs as 131.22 ±0.54 and 63.70 ± 0.35 mm Hg respectively. Systolic blood pressure more than 160 and less than 100 mm Hg have been considered as hypertension and hypotension respectively in dogs.

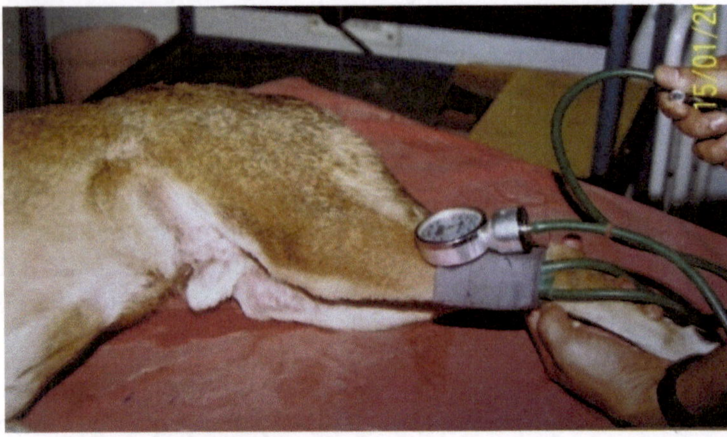

Fig. 108: Blood pressure measurement by indirect technique using a small Velcro cuff (5 x 22 cm size) tied over left hind leg on cranial tibial artery at distal medial aspect of tibia and aneroid sphygmomanometer. The hand bulb coupled to aneroid pressure gauge calibrated in mm of mercury (Hg) is used to inflate and vary the pressure in the cuff. The systolic and diastolic arterial pressures are indicated by the appearance of palpable beat with cuff deflation or appearance of throbbing of manometer or sound and ceasing of throbbing or sound respectiuvely (*Courtesy* Varshney, J.P.2020)

Oscillometric method

Devices measuring blood pressure based on oscillometric method are automated providing estimates of heart rate; and systolic, mean, and diastolic arterial blood pressure. These devices inflate the cuff automatically until arterial blood flow is impeded. As deflation of cuff begins slowly, the oscillometer measures the mean arterial blood pressure as the value at which the pressure oscillations in the cuff have maximum amplitude and an algorithm is used to compute systolic and diastolic arterial BP values from the measured mean arterial blood pressure (Williamson and Leone, 2012).

Doppler ultrasonic flow detector

This method of measuring blood pressure is based on the similar principle of auscultatory method except that it uses a distally placed flow detector rather than stethoscope. Ultrasonic gel is liberally applied on the surface of the Doppler probe and then it is placed over a peripheral artery (hairs of the area should be clipped). An occlusive cuff is placed proximally and a sphygmomanometer is used to inflate the cuff until blood flow, and therefore systolic arterial BP, is impeded, causing the audible signal to disappear (Clarke *et al.*, 2014). The cuff is slowly deflated and the cuff pressure at which the first audible sound returns suggests the value of systolic arterial blood pressure. The technique is minimally invasive, cheap, and easy to use.

Electrocardiography

Electrocardiography is a non-invasive diagnostic technique being routinely used to examine the electrical activity of the heart in health and diseases in humans and animals. ECG measures and record varying potential difference occurring over the surface of the body owing to electrical activity within the heart during systole and diastole. It is an important tool for the clinical evaluation of the heart. The technique provides important information regarding heart rate, heart rhythm, impulse formation and its conduction. Heart size can also be assessed to some extent in humans and dogs. The process of electrocardiography results in to generation of an electrocardiogram. The information on electrocardiography is by no means a complete information to make readers expert but it aims to sensitize and make familiar with the technique and its application so that cases with compromised cardiac function are referred timely for electrocardiographic investigation.

Electrocardiograph or Electrocardiographic machine

Electrocardiograph (Fig.109) is a machine used to examine the electrical activity of the heart and produces a visible record popularly known as an electrocardiogram (Fig.110). In simple terms, electrocardiograph is a machine consisting of a voltmeter, amplifier, galvanometer, display screen, control panel with an attached recording device. The sensitive galvanometer of electrocardiograph is capable of recording difference in electric potential up to 1 mV (1/1000 of a volt). Various basic features viz. running (paper) speed of the machine, sensitivity, lead selector, off/on switch, filter etc are displayed on the screen. Electrocardiographic machine operates at the paper speed of 25 or 50 mm per second as per selection. The sensitivity switch controls amplitude of the waves (as 0.5, 1. or 2). The sensitivity of 1 means that 1 mV input causes deflection of 1 cm or 10 mm on the ECG paper; the sensitivity of 2 means that 1 mV input causes deflection of 2 cm or 20 mm on the ECG paper; and the sensitivity of 0.5 means that 1 mV input causes deflection of 0.5 cm or 5 mm on the ECG paper. Lead selector switch/ touch spot is used to select the leads in single channel ECG machines. In automatic six channel machine or twelve channel ECG machines, lead selection is automatic. These are minimal basic features comes with every ECG machine, however, they may vary with machine to machine and depend on the model. Hence, manufacturers' specifications or instructions given in the user manual should be followed before using any machine for better results.

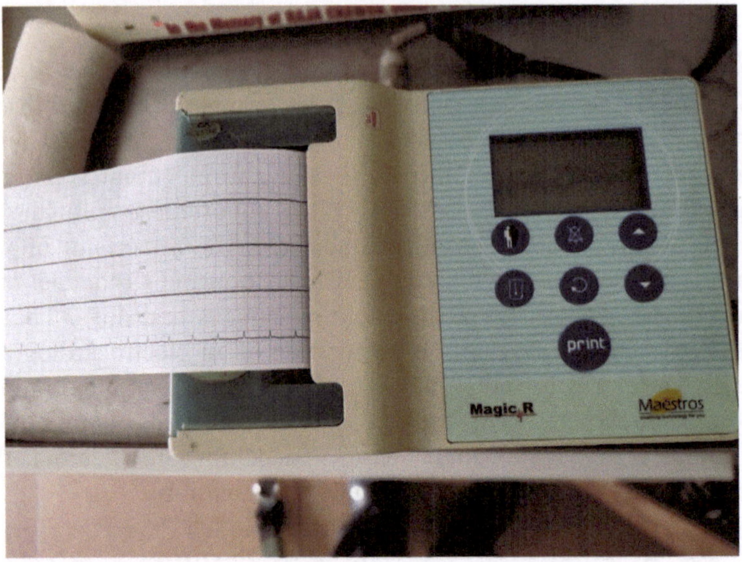

Fig. 109: Six channel Electrocardiographic machine

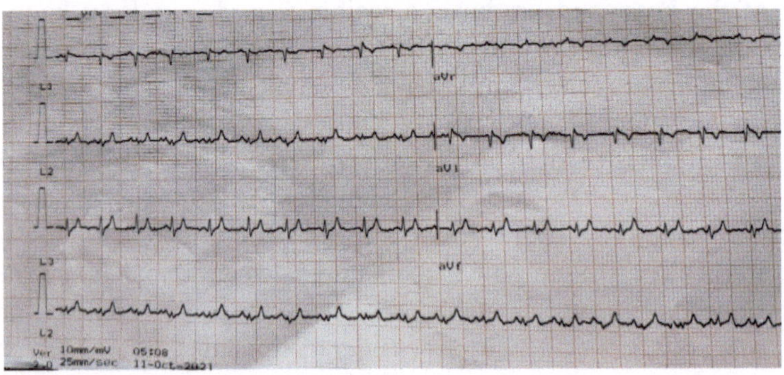

Fig. 110: Electrocardiogram of a dog in lead I, II, III, aVR, aVL and aVF leads

Restrain and positioning of animals for electrocardiography

Dogs and cats are restrained manually and placed in right lateral recumbency for electrocardiography on a wooden table or a table covered with a foam mattress and rubber sheet. The pets are restrained by an attendant in the manner shown in the Fig.111. Both left and right limbs should remain separated without touching each other and the forelimbs should be kept perpendicular to the long axis of the body.

Usually no drug is given for the restrain. Owner's presence at the electrocardiographic table during the process of electrocardiography is usually sufficient to comfort the pet. Uncooperative pets may be dosed with diazepam (1 mg/Kg orally 1 hour before examination) or acetylpromazine (2 mg/Kg orally 1 hour before the examination or 0.5 mg/Kg intramuscularly or intravenously 15 minute before examination) without significantly affecting the cardio-pulmonary functions.

Electrocardiogram in ruminants (Cow, Buffalo, Goat, and Sheep) and equines (horses and donkeys) is generally recorded in standing position. The animals are kept standing on a rubber mat.

Sites for placing electrodes

The sites for placing the electrodes are prepared by shaving, cleaning and applying electrocardiographic gel. Electrocardiographic gel is also applied on the electrode clips to facilitate proper contact.

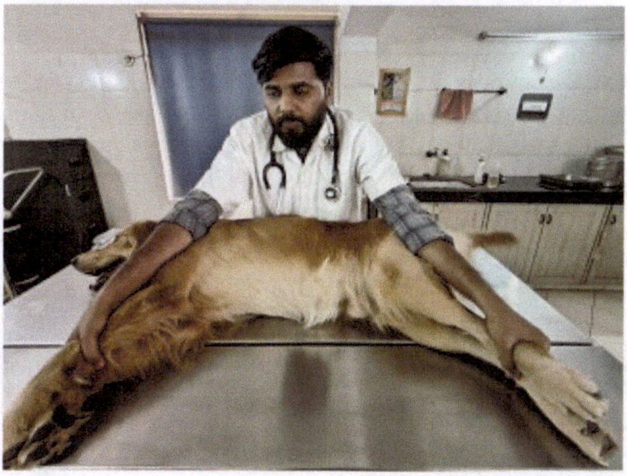

Fig. 111: Positioning of a dog for electrocardiography. The dog is positioned in right lateral recumbency for electrocardiography and is restrained as shown in the picture (Courtesy Varshney, J.P. 2025)

Electrodes clips are applied directly on the skin of the sites in case of dogs (Fig.112) and cats. RA and LA electrodes are attached directly on the skin proximal to the olecranon on the caudal aspect of respective fore limbs; and RL and LL electrodes are attached over patellar ligament on the anterior aspect of the respective hind limbs. V electrodes or chest electrodes are attached (Fig.113) on skin of the chest (CV5RL at 5th right intercostal space near edge of sternum. CV6LL at 6th left intercostal space near edge of sternum, CV6LU at 6th left intercostal space at costo-chondral junction and V10 over dorsal spinous process of 7th thoracic vertebra). Electrocardiogram is recorded in a calm and quite surroundings using electrocardiographic machine at a paper speed of 25 mm per second and amplitude of 1 mV = 10 mm or 1 cm. The recording is done in lead II in single channel machine. While automatic machines records 3, 6 or 12 channels automatically as per machines specifications.

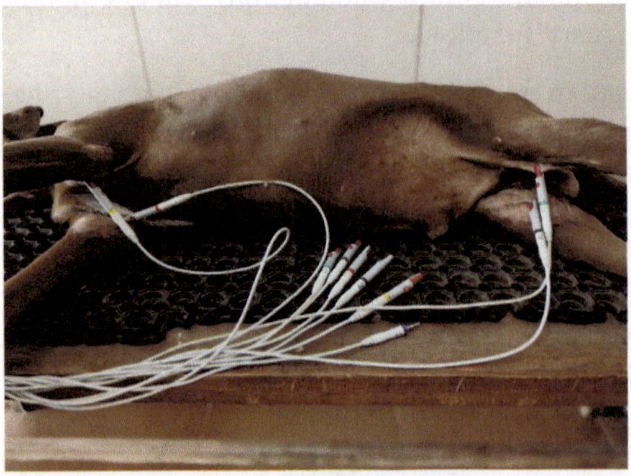

Fig. 112: Placement of electrodes in standard limb leads. LA (yellow) and RA (red) electrodes are attached to the skin proximal to olecranon on the caudal aspect of the left and right fore limb respectively. LL (green) and RL (black) electrodes are attached to the skin over patellar ligament on the anterior aspect of the left and, right hind limb respectively.

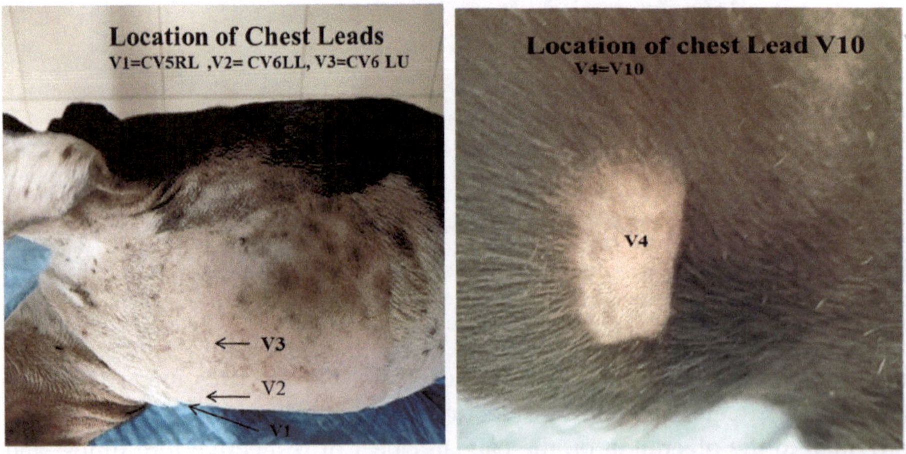

Fig. 113: Sites for placement of chest electrodes. V1 = CV5RL, V2 = CV6LL, V3 = CV6LU, V4 = V10 .CV5RL electrode is attached at 5th right intercostal space near edge of sternum. CV6LL electrode is attached at 6th left intercostal space near edge of sternum, CV6LU electrode is attached at 6th left intercostal space at costo-chondral junction and V10 electrode is attached over dorsal spinous process of 7th thoracic vertebra (Courtesy Varshney et al. 2024)

In Ruminants and horses, base-apex lead system is commonly used for electrocardiography. Recording is done with the animal standing square or with the left fore limb set slightly advance of right fore limb. The positive electrode (LA-yellow) is placed over the area of the heart apex beat on thorax behind the left elbow. The negative electrode (RA- red) is placed at the left jugular furrow two third of the way down the neck. The ground electrode (RL- black) is placed at a place remote from the heart usually at wither (Fig.114). Electrocardiogram is recorded in a calm and quite surroundings using electrocardiographic machine at a paper speed of 25 mm per second and amplitude of 1 mV = 10 mm or 1 cm. The recording is done in lead 1.With electrolyte imbalances, changes in QT intervals have been observed. Q-T interval prolongs in hypocalcaemia and shortens in hypercalcaemia.

Fig. 114: Placement of electrodes in a cow. The positive electrode (LA-yellow) is placed over the area of the heart apex beat on thorax behind the left elbow; the negative electrode (RA- red) is placed at the left jugular furrow two third of the way down the neck ; and the ground electrode (RL- black) is placed at a place remote from the heart usually at wither.

The information obtained from an electrocardiogram provides an understanding regarding the activities of the heart. An abnormal electrocardiogram may reveal the side of the heart affected and abnormalities related to rate and/or rhythm of the heart. A normal electrocardiogram does not always exclude the possibility of the cardiac disease. Therefore an electrocardiogram should not be read in isolation but it should always be interpreted in conjunction with the result of all other examinations. An electrocardiogram is popularly referred or abbreviated as 'EKG'/ 'ECG'.

Morphology of an electrocardiogram

Morphologically an electrocardiogram consists of waves (P, QRS, and T), intervals (P-R, Q-T, R-R) and segments (P-R, S-T, T-P) as shown in the Fig.115. QRS in canine and feline ECG is predominantly positive (Fig.115) while it is predominantly negative in ruminants (Fig. 116) and equine ECGs.

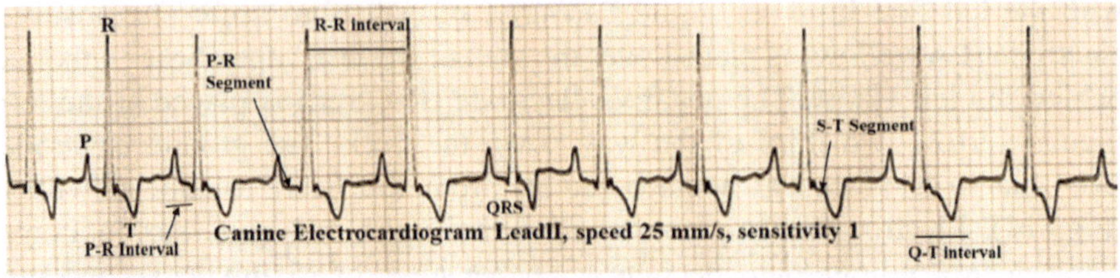

Fig. 115: Electrocardiogram of a dog showing waves (marked as P, ,R, T) , intervals (P-R, R-R, Q-T) and segments (P-R ,S-T, R-R) marked on the ECG.

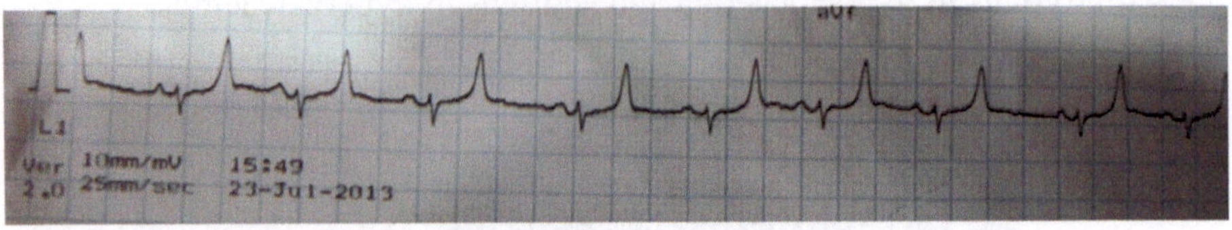

Fig. 116: Electrocardiogram of a goat showing a negative 'QRS'.

Indications of electrocardiography

Electrocardiography should be undertaken whenever clinical examination reveals cardiac murmurs and/or slow / fast /irregular heart. Broadly it is indicated for following conditions.

- For the diagnosis of disturbances of heart rate and or rhythm (arrhythmias).
- Heart monitoring during anesthesia and/or surgery.
- Heart evaluating in pets with respiratory distress / seizures/syncope/shock.
- For evaluating heart in suspected cases of cardiac diseases.
- Heart evaluation in cases presented in emergency with hypothermia, hyperthermia, snake bites or electrocution.
- Heart evaluation in pets with multi organ failure.
- For constant cardiac monitoring in cases in intensive care units.
- For monitoring heart status of geriatric patients with acute or chronic illness.
- Heart monitoring for differential diagnosis of ascites.
- For health checkup of pets engaged in sports/snuffing activities.
- Heart monitoring of pets with trauma/ accidents.
- For regular cardiac monitoring of pets with heart diseases.
- Heart monitoring in animals with electrolyte disorders or severe dehydration.
- For regular cardiac monitoring of animals on cardiac drugs.
- For assessing disease progression in animals with cardiac disease.

- For generating database for differentiating nonspecific causes of weakness, fatigue, lethargy, syncope, collapse, seizure/epilepsy and fever.
- For routine pre -surgical examination of the heart.
- For routine health checkup of pets.
- In pets, ECG can be used to assess the status of myocardium and the presence of hypertrophy. But this purpose is defeated in ruminants and horses as the nature of depolarization is different, Purkinje fibers penetrate much more deeply and depolarization occurring over multiple minor fronts simultaneously. Therefore ECG in ruminants and horses is usually indicated for the diagnosis of arrhythmias and conduction disturbances.

Benefits of the electrocardiography

- The technique is simple, handy, non-invasive and easy to use in practice.
- The technique provides a visual graph instantly for reading and making interpretation.
- The technique can be used on ambulatory, indoor and outdoor patients as well as on patients at owner's place.

Limitations of the electrocardiography

- As with other techniques, electrocardiography does have limitations.
- Definite diagnosis of congestive heart failure is not possible.
- Prognostication during anesthesia or surgery is difficult.
- Electrocardiography is unsuitable for making definite diagnosis of the diseases of cardiac valves (mitral, aortic, pulmonic and tricuspid), endocardium, and pericardium; wall thickness of cardiac chambers and obstructive blood flow diseases.
- Because of limitations, electrocardiogram should always be interpreted in conjunction with detailed observations and investigations by a technically competent veterinarian as casual reading may lead to erroneous interpretation.

Echocardiography

Echocardiography (Fig.117), a non-invasive technique, is the gold standard for assessing structure and function of the heart. It is cardiac ultrasonography. It is becoming an indispensible diagnostic technique for diagnosing congenital heart diseases, shunts, complex malformations, valvular heart diseases, cardiomyopathies, endocarditis, pericardial effusion, corpulmonale and cardiac tumor. The diagnosis of these diseases was otherwise difficult before the advent of echocardiography. M -mode and color Doppler imaging is further improving our diagnostic skill. Evaluation of the heart involves real time

B-mode, M-mode and Doppler echocardiography for better results. Doppler echocardiography is used for the diagnosis of blood flow disturbances in the heart and vessels. Measurements of chamber size can provide information regarding chamber dilatation. Increase in wall thickness can assist in the diagnosis of hypertrophy. Abnormal valve motion, incompetent valve orifice, mass associated with valve are the important observations in cases of valvular defects and endocarditis. Congenital cardiac defects can also be diagnosed by echocardiography. Echocardiography can be used in the diagnosis of pleural and pericardial effusion.

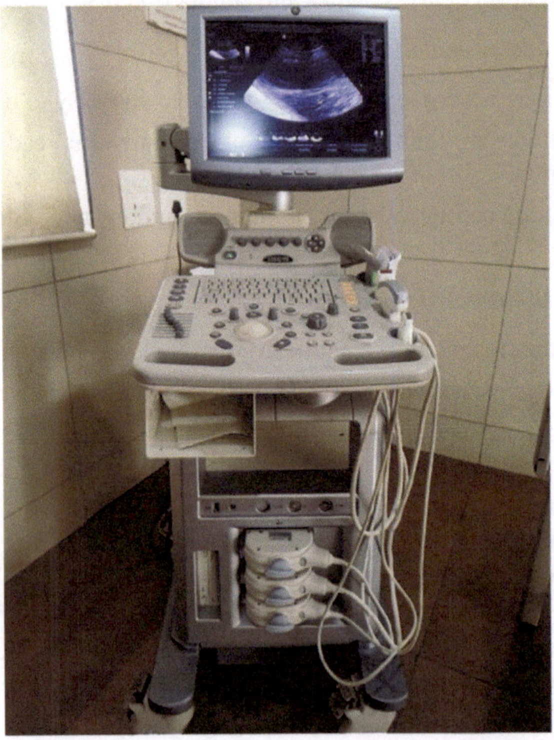

Fig. 117: An ultra sound machine used for echocardiography.

Basically three modes (real time B-mode, M-mode and Doppler) of echocardiography are used to evaluate heart in routine practice. M-mode measurements and all standard real time B-mode imaging planes from right and left side of thorax are obtained.

Real-time B- mode cardiac ultrasound is used to examine heart, blood vessels, soft tissues and thoracic organs. It can detect chamber enlargement, hypertrophy, abnormal valve, dilation of great vessels.

M-mode cardiac ultrasound is used for dimensional analysis of heart chambers, thickness of the ventricular wall, motion of the heart and valves.

Doppler ultrasound has revolutionized cardiac imaging. It provides information about blood flow through heart chambers, along vessels, across valves and its disturbances as well as velocity of blood flow.

References and Further Readings

Clarke, K.W., Trim, C.M. and Hall, L.W. (2014) Patient monitoring and clinical measurement. In: Veterinary Anesthesia.Clarke, K.W.,Trim,C.M. and Hall, L.W.(Eds.). 11th ed. St. Louis, Missouri: Elsevier; 2014. pp. 30–38. (Google Scholar).

Gaur, T. and Varshney,J.P.(2002). Non-invasive measurement of blood pressure and prevalence of blood pressure anomalies in dogs. Indian J. Anim.Sci.72:68-69.

Reef, V.B. (1998). Cardiovascular system examination. In: Manual of Equine Emergencies, Treatment and Procedures. Orsini, J.A. and Divers,T.J. (eds) WB Saunders Co., Philadelphia. pp 94-156.

Varshney, J.P.(1997). A preliminary study on indirect arterial blood pressure in horses using an electronic sphygmomanometer. Remount Vet. Corps 36: 79-82.

Varshney, J.P. (2020). Electrocardiography in Veterinary Medicine. Springer Nature, Singapore, Pte, Ltd.

Varshney, J.P. (2023). Heart Failure in Animals: Current Concepts.NIPA® Genx Electronic Resources & Solutions P. LTD. New Delhi-110 034.

Varshney,J.P. (2025). A Fully Illustrated Handbook on Clinical Diagnostic Processes in Canine Medicine. NIPA® Genx Electronic Resources & SolutionS P. LTD. New Delhi-110 034.

Varshney,J.P., Saini, Neetu and Kumar, K.S. (2024).Electrocardiogram Made Easy for Veterinary Students and Practitioners.NIPA® Genx Electronic Resources & Solutions P. LTD. New Delhi-110 034.

Williamson, J.A. and Leone, S. (2012). Non-invasive arterial blood pressure monitoring. In:Advanced Monitoring and Procedures for Small Animal Emergency and Critical Care. Jamie M., Burkitt Creedon, and Harold Davis (Eds.). Online ISBN: 9781118997246 |DOI:10.1002/9781118997246. John Wiley & Sons, Ltd

14

Neurological Examination of Farm Animals

Large Ruminants

Neurological examination of farm animals is an essential part of the physical examination, especially when there's a suspicion of neurological disease. The examination begins with observing the animal's gait and how it interacts with its surroundings. The main goal is to determine whether the neurological issue is primary (originating in the nervous system) or secondary (caused by something else) and to pinpoint where the problem is in the nervous system (central nervous system -CNS or the peripheral nervous system –PNS).

When examining an animal, it is best to start with procedures that cause the least distress. Painful tests or more invasive checks should be done at the end of the examination.

Etiology of neurological diseases in farm animals varies from bacteria, viruses, toxins, nutritional problems, to birth defects leading to widespread neurological issues with similar symptoms. Diseases like polioencephalomalacia (PEM), listeriosis, and nervous ketosis can be treated or prevented, while others, such as Bovine Spongiform Encephalopathy (BSE), is untreatable and of global concern.

The field of large-animal neurology is not as advance as of small animals and humans. Equine neurology has seen some significant advances. Diagnosing neurological conditions in large animals using imaging techniques, like X-rays or myelography, is challenging owing to animal size and difficulty in handling. Performing neurological examination on comatose or sedated large animals is difficult.

Conducting a neurological examination in farm animals involves a systematic approach to assess the nervous system, including the brain, spinal cord, peripheral nerves, and muscles.

Basic Structure and Function of Nervous System

Animals have a complex nervous system. It receives and interprets the signals from the surroundings, and sends the messages to the brain. The structure of the nervous system depends upon the body plan of an organism e.g. animals that do not have a defined head or tail, they have web-like arrangements of nerve cells throughout the body. Animals with a defined head have the nervous system divided into three parts: the central nervous system, peripheral nervous system, and autonomous nervous system. Central nervous system comprises of the brain and neurons located in the head and continues along the back (spinal cord). Peripheral nervous system includes all the nerves continuing from the central nervous system to the entire body. Autonomous nervous system consists of sympathetic and parasympathetic nerves.

Equipment for Neurological Examination

The following equipment are used in neurological examination.

- Reflex Hammer: To test deep tendon reflexes.
- Penlight or Flashlight: Helps examine pupillary light reflexes.
- Stethoscope.
- Cotton Swabs or Gauze: To test sensory responses.
- Hoof Testers: Specifically for hoofed animals, to test for pain or sensitivity in the hooves.
- Pen or Blunt Object: To assess cutaneous sensation by gently pricking the skin.
- Tape Measure or Ruler: For measuring any muscle atrophy or asymmetry.
- Tongue Depressor: Useful for examining oral and throat reflexes.

Examination of Nervous System

The neurological examination involves observing, palpating, and testing reflexes and responses to assess nervous system functioning. It consists of key areas such as history, behavior, mental status, gait and posture analysis, cranial nerve function, postural reactions, and spinal and withdrawal reflexes.

History –Taking an accurate history is important. Traumatic injuries are sudden in onset. Neoplastic diseases are progressive and infectious diseases may have acute or chronic onset. Pattern of the disease is also important. Convulsions associated with pallor or cyanosis may have their cardiac origin rather than nervous system.

Behavior- Information about behavior is also important. Bellowing, yawning, licking, mania, convulsions, aggression, wandering, head pressing, head shaking and compulsive walking are abnormal behavior.

Mental status evaluation- It assists in determining the probable location of the lesion (the brain, spinal cord, or peripheral nerves). Animals with spinal cord or peripheral nerve lesions typically have a normal mental state, though they may become dull or depressed if in pain. Brain lesions can cause signs like dullness, coma, abnormal behavior, head pressing, aggression, seizures, circling, blindness, and facial asymmetry. Eyelid spasms and excessive blinking indicate pain or peripheral nerve involvement, while prolapse of the nictitating membrane is a sign of central nervous system issues, such as tetanus. In routine veterinary practice, components of a neurological examination are often included in a general clinical examination. Most diagnoses are made from the history and clinical findings, but if uncertain, a full neurological examination is conducted.

Mental status and behaviour: When an animal separates itself from the group, it is often a sign of illness. The animal's behaviour also reflects its health status. If it reacts appropriately to external stimuli like sound and movement, it is considered bright. Conversely, if its responses are slow and it shows little interest in typical stimuli, it is described as dull or apathetic. Cattle with carbohydrate engorgement are commonly reluctant to move unless coaxed.

Dummy syndrome- A pronounced state of indifference in which the animal remains standing and can move but does not respond at all to external stimuli. This occurs in sub-acute lead poisoning, listeriosis, and in some cases of acetonemia in cattle, and in encephalomyelitis and hepatic cirrhosis in horses.

Coma is a state in which the animal is unconscious and cannot be roused. The terminal stage of apathy or depression.

Abnormal mentation is manifested in ruminants as inappropriate response to stimuli, social and grazing behavior abnormalities, hyperactivity or excitation, abnormal head carriage or movement or stereotypic behaviors.

- Inappropriate response to stimuli- Assessing whether the animal responds appropriately to environmental stimuli is the key. Normal behavior reflects alertness, while sluggish or indifferent reactions may signal illness. Observing animals in their group context is crucial, as social withdrawal can indicate poor health.
- Social and grazing behavior abnormalities-Separation or aggression within the group may suggest conditions like Bovine Spongiform Encephalopathy

(BSE). Deviations in grazing behavior, such as reluctance to eat or chew, might point to metabolic disorders like ketosis.

- Signs of hyperactivity or excitation- Hyperesthesia (heightened sensitivity) is common in hypomagnesemia, while frenzied behavior can occur in cases of rabies, acute lead poisoning, or nervous acetonemia. Unique behaviors like soundless bellowing or excessive yawning might indicate rabies or liver disease.
- Neurological abnormalities in movement or head carriage-Abnormal head tilt often indicates otitis or vestibular damage, while head aversion may suggest cerebral issues like cerebrocortical necrosis. Tremors, especially head tremors, are associated with cerebellar hypoplasia, often accompanied by ataxia and a wide-based stance.
- Compulsive or stereotypic behaviors- Actions such as self-licking may indicate nervous ketosis. Compulsive circling can signify listeriosis or brainstem lesions. In recumbent animals, attempts to rise may reveal issues like paresis or ataxia.

Posture and gait evaluation- It provides key insights into neurological health of ruminants. Voluntary locomotion is controlled by the cerebral cortex, extrapyramidal basal nuclei, and red nuclei, coordinating flexor and extensor muscle activity.

- Abnormal gait, often caused by cerebellar, brainstem, spinal cord, or peripheral nerve lesions, manifests as ataxia (incoordination) with or without weakness. Clinical assessments focus on coordination and strength, using tools like tight circling, walking over obstacles, and tail-pulling to detect hypermetria, truncal sway, weakness, and proprioceptive deficits.
- Posture, maintained by trunk and limb extensor muscles with input from vestibulospinal and reticulospinal tracts, depends on sensory feedback from pressure receptors, the vestibular system, and vision, processed by the brainstem, cerebellum, and cerebrum. Abnormal head tilt (Fig.118) signals vestibular disease, while trunk deviations (scoliosis, lordosis, or kyphosis) and abnormal limb positions (weakness, knuckling, or a wide stance) indicate upper or lower motor neuron involvement. Slow movement may highlight spasticity, tremors, or incoordination linked to conditions like milk fever or tetanus, while rapid movement can reveal hypermetria in BSE or knuckling due to trauma. Proper restraint and systematic evaluation are essential to identify these neurological abnormalities effectively.

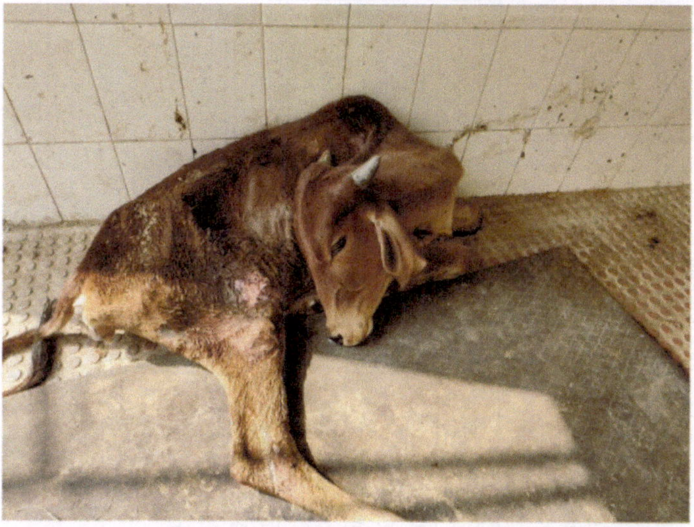

Fig. 118: Abnormal head tilt in a heifer.

Clinical signs associated with lesions of the cerebrum, cerebellum and vestibular disease are summarized in the table 4.

Table 4: .Clinical signs associated with lesions at different location

	Cerebral Disease	**Cerebellar Disease**	**Vestibular Disease**
Mentation	Altered mentation in symmetric lesions; normal in rare early cases.	Normal mentation.	Typically, alert and aware of surroundings.
Gait	Normal gait in symmetric lesions; circling toward the lesion side in asymmetric lesions.	Ataxia with a high stepping gait, truncal sway, and hypermetria.	Ipsilateral head tilt, circling, falling toward the affected side, nystagmusand a staggering gait.
Proprioception	May show deficits in severe cases.	Normal proprioception.	Deficits in balance and coordination (e.g., poor control over head and body movement).
Reflexes	Generally intact; not a prominent diagnostic feature.	Absent menace response; normal spinal reflexes.	Presence of nystagmus, abnormal balance reactions.
Nystagmus	Rarely seen; not a typical feature.	Rare; less rhythmic and consistent compared to vestibular disease.	Consistent, rhythmic nystagmus that is ipsilateral to the head tilt.
Behavioral Signs	Misinterpreted as "head pressing"; more accurately slow forward locomotion interrupted by obstacles. Aversion of the head is a lateral, upward or downward deviation	Intention tremors are often present. Compulsive circling in an unrestricted space	Horizontal or rotary nystagmus. A head tilt is rotation around the long axis of the midline
Common causes	Metabolic imbalances (e.g., dehydration, electrolyte, and acid-base disturbances like hypokalemia). Signs may be caused by diffuse (e.g. cerebrocortical necrosis, hypomagnesaemia, nervous ketosis, lead poisoning), localized (e.g. traumatic injury) or space occupying lesions (e.g. brain abscess or tumor).	Structural or functional cerebellar pathology.	Otitis media/interna, listeriosis (most common), other infections, trauma.

Spinal and withdrawal reflexes- In ruminants, four spinal reflexes (the panniculus, patellar, perineal and withdrawal reflexes for both front and hind limbs) are routinely evaluated sequentially to maintain the animal's cooperation during the examination. Other reflexes, such as the triceps, biceps, cranial, tibial, and gastrocnemius, are not very much reliable. The animal should be relaxed during evaluation of spinal reflexes. Muscle tone is assessed through passive manipulation of each limb, and classified as absent, hypotonic, normal, or hypertonic. Hypertonicity, or spasticity, indicates an upper motor neuron lesion, where the limb shows increased tone that suddenly gives way, allowing complete flexion with minimal resistance.

Panniculus reflex: This reflex (Fig. 119) is tested by lightly touching the skin on the lateral abdomen with a ballpoint pen. It is mediated by the lateral thoracic nerve and spinal segments C8, T1, and occasionally T2 in cattle, and C8 and T1 in sheep and goats. The normal response is contraction of the cutaneous trunci muscle. The reflex is often absent in dermatomes caudal to L3 in ruminants.The panniculus reflex is best examined in lateral recumbency.

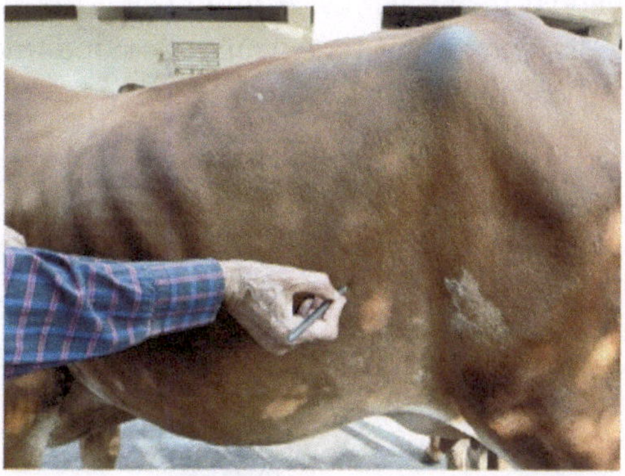

Fig. 119: Panniculus reflex is tested by touching the skin on lateral abdomen with a ball point or finger. The response to this test in normal animal is contraction of cutaneous trunci muscle.

Patellar reflex: It (Fig.120) evaluates the motor and sensory components of the femoral nerve.The patellar reflex can only be tested when the animal is in a laterally recumbent position. To test this reflex, tap the patellar ligament in a laterally recumbent ruminant with a small heavy object (e.g., wrench). The normal response is stifle extension. It requires intact spinal segments L4 and L5 and the femoral nerve. The reflex may be exaggerated with sciatic nerve damage or UMN lesions, and decreased with myonecrosis or femoral nerve paralysis, such as in calves with hip lock during birth.

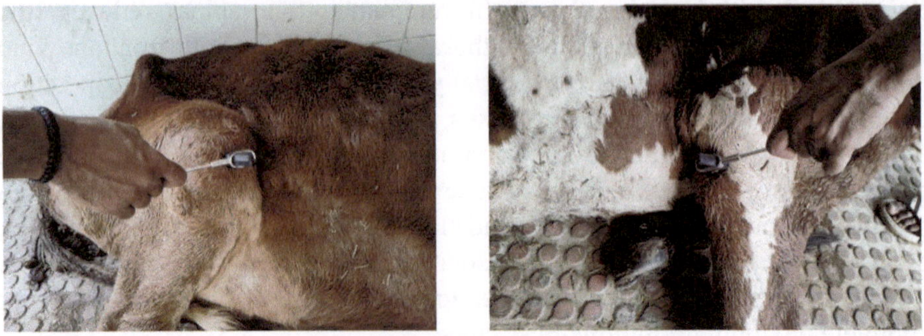

Fig. 120: Patellar reflex test is conducted in recumbent position by tapping the patellar ligament with reflex hammer. Normal response is stiffele extension.The test evaluate motor and sensory components of femoral nerve.

Perineal reflex: The perineal reflex (Fig. 121) can be tested while the animal is either standing or in lateral recumbency.This reflex is assessed by pinching the skin in the perineal area with a haemostat. A normal response is contraction of the anal sphincter and tail ventroflexion. It requires spinal segments S2-S4 and the pudendal and caudal rectal nerves. A diminished reflex can be normal in periparturient cattle, but an absent reflex suggests conditions like a lidocaine epidural, broken tail, or spinal lymphosarcoma.

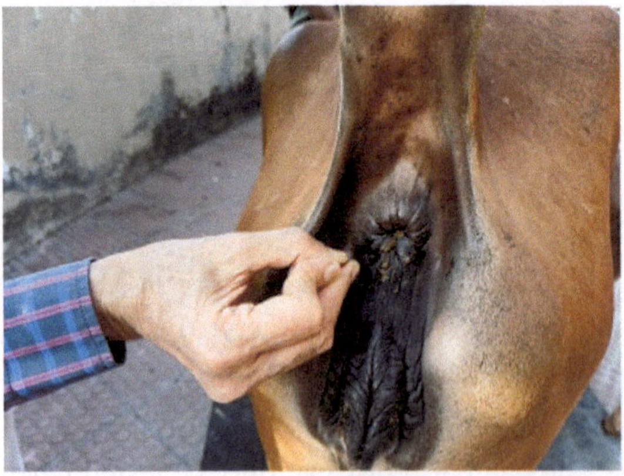

Fig. 121: The perineal reflex is tested while the animal is standing by pinching or touching the skin with needle in the perineal area. Response in healthy normal animal is associated with contraction of anal sphincter and tail ventroflexion.

Withdrawal reflex: The withdrawal reflex (Fig.122) can be tested while the animal is either standing or in lateral recumbency. A firm pressure is applied above the coronary band in standing ruminants or between the claws in recumbent animals. A normal response is limb withdrawal, and awareness of the animal by looking at the stimulus. The reflex requires spinal segments

C6-T2 for front legs and L5-S1 for hind legs, along with the radial and sciatic nerves. A crossed extensor reflex, which causes the opposite leg to extend while the tested leg withdraws, is normal in neonates but abnormal in adults, indicating UMN pathway issues.

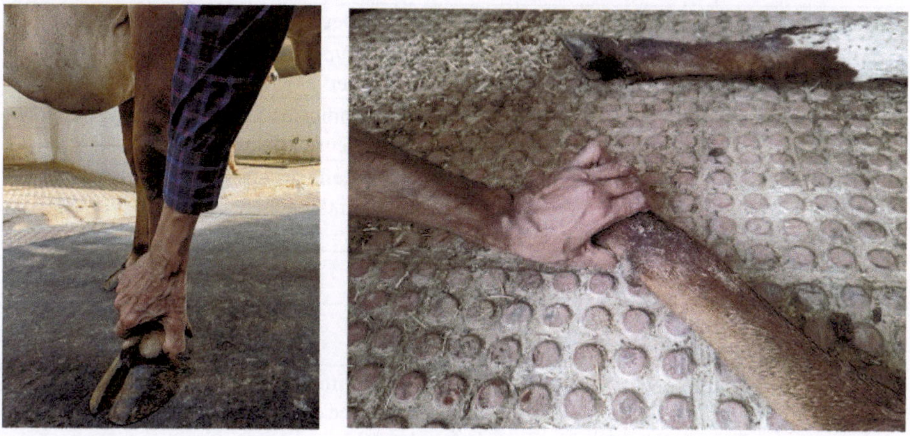

A B

Fig. 122: The withdrawal reflex is tested both in standing (A) or in lateral recumbency (B) by applying firm pressure above the coronary band. Limb withdrawal is the normal response in healthy animal.

Spinal cord lesion localization

- C1 to C6 Lesions: Cause upper motor neuron (UMN) deficits in both the forelimbs and hind limbs.
- C6 to T2 Lesions: Result in lower motor neuron (LMN) deficits in the forelimbs and UMN deficits in the hind limbs, causing forelimb weakness or paralysis with diminished or absent reflexes and milder hindlimb weakness.
- T3 to L3 Lesions: Lead to normal forelimbs but weakness or paralysis in the hind limbs, often with normal or exaggerated reflexes. Affected animals may sit like a dog, relying on their forelimbs for support.
- L4 to S2 Lesions: Cause LMN deficits in the hind limbs, potentially leading to a dog-sitting posture. Additional signs include reduced anal tone, bladder paralysis, and decreased perineal sensation.
- S1 to S3 Lesions: Do not affect the limbs but cause anal sphincter and bladder paralysis.
- Differences in the features of UMN and LMN Lesions is given in the table 5.
- Signs and differentiating features of peripheral nerve disease /injuries are summarized in the table 6.

Table 5: Differences in features of UMN and LMN spinal cord lesion

Feature	UMN Lesions	LMN Lesions
Gait Abnormalities	Proprioceptive deficits, weakness, paralysis, increased extensor tone	Decreased muscle tone (paresis) or absent tone (paralysis)
Muscle Tone	Increased extensor tone	Decreased or absent muscle tone
Spinal Reflexes	Exaggerated reflexes	Decreased or absent reflexes
Proprioception	Impaired (abnormal awareness of limb position in space)	Typically, unaffected unless severe nerve damage occurs
Interpretation	Indicates lesion in the upper motor neuron pathway	Indicates lesion in the lower motor neuron pathway
Other Indicators	Exaggerated spinal reflexes may also suggest LMN lesion in the antagonistic muscle group	Reflexes always decreased or absent in the affected muscles

Table 6: Peripheral nerve disease/injuries

Nerve	Spinal Cord Origin	Key Clinical Signs	Differentiating Features
Brachial Plexus	C6, C7, C8, T1, T2 (cattle); C6, C7, C8, T1 (sheep/goats)	Inability to bear weight; limb appears to "hang"; reduced skin sensation on lateral scapulohumeral joint.	Unable to support weight, differentiating from radial nerve injury.
Suprascapular	C6, C7	Mild abduction at rest; stiff, circumducted limb movement; muscle atrophy without weight-bearing issues.	Weight-bearing unaffected; no scapulohumeral joint laxity despite muscle atrophy.
Radial	T1 (main), C7, C8	Limb dragging, inability to advance limb; reduced skin sensation on metacarpus/ digit.	Weight-bearing possible when limb is positioned under the body, unlike brachial plexus.
Femoral	L4, L5	Inability to bear weight; quadriceps atrophy; absent patellar reflex; difficulty advancing limb.	Patellar luxation possible in severe atrophy; often occurs during difficult parturition. The affected calf may be recumbent or stand with a frog-like posture.
Obturator	L4, L5, L6	Wide-based hind-limb stance (abduction); difficulty correcting posture in extreme abduction.	Absence of fetlock knuckling differentiates it from sciatic nerve damage.
Sciatic	L5, L6, S1, S2	Hind-limb weakness; fetlock knuckling; normal/increased patellar reflex.	Knuckling differentiates it from obturator damage; often occurs during calving. Most commonly seen in a young animal following incorrectly placed intramuscular injections into the gluteal or biceps femoris

Nerve	Spinal Cord Origin	Key Clinical Signs	Differentiating Features
Peroneal	L6, S1 (main), L5, S2	Knuckling of fetlock; dragging digit; loss of skin sensation on anterior metatarsus/digit.	Full weight-bearing possible when digit is manually positioned.
Tibial	L6, S1 (main), L5, S2	Hyperflexion of hock; forward knuckling of fetlock.	Rarely injured due to its anatomical location.

Nerve damage related to calving commonly involves the sciatic, obturator, or peroneal nerves. Of these, the sciatic nerve is the most frequently affected. Therefore, the term "calving paralysis" is preferred over the narrower and often inaccurate term "obturator nerve paralysis" to describe these cases.

Evaluation of Cranial Nerves Function- Evaluation of cranial nerves' function assists in localizing a lesion near or within the brainstem. There are 12 pairs of cranial nerves numbered as CN I, CN II, CN III, CN IV, CN V, CN VI, CN VII, CN VIII, CN IX, CNX, CN XI and CN XII. It is challenging to assess the functioning of CN I (olfactory nerve) and CN XI (accessory nerve) in large animals. Location of the nucleus for cranial nerves and clinical signs associated with their abnormalities are summarized in the table 7.

Table 7: Location of nucleus for cranial nerves and clinical signs associated with their abnormality

Nucleus location	Cranial nerve	Clinical signs
Thalamocortex	I: Olfactory nerve	Loss of smell
	II: Optic nerve	Loss of vision
Midbrain	III: Oculomotor nerve	Pupil dilatation and ventrolateral strabismus
	IV: Trochlear nerve	Dorsolateral strabismus
Pons	V: Trigeminal motor nerve	Loss of facial sensation and motor function of the muscles of mastication
Rostral medulla	V: Trigeminal sensory nerve	
	VI: Abducent nerve	Medial strabismus and unable to withdraw the eyeball deeper into the orbit
	VII: Facial nerve	Loss of motor function to the muscles of facial expression
	VIII: Vestibulo-cochlear nerve	Loss of hearing (cochlear) and/or nystagmus and head tilt
Caudal Medulla	IX: Glossopharyngeal nerve	Dysphagia
	X: Vagal nerve	
	XI: Accessory nerve	
	XII: Hypoglossal nerve	Tongue paresis or tongue atrophy

Determining whether deficits are unilateral, bilateral, or multifocal is crucial, as this can help prioritize the differential diagnosis list.

CN I (Olfactory Nerve): Responsible for smell; difficult to assess in animals as it relies on their reaction to odors.

CN II (Optic Nerve): Optic nerve is related to vision. Eye preservation tests are conducted for vision acuity. The primary tests for visual acuity in animals are menace response and way finding an obstacle course.

The menace response test- It evaluates cranial nerves II and VII, along with the entire visual pathway, including the retina, optic nerves, chiasm, tracts, radiation, and occipital cortex. It also requires an intact cerebellum and CN VII. This response, characterized by blinking and head withdrawal, is not innate but develops within the first week in farm animals. It is absent in neonates, non-visual ruminants, and those with advanced cerebellar disease. The test (Fig.123) involves a single finger moved toward the eye (30-50 cm away) to avoid air currents influencing the response, with transparent plastic interposed if necessary to ensure accuracy. In cows, covering the non-tested eye is unnecessary due to lateral eye positioning. Additional methods, such as dropping light objects like feathers or handkerchiefs to assess gaze tracking, can be used in young ruminants before the menace reflex develops. Throwing cotton wool balls is considered too slow and imprecise for assessing vision in ruminants. Finally, ophthalmoscopic examination is essential for directly evaluating the optic nerve.

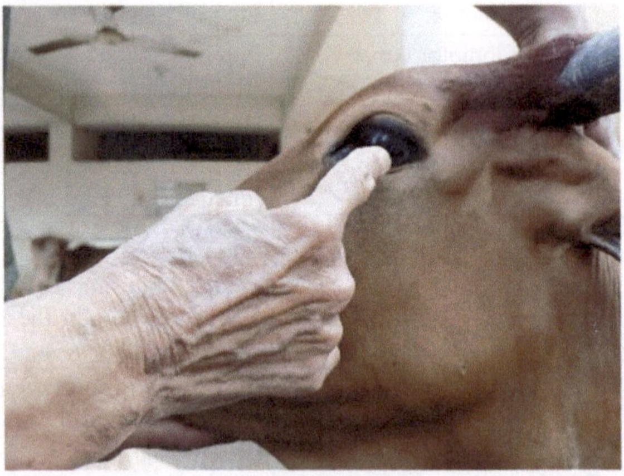

Fig. 123: Menace test in a cattle. It is conducted by taking a single finger towards eye (30 cm away from eye) without creating an air current. Blinking of eye and head withdrawal is a normal response.

Fixation reflex - Fixation reflex involves the head and eyes tracking a moving object, such as a falling handkerchief. This reflex pathway includes the retina, optic tract, rostral brain, posterior cerebrum, and cranial nerves II, IV, VI, and/or cervical nerves.

CN III (Oculomotor nerve): Abnormalities of oculomotor nerve results in papillary dilatation and defective papillary constriction (on increased light intensity), defective eye wall movement or position (ventrolateral deviation) and palpebral ptosis. The functioning of CN III can be evaluated by menace reflex (described above), and papillary light reflex tests.

Pupillary light reflex - The animal should first be kept in a dark environment for a few minutes. A bright light is then directed into one eye (Fig. 124), eliciting a direct response (constriction of the illuminated pupil) and an indirect or consensual response (constriction of the opposite pupil). Blindness caused by occipital cortex damage does not affect this reflex, as the reflex pathways remain intact. However, blindness with dilated, non-responsive pupils suggests a lesion in the retina, optic tract, or brainstem. In cows with severe milk fever, dilated pupils that do not respond to light result from flaccid iris paralysis. The test assesses cranial nerves II and III and relies on bright light and dilated pupils for accuracy. Ruminants have a slower reflex response compared to other domestic animals, and the lateral placement of their eyes makes evaluating the consensual reflex difficult with a single observer.

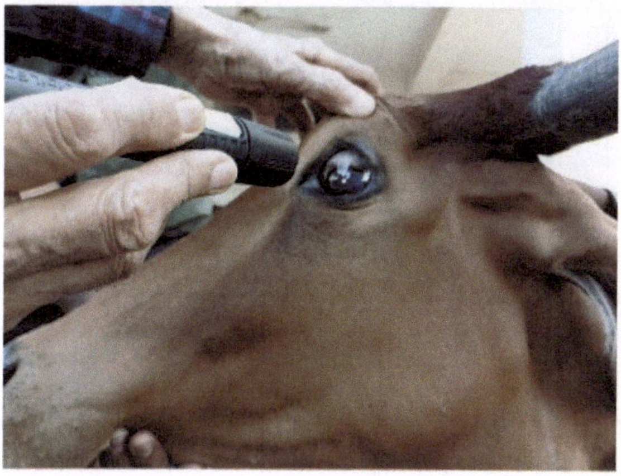

Fig. 124: Pupillary light reflex is conducted by directing a bright light into one eye. Constriction of pupil is a normal response.

CN IV (Trochlear nerve): It supplies the dorsal oblique muscle of the eye. Its injury causes external movement and position of the eyeball (dorsolateral fixation). In polioencephalomalacia in cattle there is dorsomedial fixation of eye ball.

CN V (Trigeminal nerve): Its sensory fibers supplies to face and its functioning can be examined by palpebral reflex test and sensitivity of the face. Its motor fibers supplies the muscles of mastication. Its abnormality results in to chewing defect, abnormal jaw movement and asymmetrical muscle contraction.

Palpebral reflex test - It is assessed by gently touching the skin around the eye without allowing the animal to see the finger's approach. In ruminants with intact CN V, CN VI, and CN VII, the normal response includes eyelid closure, eye retraction, and head aversion. However, this reflex can be challenging to evaluate in depressed animals or those with CN VII damage.

Corneal reflex test - The corneal reflex test (Fig. 125) provides an assessment of functioning of cranial nerves V and VII. It is conducted by gently touching the cornea after slowly approaching the eye. A normal response, indicating intact CN V, CN VI, and CN VII, includes eyelid closure, eye retraction, and head aversion. Touching the cornea near the medial canthus evaluates the ophthalmic branch of CN V, while touching near the lateral canthus tests the maxillary branch. This reflex can be challenging to assess in depressed animals or those with CN VII dysfunction.

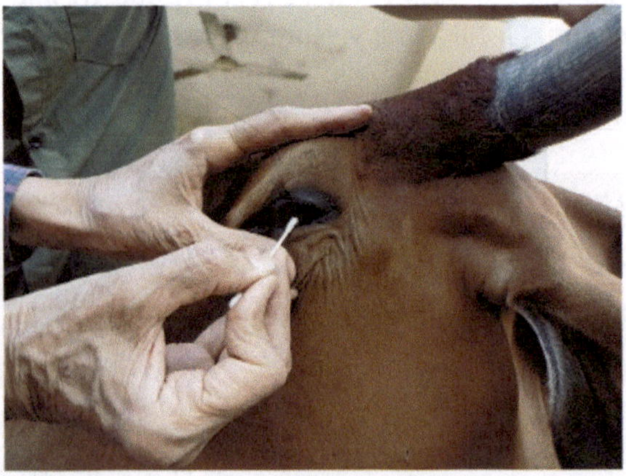

Fig. 125: Corneal reflex test is conducted by gently touching the cornea by a cotton bud. Eye lid closure is a normal response.

CN VI (Abducent nerve): Its motor fibers supply to the retractor and lateral rectus muscles of the eye ball. Injury to CN VI is reflected as protrusion and medial deviation of the globe.

CN VII (Facial nerve): CN VII supplies motor fibers to ear, eye lid, lips, and nostrils and also to menace, palpebral and corneal reflexes. Therefore abnormalities of these may give an indication of CN VII injury. Facial paralysis is one of the indication of CN VII injury.

Photomotor reflex- The photo motor reflex (Fig.126) causes the eyelids to close when a bright light is directed into the eye. This reflex pathway involves the retina, optic tract, anterior brainstem, posterior cerebrum, and cranial nerve VII.

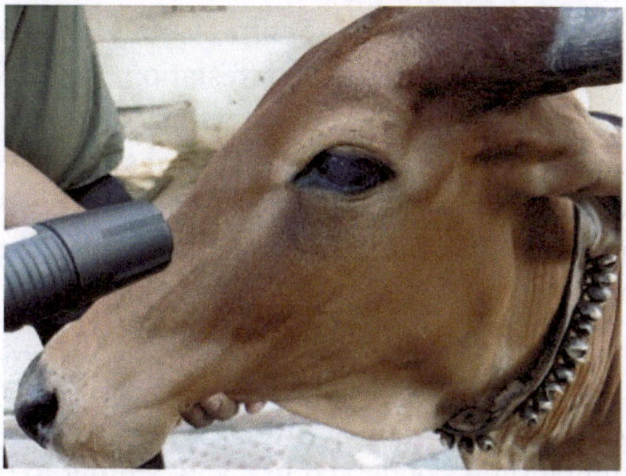

Fig. 126: Photomotor reflex test is conducted by directing a bright light into the eye. Closure of the eye lid is a normal response.

CN VIII (Vestibulocochlear nerve): Abnormality of vestibular part of the nerve is reflected in head rotation around long axis. Cochlear part abnormality is difficult to assess by clinical examination. Failure to respond sudden sharp sound, without creating air current and out of site, indicates deafness.

CN IX and X (Glossopharyngeal nerve and Vagus nerve): Dysfunction of glossopharyngeal and vagus nerve causes paralysis of pharynx and larynx accompanied with dysphagia, regurgitation through nostrils, voice abnormality and respiratory interference.

CN XI (Accessory Nerve): It controls neck and shoulder movement; hard to test in animals due to subtle and overlapping muscle actions. Injury to this nerve is infrequent.

CN XII (Hypoglossal nerve): The nerve supplies motor fiber to the tongue. Therefore CN XII dysfunction is reflected in the motor abnormality of the tongue (tongue protrusion, deviation, fibrillation leading to difficult food prehension and water uptake).

Pigs

The neurological examination of pigs is conducted in a way similar to other animals except some differences in approach owing to anatomical and behavioral characteristics. Change in the behavior of pigs (aggression, lethargy, or hyper excitability) can be easily appreciable. Frequent head shaking in pigs is mostly due to aural hematomas or mange infestations (*Sarcoptes scabiei*). Lateral deviation of spine and swelling of longissimus dorsi muscle may be seen in cases of acute myopathy due to deficiency of vitamin E or selenium. Twisting and nose deviation is also seen in chronic atrophic rhinitis. Hereditary humpy-backed syndrome is manifested by abnormal body shapes. Excessive self-rubbing may also be due to mange or lice infestation. Head tilt may also be due to vestibular disease. Seizures accompanied with nystagmus and opisthotonos, is seen in meningitis, water deprivation or classical swine fever. Incoordination in pigs may also occur in bowel edema. Spinal injury or abscessation may also lead to ataxia. Muscle tremors during sleep and disappearing on awakening of the pigs are normal

Sheep

Neurological disorders of diversified etiology have been reported in sheep. Neurological examination is conducted in a way similar to other species.

Asymmetric head conformation in sheep can also be ascribed to neurological dysfunction. Unilateral cheek bulging (due to cud retention), lingual paralysis, and drooping of the ear or eyelid is the result of facial paralysis indicating cranial nerve impairment. Nystagmus, an oscillatory eye movement, is often associated with lesions in the vestibular apparatus or cerebellum. Prehension loss and excessive drooling is seen in flaccid paralysis. Muscle tone for spasticity (as seen in tetanus) or flaccidity (such as in botulism or hypocalcaemia) needs to be assessed. In cases of tetra paresis and paraparesis, evaluation of spinal reflexes and motor response should be done. Tests such as proprioception, skin sensation, panniculus response, deep pain, and reflexes (pedal, triceps, and patellar) are recommended for localizing the lesion. The wheelbarrow and hemi walking tests help detect proprioceptive or motor deficits, with a wide-based stance suggesting proprioceptive abnormalities or cerebellar lesions.

Some neurological signs can assist in identification of lesion sites. Cerebellar lesions often associated with head tremors, head tilt, dysmetria, wide-based stance, bilateral blindness, and nystagmus, while cerebral lesions are associated with head aversion, wide circling (ipsilateral), tight circling (contralateral), depression or excitability, and unilateral blindness (contralateral). *Taenia multiceps* (*Coenurus cerebralis*) cysts, cause of gid in sheep and goats, typically affects the cerebrum or cerebellum. Its infestation is associated with neurological symptoms. Scrapie or *Psoroptes ovis* infections cause intense rubbing along the back near the tail base and trigger a nibbling reflex.

Goats

Neurological examination is conducted in almost same way as in other species. Ataxia and occasionally coma is seen in cases of ruminal acidosis. Copper deficiency may also lead to swayback, ataxia and hind limb weakness. Around kidding, hypocalcemia may lead to muscle tremors, ataxia, and recumbency; and pregnancy toxemia may be marked by depression, ataxia, and coma. Floppy kid syndrome may also be associated with severe muscle weakness, lowered consciousness and diminished reflexes because of metabolic acidosis.

Caprine arthritis encephalitis virus infection in kids is also manifested with neurological signs (ataxia, hemiplegia or tetraplegia, blindness and coma).

Adult goats infected with CAEV encephalitis may show limb incoordination, progressing to paralysis. Kids affected with swayback develop progressive hind limb ataxia due to demyelination from copper deficiency, though they remain alert and show interest in eating. Scrapie affects goats around 3 to 4 years age. Initial symptoms are mild and non-specific, such as slight temperament changes, weight loss, and reduced milk production. Later, distinct signs such as pruritus, skin nibbling or scratching and neurological symptoms like incoordination, high-stepping gait, muscle tremors and changes in temperament develop.

References and Further Readings

Constable, P. D., Hinchcliff, K. W., Done, S. H., & Gruenberg, W. (2016). Veterinary Medicine: A Textbook of the Diseases of Cattle, Horses, Sheep, Pigs and Goats - Two-Volume Set. Saunders Limited.

Jackson, P. and Cockcroft, P.(2002). Clinical Examination of Farm animals-Vet Books. Agriculture and Food Sciences, Medicine

15

Neurological Examination of Pet Animals

A neurological examination is an advanced and specialized clinical examination that is performed when there is suspicion of potential nervous system involvement. Before performing neurological examination a detailed history is essential. Neurological examination along with history not only helps in confirming whether the neurological problem is of primary or secondary origin but also aids in accurately identifying the location of the lesion.

Common Terminology used in Neurology

The following terms are commonly used in neurology. For a better understanding, common neurological terms with their definition are given below:

- Ataxia - An abnormal gait marked by lack of coordination, but without spasticity, weakness, or involuntary movements. Ataxia results from damage to the proprioceptive pathways, the vestibular system, or the cerebellum.
- Circumduction - It refers to the movement of a limb where the distal part traces a circular path while the proximal part remains stationary.
- Coma - an unconscious state where vigorous application of a noxious stimuli does not induce arousal. Spinal reflexes might be normal in such cases.
- Dysmetria - A condition in which the ability to control the range, force, or distance of a movement is impaired, leading to movements that are either too short (hypometria) or too exaggerated (hypermetria). It is often linked to cerebellar dysfunction.
- Hypermetria - Exaggerated movements of the limbs during locomotion, more easily observed in the front legs during their voluntary advancement.
- Nystagmus - Inappropriate involuntary, rapid, and repetitive eye movements which can be horizontal, vertical, or in a rotary pattern. It may be associated with problems of various underlying issues (vestibular disorders, neurological problems, or certain medications).

- Opisthotonous: - Dorsiflexion of the head and neck.
- Proprioception - It refers to the ability to sense the position and movement of body parts in space. It involves the integration of muscles, joints, and tendons, allowing dogs to maintain balance, coordinate movements, and navigate their environment effectively.
- Seizure - Involuntary episode of muscular activity that may be generalized or localized.
- Stupor - Animal tends to sleep when undisturbed, but vigorous application of a noxious stimulus will induce arousal.
- Tremor - Small and rapid alternating contractions of muscle groups that occur at rest or during movement.
- Truncal ataxia - It is swaying of the body (trunk) from side to side as it attempts to walk forward.
- Weakness - Loss of muscle strength.
- Non ambulatory animal - Capable of walking on all four limbs, supporting its body weight, and advancing without assistance.
- Ambulatory animal - Unable to support its weight (on all four limbs or specifically on hind limbs) or walk.

Neurologic Signs Caused by Systemic Diseases

Neurologic signs in systemic diseases typically arise in the cerebral cortex or the peripheral nervous system (PNS). Changes in white matter and signs associated with the brainstem and cerebellum, have also been observed as secondary effects to systemic conditions. The following systemic diseases are associated with neurological signs.

- Renal encephalopathy - Toxic substances not excreted due to kidney failure can cause encephalopathy manifesting many neurological signs.
- Hypothyroidism – Acute or chronic progressive central vestibular signs may be observed in hypothyroidism.
- Hyperthyroidism – In cats hyperthyroidism causes change in sleep-wake cycle, aggression, or obtundation.
- Pituitary dependent hyperadrenocorticism (PDH) - Its clinical manifestations include inappetence, mild obtundation, disorientation, and pacing in dogs.
- Hypoglycaemia - Persistent hypoglycaemia (caused by poor nutrition, insulinoma, liver failure, or insulin overdose), can lead neurologic signs such as obtundation, weakness, disorientation, tremors, seizures, blindness, and coma.

- Thiamine deficiency – It may cause lesions in cerebrum and vestibular nuclei giving rise to many neurologic signs.
- Hypokalaemia - By altering resting muscle cell membrane potential it can lead to generalized muscular weakness in dogs and cats and ventroflexion of the neck in cats.
- Hypocalcaemia – Tetany and seizures have been observed in pets with severe hypocalcaemia (lactating pets, pets with CKD or hypoparathyroidism) because of increase in membrane excitability.
- Neoplasia - Neoplasia can cause neurologic signs from direct invasion or metastasis, or secondary to ischemic and hemorrhagic infarcts.
- Hyperthermia – Damage to cerebral cortex and hypothalamus in dogs with hyperthermia leads to neurological signs such as seizures, disorientation, and coma.
- Hyperparathyroidism - Weakness may result from renal secondary hyperparathyroidism leading to peripheral neuropathy and myopathy.
- Diabetes Mellitus (DM) – Sensory motor polyneuropathy, a late-stage complication of diabetes mellitus, can result from metabolic imbalances or vascular alterations. It is common in diabetic cats. The condition primarily affects the pelvic limbs, causing a plantigrade stance, jumping difficulties, weak postural reactions, reduced tendon reflexes, and muscle atrophy.
- Hyperadrenocorticism - It often leads to muscle weakness in dogs, occasionally presenting with fibrotic myopathy or polyneuropathy.
- Hypoadrenocorticism – It may cause lethargy, weakness, tremors, and rare muscle cramps, sometimes accompanied by hypoglycaemia.
- Systemic Hypertension - Rapid increase in blood pressure may cause hypertensive encephalopathy with manifestation of seizures, ataxia, stupor, and blindness. Hypertensive cats may develop thrombosis, causing myelopathy or encephalopathy.
- Hepatic encephalopathy - Hepatic encephalopathy (HE) in dogs (from porto systemic shunt -PSS or liver failure) is associated with nervous symptoms ranging from cognitive issues to seizures.

Equipment/Instruments for Neurological Examination

- Reflex hammer.
- Cotton tip applicator.
- Pen torch light or transilluminator.
- Cotton balls.

- Haemostats.
- Lens.

Anesthesia/ Sedative

- Generally no anesthetic/ sedative is needed for neurological examination.

Place of neurological examination

- Neurological examination should be conducted in a calm and quite place.
- Non-slippery rough surface helps in detecting subtle defects.

Examination procedure

Neurological examination is conducted to evaluate mentation, posture and gait, cranial nerves, spinal reflexes, postural reaction, palpation of spine and muscle, and assessing pain perception.

(a). Evaluation of Mentation- It is a visual assessment of the patient (Fig.155). Whether, the animal is normal (bright, alert, responsive ,calm), dull or obtund (drowsy, inattentive but reactive to auditory stimuli such as calling its name, noise, clapping), stuporous (sleepy needing strong stimulation to show reaction), comatose (unconscious) or showing behavioral changes (agitation, aggression, dementia or obsessive compulsive behavior). Different mental states, their clinical characters and association with involvement of brain parts is summarized in the table 8.

Table 8: Mental states, their clinical characters and association with involvement of different brain parts.

Mental State	Clinical Characteristics	Associated part of Brain involved
Normal	The pet is alert and responds appropriately to environmental stimuli	
Depressed	The pet is dull, drowsy, and baffling but responds to environmental stimuli	Metabolic disturbance, systemic illness, or damage or disease affecting the cerebrum or brainstem
Delirious	The pet is alert, agitated or confused. Response to stimuli is inappropriate.	Cerebral cortical disease or a metabolic encephalopathy.
Stuporous	The pet is unconscious but aroused by strong painful stimuli.	Metabolic disturbance, systemic illness, or damage or disease affecting the cerebrum or brainstem
Comatose	The pet is in the state of deep unconsciousness and cannot be aroused even with noxious stimuli.	Brain stem lesion.

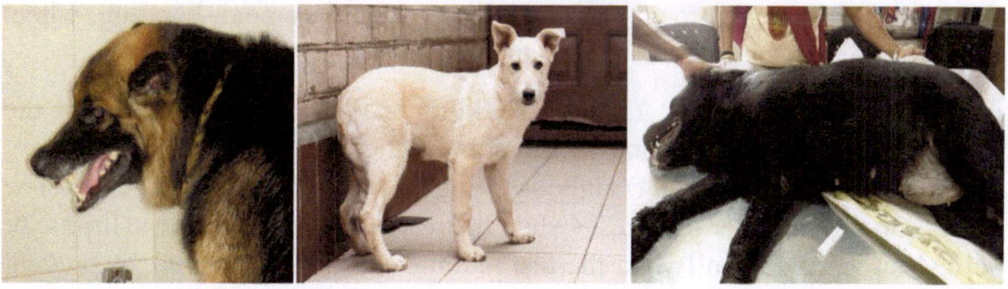

A B C

Fig. 127: Different mentation in dogs. **A.** Aggression. **B.** Alertness. **C.** Coma. Figures are with the Courtesy of Varshney, J.P. (2025).

(b) Assessing posture - The orientation of the animal is to be noted. Whether the spine is normal, curved dorsally (kyphosis), curved ventrally (lordosis), deviated laterally (scoliosis), twisting of neck (torticollis), head tilt, head turn, wide base stance or dorso flexion of head and neck (opithotonus). Whether the head is held normally or tilted. Neck and head is aligned normally or turned. The dog is assessed in standing position as standing normally (Fig.128 A) or wide based stance.Upright position is the normal posture. Abnormality in posture is an indication of a lesion anywhere along ascending or descending pathways in the peripheral or central nervous system. Marked abnormalities in postural reactions without any change in gait is suggestive of a lesion in cerebral cortex. Head tilt (Fig.128 B) is suggestive of disease affecting the vestibular system. Fore brain lesions may lead to head turn. Postures such as swaying, leaning on objects or wide based stance suggests cerebellar or vestibular system dysfunction.

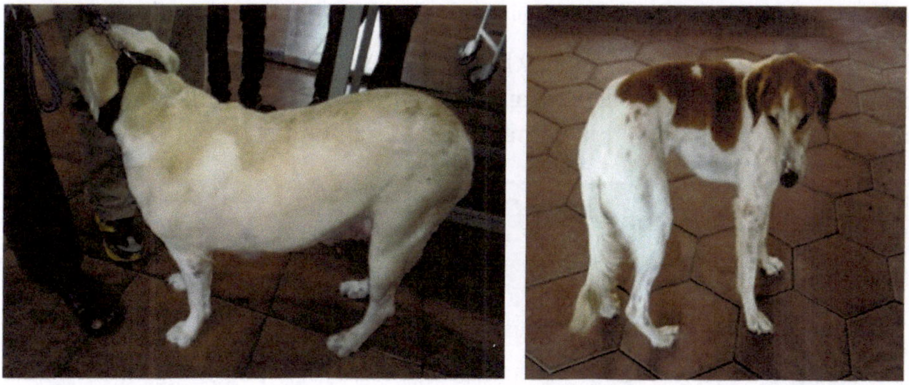

A B

Fig. 128: Different postures in dogs. **A.** Normal posture. **B.** Head tilt posture. Figures are with the Courtesy of Varshney, J.P. (2025)

Postural reaction assessment- It is done by paw replacement, hopping, wheel barrowing, extensor postural thrust, and hemi walking.

Paw replacement test (Fig. 129) or knuckling test is conducted by flexing the paw with the help of few fingers (dorsum of the paw is on the non-slippery floor without bearing weight) and observing its return to normal position. The dog is supported under the chest (when the test is conducted for fore limbs) or pelvis/caudal abdomen (when the test is done for the hind limbs), to prevent loss of balance at the time of paw flexion. A dog with normal proprioception immediately returns the paw to its normal position. Delayed or absent correction (knuckling) indicates a potential deficit in proprioception. This test helps localize lesions affecting motor coordination and sensory pathways.

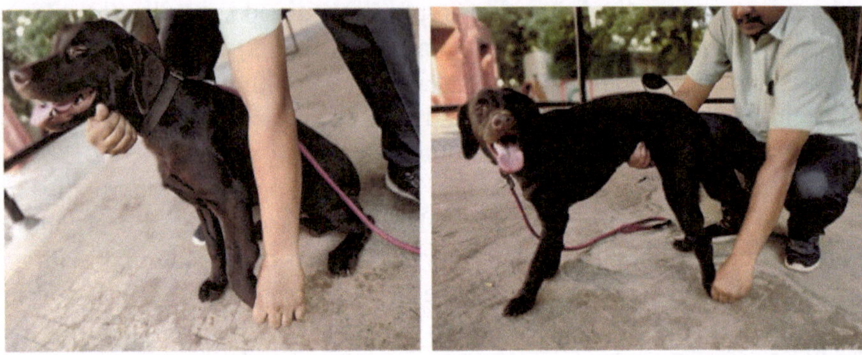

A B

Fig. 129: Paw replacement test in dogs. For paw replacement test in thoracic limb the dog is supported under the chest and the toe is flexed (A).For paw replacement test in pelvic limb the dog is supported under the caudal abdomen or pelvis and then the toe is flexed (B) . Returning of paw to normal position is noticed. Figures are with the Courtesy of Varshney, J.P. (2025).

Hopping test (Fig.130) is done by holding the patient in such a manner that all its weight is on one limb and the animal is moved forward or laterally. Normal animal hops on the limb. Each limb is tested individually.

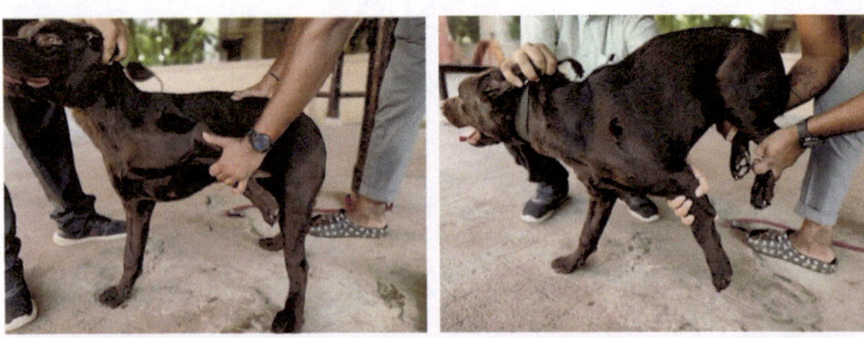

Fig. 130: Hopping Test in dogs. Figures are with the Courtesy of Varshney, J.P. (2025).

Wheel barrowing test (Fig.131) is conducted by supporting the dog under abdomen and lifting the hind limbs from the ground, allowing the animal to bear weight on its fore limbs (thoracic limbs) and then move the animal forward in the manner as wheel barrow is pushed.Normal dogs walk forward with coordinated movement.

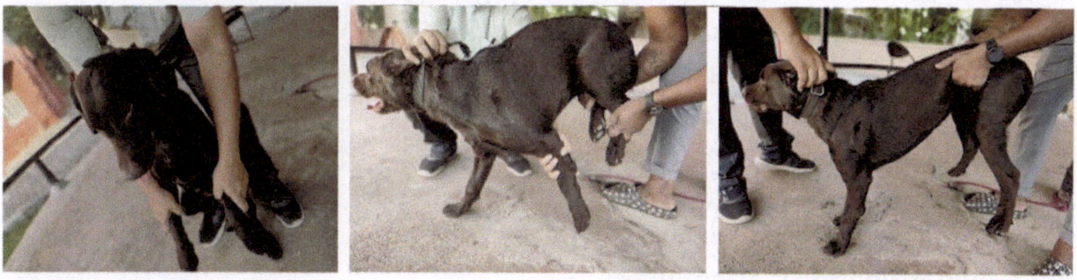

Fig. 131: Wheel –barrow test in dogs. Figures are with the Courtesy of Dr.J.P.Varshney (2025).

Extensor postural thrust test is conducted by supporting the dog under the thorax to lift fore limbs from the ground and allowing weight bearing on pelvic (hind) limbs. Then the animalis moved in symmetric walking movements to achieve a position of support. The dog should not be walked backwards (i.e., reverse wheelbarrowing).

Hemi-walking test is conducted by lifting the front and rear limbs of one side and all weight bearing on the limbs of opposite side (if fore and hind limbs of left side are lifted then weight bearing is on the fore and hind limbs of right side). The dog is forced to hop with the limbs on the ground and the lateral walking movement is evaluated.

Postures such as ophisthotonus (dorso reflection of head and neck), decerebrate rigidity (all four limbs rigidly extended, stupor or coma and dorsal extension of head and neck indicating severe rostral brain stem lesion), decerebellate rigidity (extended thoracic limbs with flexed pelvic limbs, opisthotonus with normal mentation owing to an acute lesion in the rostral cerebellum) or Schiff-Sherrington posture (increased tone in thoracic limbs and hind limb paralysis. It may be due to acute thoracic or cranial lumbar spinal cord trauma) are observed rarely.

(c) Assessment of the gait- Gait refers to the movement pattern of an animal's limbs. It is to be observed before a hands-on examination. The gait should be assessed on a flat, non-slippery surface, where the pet can walk, make frequent turns, and move in circles. Pets requiring assistance should be supported with a harness or sling to better assess their voluntary movement and gait. Changes in gait are sometimes secondary to paresis, ataxia, or pain. The gait should be observed for signs of weakness (paresis) or paralysis (plegia),

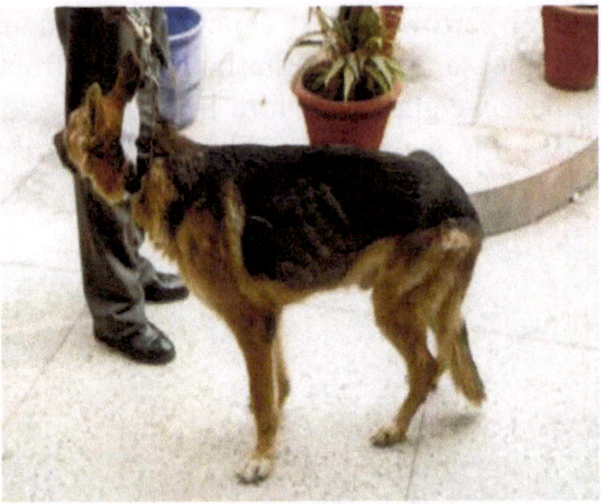

Fig. 132: Hind limb ataxia in a German shepherd. Figure is with the Courtesy of Varshney, J.P. (2025).

ataxia, or lameness. Gait is to be observed from the front and back while the animal is walking in a straight line. For detecting subtle gait defects, the animal is to be observed while going up and down the stairs. Gait abnormalities are described as mix of weakness, ataxia, paresis/plagia or lameness. Ataxia (Fig.132) is classified as proprioceptive (symmetric and mild incoordination), vestibular (asymmetric, falling or drifting to one side of mid line), or cerebellar (symmetric, bouncy gait with hypermetria without loss of strength in limbs). Paresis is characterized by reduced voluntary motor function. It is associated with certain degree of weakness. It may be neurogenic or muscular. While plegia (Fig.133) is characterized by loss of motor function. It may be monoplegia (single limb paralysis), hemiplegia (one side body paralysis), paraplegia (hind limb paralysis) or tetraplegia (all four limbs paralysis). Lameness can be either of orthopedic or neurologic origin. Motor function should also be assessed. Tetraparetic/plegic dogs or cats require support under the pelvis. Animals are encouraged to move forward to see their voluntary movement. Motor function can be assessed by observing response on calling its name and walking with the animal. Response on touching is a reflex action and should not be considered voluntary movement.

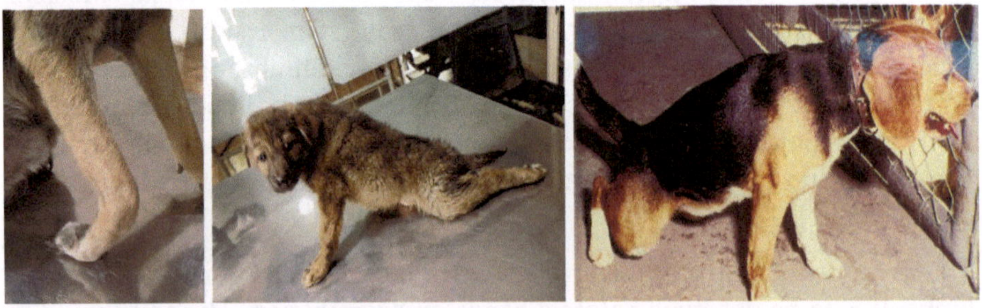

A B C

Fig. 133: Plegia in dogs. **(A)**. Right fore limb paralysis in a male German shepherd. Note the paw is flexed and the dog is not able to keep the paw in right position. **(B)** Paraplegia in a three month old non-descript dog caught by other dog. **(C)** Paraplegia in another dog. Note there is complete loss of motor function in the hind limbs. Figures are with the Courtesy of Varshney, J.P. (2025).

Animals with pain in all limbs typically show a stiff, short-stride gait (lameness), commonly observed in conditions such as polyarthritis. When only one limb is affected, that limb has a shortened weight-bearing phase, while the opposite limb bears weight for longer than usual. Single limb involvement is often linked to orthopedic diseases, spinal nerve or nerve root entrapment due to lateralized disk extrusion, or nerve root tumors. Painful limbs may also be lifted, elevated, or carried by the animal.

Circling gait is linked to lesions in the forebrain or vestibular system. In cases of unilateral forebrain lesions, the affected animal typically walks or paces in wide circles toward the side of the lesion. In contrast, vestibular disorders often lead to tight circling toward the lesion side, accompanied by ataxia, loss of balance, head tilt, and nystagmus.

(d) Cranial Nerve Evaluation-Cranial nerves are peripheral nerves originate from brainstem and perform both sensory and motor functions. There are 12 cranial nerves namely olfactory (CN I), optic (CN II), oculomotor (CN III), trochlear (CN IV), trigeminal (CN V), abducent (CN VI), facial (CN VII), vestibulocochlear (CN VIII), glossopharyngeal (CN IX), vagus (CN X), spinal accessory (CN XI) and hypoglossal (CN XII) nerve. Cranial nerves evaluation is done through reflexes (cranial nerve reflex and spinal reflex) and/ or reactions (sensory pathway, motor pathway and cerebrum). Their function can be assessed through standard tests.

CN I (olfactory) is evaluated to ascertain patency of nares for smell in sniffers, detective or hunter dogs. The test is conducted by covering the eyes and presenting the food beneath the nose. The sniffing behavior is observed. Primary neurological problems related to CN I are rarely seen.

Fig. 134: Menace response test in a dog. The test is performed by waving a hand towards patient's face in a threatening manner. Normally it is associated with eye blinking. The image is with the Courtesy of Varshney, J.P. (2025)

CN II (optic nerve) is evaluated by visual following, menace response, papillary light response and fundus examination.The visual following test is conducted by dropping cotton balls, moving a toy or ball in front of the patient and to see whether eyes and head follow the object or not. Menace response test (Fig.134) is performed by waving a hand towards patient's face as a threatening gesture. Normally it is associated with eye blinking. It is used to assess the integrity of optic nerve, facial nerve, forebrain and cerebellum.

CN III (Oculomotor) nerve's motor functioning is evaluated by pupil light response, eye position, and pupil size and observing for physiological nystagmus when turning the head. Pupillary light reflex test (Fig. 135) is done to assess the integrity of optic and oculomotor nerve. The test is done by shining a bright light into one eye and noting the constriction of pupil of the eye being tested as well as other eye also. CN III dysfunction causes ventrolateral strabismus, ptosis, mydriasis, unresponsive pupil, reduced or asymmetrical eyeball movement.

Fig. 135: Pupil light response test in a dog. The test is performed by shining a bright light into one eye and noting the constriction of pupil of the eye being tested as well as other eye also. The image is with the Courtesy of Varshney,J.P.(2025).

CN IV dysfunction leads to lateral deviation of the superficial retinal vein on ophthalmoscopic examination in the dog (round pupil). Motor functioning of CN IV nerve is evaluated by observing for dorsolateral rotation of the pupil and physiological nystagmus.

CN V (trigeminal) nerve has three branches (ophthalmic, maxillary and mandibular). Its sensory fibers innervate to face and motor fibers innervate to muscles of mastication (temporalis and masseter muscles). Sensation over all three branches should be examined. Eliciting a blink response on touching canthus of the eye (palpebral reflex – Fig.136) indicates the integrity of the ophthalmic branch of trigeminal nerve. Wrinkling of face and a blink on touching or pinching the upper lip (Fig.137) lateral to canine tooth indicates the integrity of the maxillary branch of trigeminal nerve. These responses also depends on the motor supply of the facial nerve.The patency of the maxillary branch is tested by pinching or touching the upper lip lateral to canine tooth. The patency of the mandibular branch is tested by pinching or touching the lower lip (Fig.138) lateral to the canine tooth. Neuropathy of mandibular branch of trigeminal nerve is manifested in the form of dropped lower jaw (inability to close mouth), difficulty/inability in eating and drinking and excessive drooling (Fig. 139). Sensory testing of CN V nerve is done by evaluating nasal mucosa response, corneal reflex, palpebral reflex, trigeminofacial reflex (touching area of maxillary vibrissae and observing eye blink). Motor dysfunction of CN V nerve leads to atrophy of the muscles of mastication (temporalis and masseter muscles – Fig.140) leading to inability to open mouth. Palpebral reflex test is performed by touching medial and lateral canthi and observing the blink. Blink in response to touching medial and lateral canthi indicates intact palpebral reflex. This test is used to assess the integrity of trigeminal and facial nerves. Nasal stimulation response test is performed by touching the nasal mucosa on the medial aspect of each nare with a blunted probe (Fig.141). Normally it elicits head withdrawal response. The test is used to assess the integrity of trigeminal nerve and the forebrain. In nut shell trigeminal nerve is evaluated by palpating masseter and temporal muscles for symmetry, atrophy, and pain; checking the jaw tone; looking for retraction response on touching the medial septum of the nose; observing for retraction response on touching globe; observing for blink response on touching medial canthus; and observing for snarl response on lip pinching

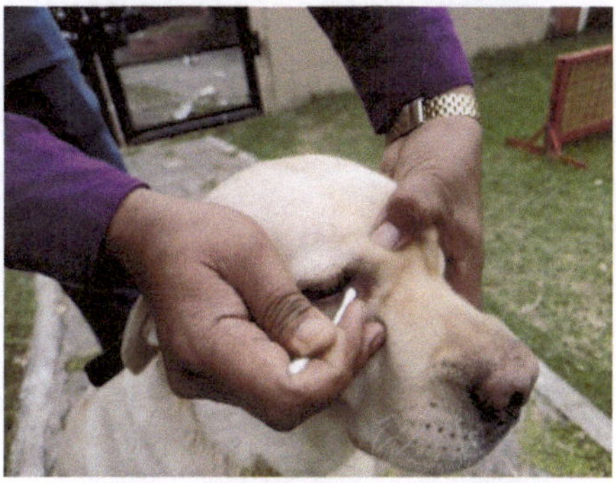

Fig. 136: Palpebral reflex in a dog. The test is conducted by touching medial or lateral canthi and observing the blink response. Figure is with the Courtesy of Varshney, J.P. (2025).

Fig. 137: Testing patency of maxillary branch of trigeminal nerve. The test is conducted By touching or pinching the upper lip and observing wrinkling of face and a blink. Figure is with the Courtesy of Varshney, J.P. (2025).

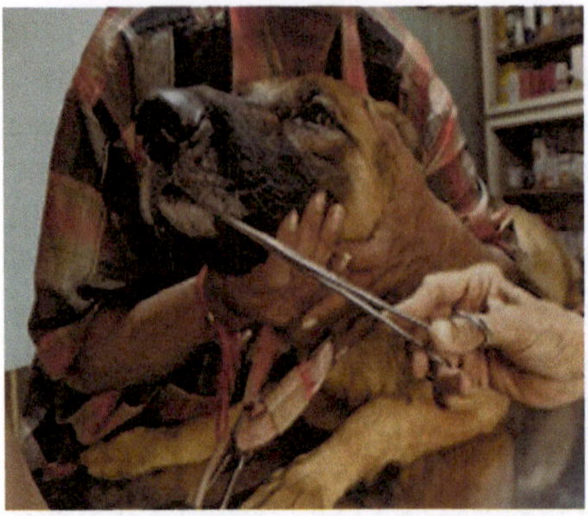

Fig. 138: Testing patency of mandibular branch of trigeminal nerve in a dog. The patency of the mandibular branch is tested by pinching or touching the lower lip lateral to the canine tooth.

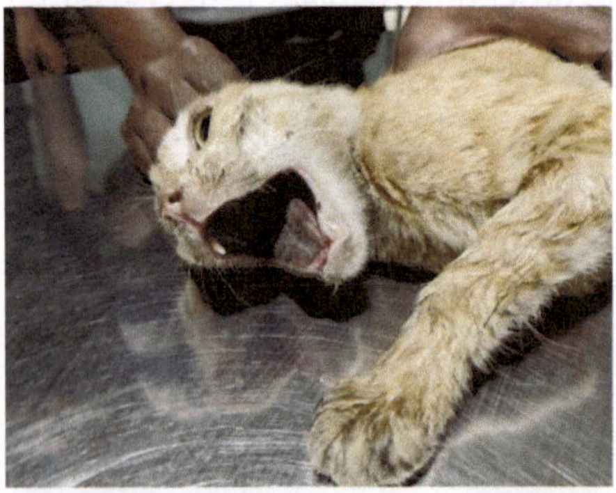

Fig. 139: Dropped jaw due to idiopathic neuritis of mandibular branch of trigeminal nerve in a cat. Radiographic examination was uncommitted in this case. The patency of the mandibular branch was tested by pinching or touching the lower lip lateral to the canine tooth. There was on response on pinching Neuropathy of mandibular branch of trigeminal nerve is manifested in the form of dropped lower jaw (inability to close mouth), inability in eating and drinking and excessive drooling

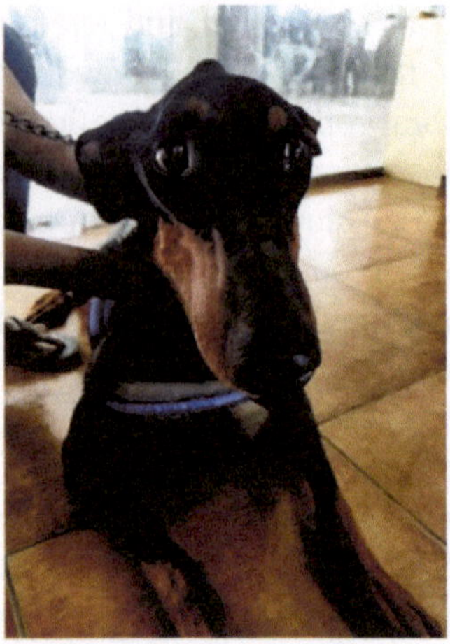

Fig. 140: Dog showing atrophy of masseter and temporalis muscles. The figure is with the Courtesy of Varshney, J.P(2025)

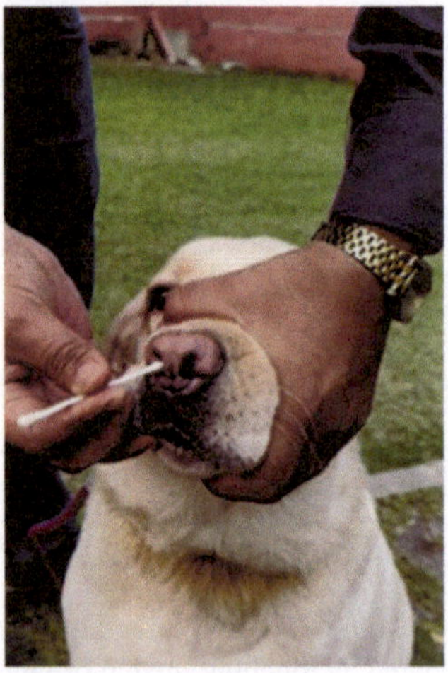

Fig. 141: Nasal stimulation response test in a dog. It is performed by touching the medial septum of the nose on each side with a cotton bud or blunted probe and observing head withdrawal response . The figure is with the Courtesy of Varshney, J.P. (2025).

CN VI nerve (efferent) can be evaluated by corneal reflex (touching cornea and observing retraction –Fig. 142), eye positioning, medial strabismus and physiological nystagmus. Dysfunction of CN VI leads to medial strabismus and reduced movement of eyeball.

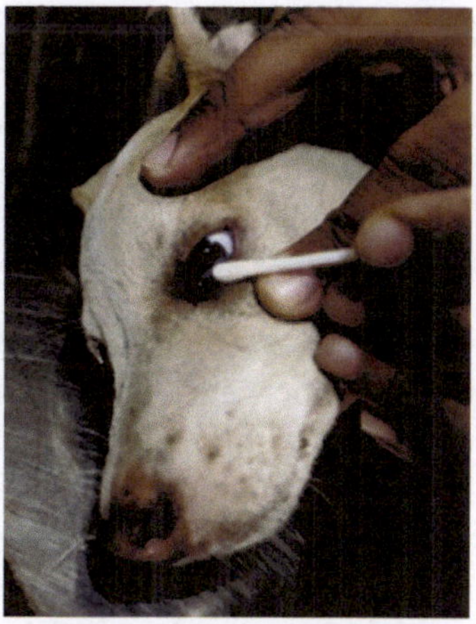

Fig. 142: Testing of corneal reflex in a dog. The test is performed by touching cornea with blunt object such as cotton buds and observing retraction. The figure is with the Courtesy of Varshney, J.P.

CN VII (facial) nerve innervates facial muscles, salivary and lacrimal glands. Its functioning is evaluated by menace response (detailed under CN II), palpebral reflex (detailed under CN V), corneal reflex (detailed Under CN VI), Schirmer tear test (Fig.143 A), ear flick response to medial pinna stimulation (Fig.143 B) and observing for facial paralysis. Asymmetrical eyelid closure, a widened palpebral fissure, spontaneous blinking or drooping ears(Fig.143 C) are suggestive of malfunctioning of the facial nerve.

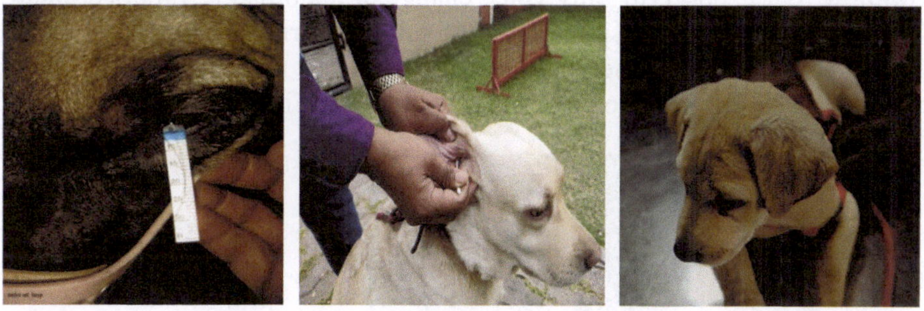

A B C

Fig. 143. Tests for assessing facial nerve in dogs. A. Schirmer tear test for lacrimal gland functioning. B. Ear flick response to medial pinna stimulation. C. Drooping ear due to malfunctioning of facial nerve. The figures at A and B are with the Courtesy of Varshney, J.P. (2025).

CN VIII (vestibulocochlear) nerve mediates hearing and vestibular function. Turning head toward loud unexpected noise is a normal response indicating the normal functioning of CN VIII nerve.Unilateral or partial deafness cannot be definitely ascertained by this crude test of response to unexpected noise. The owner might observe signs of subtle hearing loss such as no readily response to calls. Head tilt (Fig.144), nystagmus, ataxia, broad-based stance are suggestive of vestibular dysfunction.CN VIII nerve is evaluated by observing the head position and gait; and physiological nystagmus.Oculocephalic reflex test is performed by turning the head of the animal from side to side and observing the rapid movement of the eye in the direction of the head turned. The test is indicated to assess the integrity of the vestibulocochlear, oculomotor, trochlear, and abducent nerves.

Fig. 144: Head tilt in a Pomeranian dog, an indication of vestibular dysfunction (CN VIII). The figure is with the Courtesy of Varshney. J.P. (2025).

CN IX (glossopharyngeal) and CN X (vagus) nerves innervate (motor and sensory fibers) to the pharynx. Vagus nerve also controls laryngeal function.The functioning of these nerves is tested by touching (with the help of applicator stick) both left and right sides of the caudal pharyngeal wall (Fig.145 A). Elevation of the palate and pharyngeal muscle contraction (gag reflex) are the response to touching the caudal pharyngeal wall. This test is known as Gag reflex test. It is difficult to perform this test in temperamental dogs. Asymmetrical response is indicative of dysfunction. Functioning of CN IX and X nerves is evaluated by gag swallow reflex, observing any voice change or abnormal respiratory sound; and signs of coughing/ aspiration when eating or drinking. Vagus nerve can also be examined by observing laryngeal paralysis and assessing for mega esophagus and regurgitation by x-ray examination (Fig. 145 B).

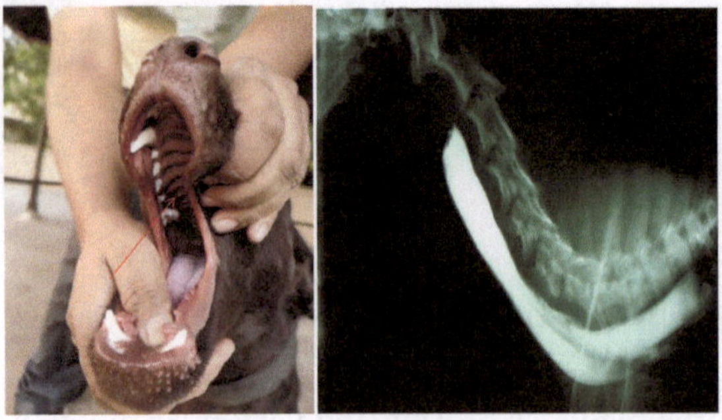

A B

Fig. 145: Tests for the functioning of glossopharyngeal and vagus nerves. A.Showing place for touching pharyngeal wall. Its stimulation results in gag reflex. B. Barium meal x-ray showing mega esophagus, an indication of malfunctioning of vagus nerve. The figures are with the Courtesy of Varshney, J.P.(2025).

CN XI (spinal accessory) nerve (motor fibers) innervates to trapezius muscle. Atrophy of trapezius muscle is suggestive of CN XI abnormality. Functioning of CN XI nerve is evaluated by examining trapezius muscle and signs of left and right muscle asymmetry.

CN XII (hypoglossal) nerve innervate tongue muscles. Asymmetry, atrophy or deviation of tongue from normal position (Fig.146) is indicative of hypoglossal nerve affection. Tongue function can be assessed by watching the dog drinking water. Changes in the symmetry of the tongue musculature and its voluntary movement are suggestive of nerve dysfunctioning.

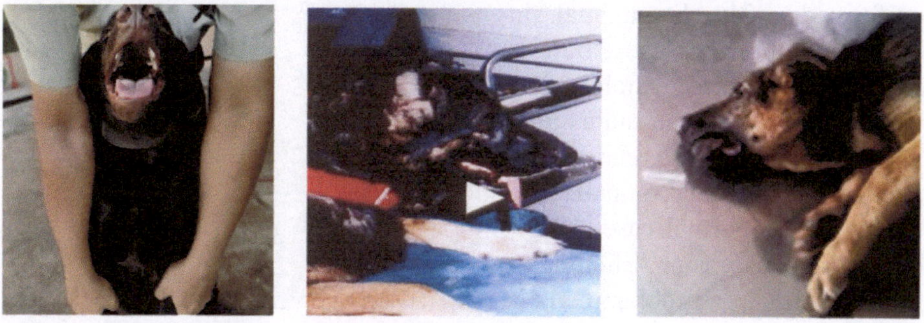

A B C

Fig. 146: Dogs showing normal and abnormal functioning of hypoglossal nerve. Normal positioning of tongue indicative of normal functioning of hypoglossal nerve **(A).** Tongue is hanging to its right side **(B)**. Tonge is hanging to its left side **(C)**. Tongue postioning and asymmetry in Fig B and C is suggestive of dysfunctioning of hypoglossal nerve (CN XII). The figures are with the Courtesy of Varshney, J.P.(2025).

(e) Spinal Reflexes - The integrity of sensory and motor components of the reflex arc and the influence of descending motor pathways on the reflex is tested by spinal reflexes. A reduction or absence of reflex is suggestive of complete or partial loss of either the sensory or motor nerves responsible for the reflex (lower motor neuron – LMN).While an exaggerated reflex is suggestive of a deformity or defect in the descending pathways from the brain and spinal cord that normally inhibit the reflex (upper motor neuron -UMN). Fore limb reflexes are less reliable for localizing lesions.

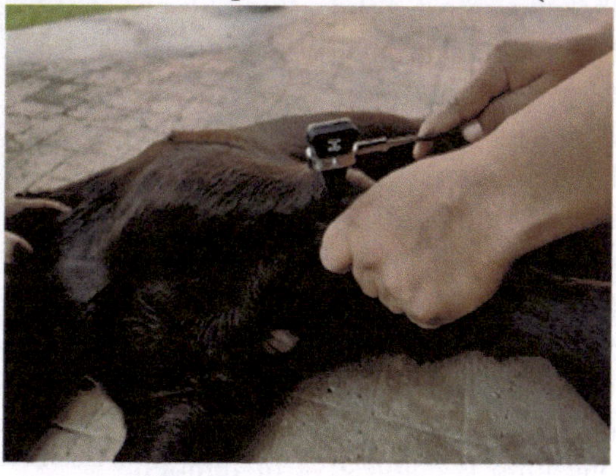

Fig. 147: Performing a biceps reflex test in a dog. The thoracic limb (fore limb) is pulled slightly caudal placing an index finger on biceps tendon and tapping it with reflex hammer.Normal biceps reflex is associated with slight flexion of the elbow. The figure is with the Courtesy of Varshney, J.P.(2025).

The Biceps reflex (Fig. 147) is performed by pulling the thoracic limb slightly caudally, placing the index finger on the biceps tendon (cranial and proximal to the elbow) and tapping the with the reflex hammer or pleximeter. In normal cases the elbow shows slight flexion. An absence or decrease in reflex is suggestive of a lesion involving spinal cord segments from C6 to T2 (LMN). While an exaggerated reflex is suggestive of a lesion cranial to spinal cord segment C6 (UMN).

The triceps reflex (Fig.148) is performed with the animal in lateral recumbency. The elbow is flexed and abducted by holding the thoracic limb over the radius ulna and tapping the triceps tendon with a reflex hammer just proximal to the olecranon. In normal cases elbow shows slight extension. Eliciting of triceps reflex in normal animals is difficult to appreciate making the interpretation of lowered or absent triceps reflex a more difficult task. While an exaggerated reflex is suggestive of a lesion cranial to C7 (UMN).

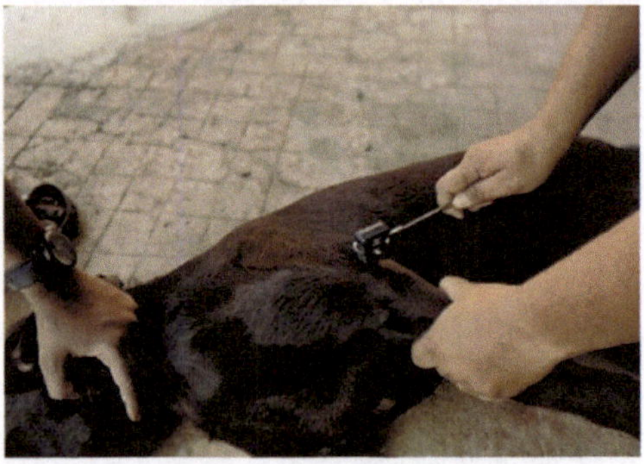

Fig. 148: Performing a triceps reflex test in a dog. The elbow is flexed and abducted by holding the thoracic limb over radius –ulna and tapping triceps tendon with reflex hammer just proximal to olecranon. A normal reflex is associated with slight extension of the elbow. The figure is with the Courtesy of Varshney, J.P. (2025).

The Patella reflex is a reliable test. It is performed with the animal in lateral recumbency supporting the uppermost pelvic (hind) limb by holding the hock with slightly flexed stifle and tapping the patellar ligament briskly with a reflex hammer (Fig.149). A single quick extension of the stifle is a normal response. An absence or reduced patellar reflex with flaccidity is suggestive of a lesion of sensory or motor component of the reflex arc (LMN). Loss of patellar reflex in one limb is suggestive of femoral nerve lesion. While loss of patellar reflex in both hind limbs indicates segmental spinal cord lesion (L4-L6). Hyperpatellar reflex (exaggeratedpatellar reflex) with spasticity (increased muscle tone) along with signs of UMN

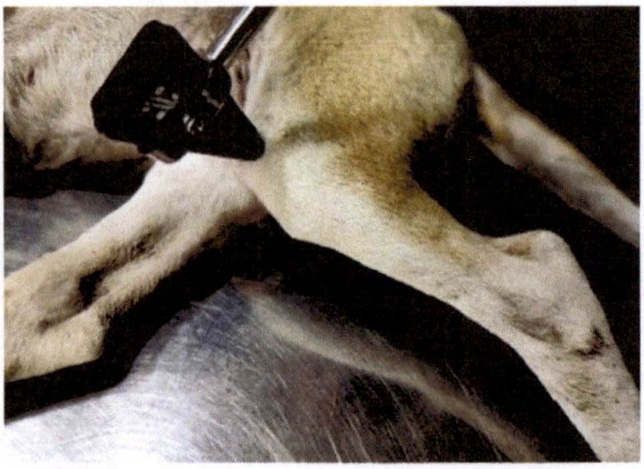

Fig. 149: Performing a patellar reflex test in a dog. The uppermost pelvic limb is supported by holding the hock with slightly flexed stifle and briskly tapping the patellar ligament with reflex hammer.A single quick extension of the stifle is a normal response. The figure is with the Courtesy of Varshney, J.P. (2025).

dysfunction is suggestive of lesion in the spinal cord cranial to L4 segment (UMN). Exaggerated patellar reflex is also seen in loss of sciatic nerve integrity.

The gastrocnemius reflex is mediated via the tibial branch of the sciatic nerve and cord segment L5 to L7 and S1. This reflex is tested with the dog in lateral recumbency supporting the uppermost pelvic limb (hind leg) by placing a hand under the metatarsal bones with slightly flexed hock and striking gastrocnemius tendon briskly with a reflex hammer (Fig.150). Extension of the hock followed by flexion is a normal reflex.

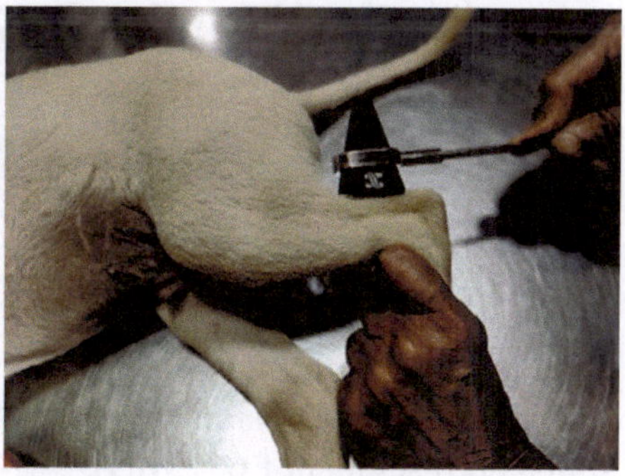

Fig. 150: Performing gastrocnemius reflex test in a dog. The reflex is tested with the dog in lateral recumbency supporting the upper most pelvic limb by placing a hand under the metatarsal bones with slightly flexed hock andstriking gastrocnemius tendon briskly with a reflex hammer. Extension of the hock followed by flexion is a normal reflex. The figure is with the Courtesy of Varshney, J.P. (2025).

The withdrawal reflex (Fig.151) should be tested in both fore and hind limbs. The dog is placed in lateral recumbency for testing the withdrawal reflex in fore (thoracic) limbs. On application of the least harmful stimulus (pinches) to the foot, flexion of the entire limb is a normal response. This reflex is primarily mediated through spinal cord segments C6-T2. Any abnormality in withdrawal reflex (absence or depression) indicates a lesion of spinal cord segments (C6 to T2) or of the peripheral nerves (LMN). Exaggerated withdrawal reflexes associated with other signs of upper motor neuron (UMN) dysfunction are suggestive of a lesion cranial to spinal cord segment C6 (UMN). The withdrawal reflex in the pelvic limbs is tested in the similar manner described for thoracic limbs. Flexion of entire hind limb in response to stimulus is a normal response. The withdrawal reflex in pelvic (hind) limbs is mediated through spinal cord segments L6 to S1 and the sciatic nerve. Any

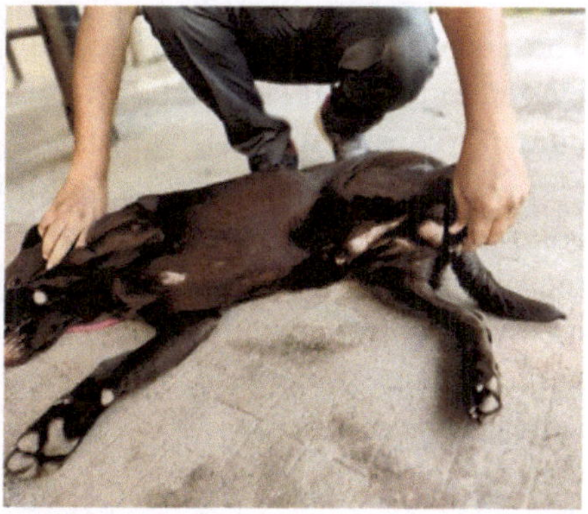

Fig. 151: Performing a withdrawal reflex test in a dog. The dog is placed in lateral recumbency and a mild stimulus is applied to the foot web (pinching). Flexion of the entire limbis normal response. The test should be conducted in both fore and hind limbs. The figure is with the Courtesy of Varshney, J.P. (2025).

abnormality in withdrawal reflex (absence or depression) indicates a lesion of spinal cord segments (L6 to S1) or of the peripheral nerves (LMN). If the withdrawal reflex is absent in only one limb, it is most likely due to a sciatic nerve lesion. Exaggerated withdrawal reflexes are suggestive of a lesion cranial to spinal cord segment L6 (UMN).

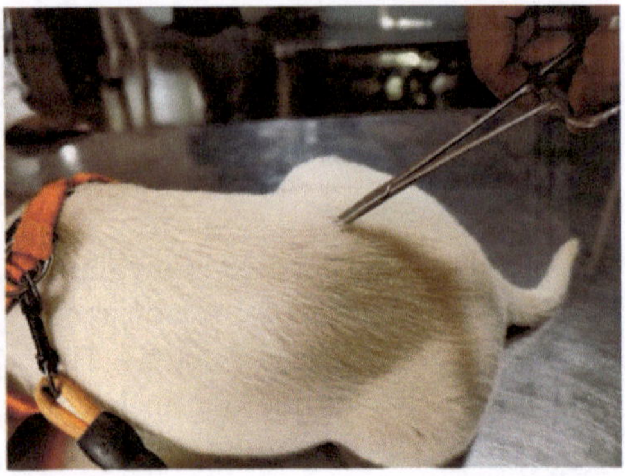

Fig. 152: Performing a cutaneous trunci reflex test in a dog. The test is performed by applying pin prick stimulus to the skin over the back starting from the lumbosacral region and continuing cranially. Twitching of the cutaneous trunci muscle on both sides of the dorsal midline, at the point of stimulation and cranially is the normal response to the pin-prick stimuli.The figure is with the Courtesy of Varshney, J.P. (2025).

The cutaneous trunci reflex is tested by applying pin-prick stimulus to the skin over the backstarting from the lumbosacral region and continuing cranially (Fig.152). Twitching of the cutaneous trunci muscle on both sides of the dorsal midline, at the point of stimulation and cranially is the normal response to the pin-prick stimuli.Consistent absence of a response occurring only on one side (avulsion side) one or two segments caudal to thespinal cord lesion may be due to brachial plexus avulsion injuries.

The anal sphincter reflex is tested by gentle perineal stimulation with blunt forceps (Fig.153). Normally perineal stimulation results into contractions of the anal sphincter muscle. Anal sphincter reflex is mediated through the pudendal nerve (perineal nerve is sensory; caudal rectal nerve is motor) and spinal cord segments S1-S3. Absence or poor contraction of anus is suggestive of a lesion of sacral spinal cord or pudendal nerve lesion (LMN). An exaggerated response is suggestive of a lesion above the S1 spinal cord segment.

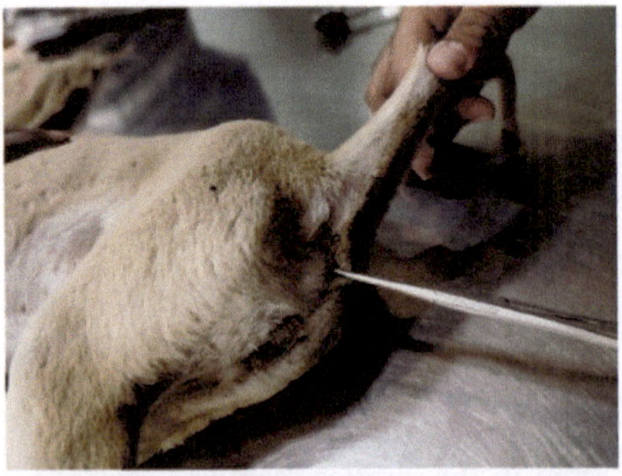

Fig. 153: Performing an anal sphincter reflex test in a dog. The test can be conducted in standing or recumbent positions. A gentle stimulation with blunt forceps is applied in perineal region. It elicits contraction of the anal sphincter muscle ishealthy dogs indicating an intact anal sphincter reflex. The figure is with the Courtesy of Varshney, J.P. (2025).

(f). Palpation - Palpation (Fig. 154) is done along the spine and muscles for pain, muscle tone and atrophy. Neck is moved in all directions to check for pain. Dog undergone surgery or suffering from atlantoaxial subluxation or suspected fracture is not subjected to neck rotation.

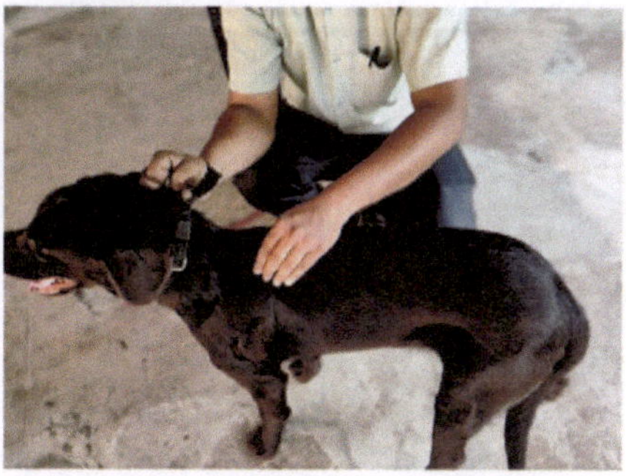

Fig. 154: Performing palpation test in a dog.Palpation is conducted along the spine and muscles to observe pain, atrophy and muscle tone. The figure is with the Courtesy of Varshney, J.P. (2025).

(g). Pain Perception - Pain perception is assessed in dogs with loss of motor function and young ones with signs of a sensory neuropathy. A conscious response (vocalizing, trying to bite, turning the head, whining, dilating pupils, increased respiratory rate) to pinching the toe web (using a hemostat) is suggestive of pain. Lack of deep pain perception carries a guarded to poor prognosis.

References and Further Readings

Carmichael, N.(1998). Nervoussystem. In: Manual of Small Animal Clinical Pathology.

Davidson, M., Else, R. and Lumsden, J. (Eds.). British Small Anim. Vet. Assoc. Cheltenham, UK, pp. 235–240.

Di Terlizzi, R. and Platt, S.R. (2009). The function, composition and analysis of cerebrospinal fluid in companion animals: Part II – Analysis. Vet J. 180: 15-32.

Nelson, R. W., and Couto, C. G. (2019). Small Animal Internal Medicine. Elsevier.

Oliver, J.E. and Lorenz, M.D.(1997). In: Handbook of Veterinary Neurology. Liner, J.E. and Lorenz, M.D. (Eds.), Saunders, Philadelphia, PA.

Varshney, J.P. (2025). Clinical Diagnostic Processes in Canine Medicine. NIPA GENX Electronic Resources and Solutions P. LTD. New Delhi- 110034

16

Peritoneal Fluid

Peritoneal Fluid

Peritoneal fluid (PTF) is an ultra-filtrate of plasma. The amount of the fluid and its constituents depends on oncotic and hydrostatic pressures, vascular permeability, mesothelial lining integrity, and lymphatic drainage. In healthy animals, the amount of peritoneal fluid is negligible making its aspiration difficult. However, in large animals, a small volume (up to 5ml) can be aspirated. Volumes of 10ml or above may indicate a pathological process in the abdominal cavity, unless the animal is in late pregnancy.

Indications for collection and analysis

The collection and analysis of peritoneal fluid is important in the diagnosis of diseases of the peritoneum and the abdominal segment of alimentary tract .Increased accumulation of peritoneal fluid may be evident in congestive heart failure, peritonitis, nephrotic syndrome, or hepatic cirrhosis, The peritoneal fluid analysis can assist in the diagnosis of ascites, abdominal neoplasm, peritonitis, uroperitoneum and hemoperitoneum etc. Abdominal paracentesis is valuable in differential diagnosis and prognosis of colic in horses and peritonitis in cattle.

Peritoneal fluid collection

Peritoneal fluid can be collected by performing abdomenocentesis. For abdomenocentesis, the animal is restrained properly. The site is prepared aseptically by hair clipping, shaving and applying antiseptic solution.

Site for collection

Cattle: The midline should be avoided in cattle, as it can puncture the rumen. Abdomenocentesis can be performed at left cranial - in the reticular-ruminal recess, 5 to 6cm cranial and medial to the left mammary vein foramen (House *et al.,* 1992); right cranial- in the region of the abomasum at the midpoint between the navel and the xiphoid appendix and to the right of the ventral midline, at the midpoint between this and the right mammary vein (Radostits *et al.,* 2007); or right caudal- medial to the fold of the right flank in a cranio-dorsal position to the udder (House *et al.,* 1992).

Horse: The site for abdomenocentesis in horses is ventral mid line, 25 cm caudal to the xiphoid (midway between the xiphoid and the umbilicus).

Dogs: Ventral anterior abdomen, midway between the xiphisternum and the umbilicus in the midline is chosen. An alternative site in small animals is in the anterior abdomen, 2-2.5 cm caudal to the xiphisternum and 2-2.5 cm left or right of the midline. Repeating the abdomenocentesis at a different site may help confirm the result.

Collection Procedure

A local anesthetic (2% lidocaine without vasoconstrictor) is infiltrated locally at the site. An appropriate sized needle (lumbar-spinal needle in case of bovines, 18 gauge needle or a teat cannula in horses and a 21-22 gauge needle in dogs and small animals) is used for peritoneal fluid collection. The needle is inserted through skin, abdominal muscle and peritoneum and fluid is collected.

In the horse, a stab incision is made through skin, subcutaneous tissues and into linea alba. A long blunt-pointed bovine teat cannula is inserted through the stab wound and moved until a cut into the linea alba is felt. The cannula is further pushed into the peritoneal cavity.

In cattle, a teat cannula or a 16 gauge 5cm hypodermic needle may be used with care. It is carefully pushed through the abdominal wall. The needle twitches when the peritoneum is punctured.

Dogs/cats are placed in lateral recumbency so that large area of the ventral abdomen is available for site selection. Proper restrain is applied. Usually no sedation is needed in most of the cases. Linea alba (down to the ventral abdomen between left and right hind limbs) usually around 4-5 cm caudal to xiphoid (Fig. 155) is the most preferred site for conducting paracentesis abdominis because of less vasculature in the this area and minimum chance of contamination. This area is swabbed with alcohol .Linea alba is carefully pierced using a sterile needle (21-22 gauge) and a syringe (5ml /10 ml) through the skin, subcutaneous tissue, and abdominal wall and into peritoneal cavity. As soon as needle pierces peritoneal cavity, fluid drops appear in the syringe. The fluid is then drawn slowly. Two to three ml content of the syringe is transferred into a well labeled vials or vacutainer tubes containing EDTA for cell count, protein analysis and cytological evaluation. Other 2- 3 ml fluid portion should be transferred in to a sterile vacutainer tube or vial without anticoagulant for bacterial culture and biochemical (bilirubin, cholesterol, triglyceride, creatinine etc.) analysis.

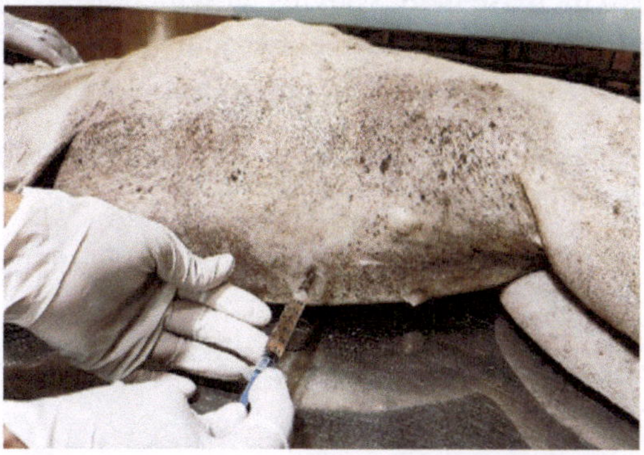

Fig. 155: Collection of peritoneal fluid in a dog. The needle is inserted usually around 4-5 cm caudal to xiphoid .The figure is with the Courtesy of Varshney, J.P. (2025).

The sample is collected into an ethylenediaminetetraacetic acid (EDTA) tube for cytology and a sterile tube/syringe/ culture vial for bacterial culture and sensitivity testing.

Examination of Peritoneal Fluid

After the sample is collected, the volume, colour, pH, viscosity, turbidity, etc. is assessed.

Transparency- Normal peritoneal fluid is crystal clear. Its turbidity indicates the presence of increased number of leukocytes and protein.

Color- Normal peritoneal fluid is straw colored. Variations in color are suggestive of pathology. Green , intense orange green, pink to red, red brown or milky color is suggestive of presence of food material ; rupture of biliary system; hemoglobin, erythrocytes, damage to vascular system or perforation; degenerated blood and hemoglobin (damaged gut wall); or lymphatic obstruction respectively. Frank blood in peritoneal fluid is suggestive of hemoperitoneum.

Particulate matter- Particulate matter in the peritoneal fluid is suggestive of either fibrin clots or gut contents.

Protein content and specific gravity- Normal peritoneal fluid of cattle has < 1.6 g protein/ dl. High protein contents and high specific gravity of the peritoneal fluid is suggestive of vascular damage as in peritonitis or mural infarction.

Volume- a continuous flow of peritoneal fluid with 10-20 ml per sample may be due to ruptured urinary bladder or ascites.

Leukocyte count- An increase in leukocytes with increased polymorphonuclear cells in peritoneal fluid is suggestive of acute inflammation. In chronic peritonitis, the number of mononuclear phagocytes increases.

Mesothelial cells- Increased number of mesothelial cells with dividing mitotic changes indicates neoplasia.

Organism- Isolation of bacteria from peritoneal fluid is suggestive of infective peritonitis.

After preliminary examination, the samples can then be sent to the laboratory for detailed analysis.

- Serum-ascites albumin gradient (SAAG): The ratio of albumin level in the serum and ascetic fluid is calculated. The serum albumin level minus the ascetic fluid albumin level is calculated to determine the gradient for differentiating between transudates and exudates. A SAAG level of 1.1 g/dL or greater suggests the presence of a transudate and less than 1.1 g/dL, an exudate.
- Peritoneal fluid glucose, amylase, bilirubin, creatinine, lactate dehydrogenase (LD), tumor markers for malignancy, can be evaluated.
- Microscopic examination of the peritoneal fluid may be performed if infection or cancer is suspected. The sample is centrifuged to obtain a sediment which has maximum concentration of the cells in the fluid. The sediment can then be smeared on a clean grease-free glass slide, stained and examined under microscope and evaluated for abnormal cells, such as malignant cells (cancer cells). The smear can also be stained for identification of bacteria or fungi, Acid Fast Bacillus (AFB) under a microscope for infectious diseases, such as mycobacterium infections.
- Adenosine deaminase test done for detecting tuberculosis in peritoneal fluid.
- Bacterial culture and antimicrobial sensitivity testing can be done to detect the presence of microbes by culturing and then knowing the antimicrobial sensitivity for antimicrobial therapy.
- Testing for viruses and parasites can also be undertaken as per requirement.

Classification of peritoneal fluid

The peritoneal fluid can be classified as transudate or exudate depending upon the transparency and cellular content of the fluid.

Transudate are fluid that pass through a membrane or squeezed through tissue or into the extracellular space. Transudates are thin, watery containing very few cells and protein. Transudates are due to portal hypertension, or hypoalbuminemia or congestive heart failure.

Exudates fluid containing cells and cellular substances that are slowly discharged from blood vessels and are usually from inflamed tissues. It may be secondary to infection, inflammation or malignancy.

The characteristics of transudate and exudates are given in the table 9 .

Table 9: Characteristics of transudate and exudate

S.No.	Contents	Transudate	Exudate
1.	Appearance	Clear or pale	Turbid or bloody
2.	Specific gravity	<1.015	>1.015
3.	Cells	Negligible/few cells	More cells
	a. Nucleated	<1500/μL	>1500/μL
	b. RBCs	<5000/μL	>5000/μL
4.	Protein content	Less than or equal to 3gm/L, does not contain large plasma proteins like fibrinogen	More than 3gm/L, contain large plasma proteins like fibrinogen.
5.	Lactate dehydrogenase	<0.6	>0.6
6.	Glucose	Equal to level in blood	Less than 60 mg/dL

References and Further Readings

House, J.K., Smith, B.P., VanMetre, D.C., Fecteau, G., Craychee, T. and Neves, J. (1992). Ancillary testes for assessment of the ruminant digestive system. Vet. Clin. N. Am., Food Anim. Pract. 8(2):203-232.

Radostits, O.M., Gay, C.C., Hinchcliff, K.W. and Constable, P.D. (2007). Diseases of the alimentary tract II, p.293-382. In: Veterinary Medicine: a textbook of the disease of cattle, horses, sheep, pigs and goats. Radostits, O.M., Gay, C.C., Hinchcliff, K.W. and Constable, P.D. (Eds). 10th edn. Saunders Elsevier, Edinburgh.

Varshney, J.P.(2025). Clinical Diagnostic Processes in Canine Medicine. NIPA GENX Electronic Resources and Solutions P. LTD. New Delhi- 110034

17

Clinical Examination for Skin Disease

The skin is the largest organ of the body that plays an important role in defending the body against minor physical injuries, providing some defense against microbial invasion, maintenance of homeostasis, water conservation, temperature conservation, supporting hair growth and heat loss. Changes in the secretion of sebaceous glands lead to compromised defence. Abnormalities of the skin may arise from specific skin disease or by the poor general health status. Therefore, a detailed clinical examination of the animal as a whole and of its skin in particular, is an essential part of the process of diagnosis. The skin has three main layers starting from outside as the epidermis, the dermis and the subcutis. The epidermis consists largely of epithelial cells and pigment. The dermis is a connective tissue layer containing blood vessels, nerves, hair follicles, sebaceous and sweat glands. The inner most subcutis contains fibrous and fatty tissues that provides insulation and support. In healthy animals, the skin has considerable elasticity. It reduces in dehydration and inflammation / injury to skin. General clinical examination should be carried out first as a routine followed by special examination of the skin. Clinical examination for skin disease diagnosis involves history, visual appraisal or inspection, manual appraisal or inspection, distribution pattern of the lesions, description of the lesion, clinical signs and symptoms and special diagnostic procedures.

History

History is the first step in diagnostic process in cases of skin disease also. Knowledge of the geographical area and the history of the herd may provide some important clues for the diagnosis of skin disease. Coat color changes are frequent in copper deficient areas. Herd history of previous skin disease and its treatment can help in the diagnosis of the present skin disease. The history of recent contacts with other is also valuable. Recent changes in diet and management should be given due share. Poor nutrition is also an important factor that can lead to a dull, dry, thin and brittle skin coat. Complaint of frequent rubbing is indicative of pruritus. Abrasive surfaces or corners of the cubicles, parlor or the floor are the cause of repeated skin injury. Overcrowding and insufficient feeding facilities to animals are a contributing factor to poor coat condition. Though, poor coat conditions and marks of excessive self-grooming or large patches of alopecia may be visualized from a distance, close examination of the skin coat should invariably be done.

Visual Appraisal or Inspection of the Skin

The skin of whole body should be methodically examined both from a distance and closely for areas of abnormal skin or hair. Lick marks may be noticed on the skin of healthy animals. Heavy lice infestation may cause intense pruritus with excessive grooming or licks marks leading to hair loss and skin thickening. Presence of any swelling with or without discharge should be noted.

Manual Appraisal or Inspection of the Skin

The skin of the body, as much as possible, should be inspected with gentle touching to detect lesions that are not visible on the visual appraisal as these may be hidden beneath the matted hairs. Manual appraisal may help in detecting enlarged lymph nodes, thickened skin, subcutaneous edema or infection. Skin turgor (flexibility and resilience) can also be assessed by picking up a skin fold of upper eye lid between finger and thumb. Thickened skin is seen in cases of sarcoptic mange and callus in the joint area subjected to repeated trauma.

Distribution Pattern of Lesions

The distribution pattern of skin lesions is to be noted owing to their diagnostic significance. Lesions of photosensitization are commonly seen in unpigmented or lightly pigmented skin area on the dorsal parts of the body directly exposed to sun light. While ringworm lesions in calves are commonly seen on head and neck but can also be seen in other parts of the body

Description of the Skin Lesions

The examination of the skin for disease should invariably include the description of the skin lesions and involvement of the tissues. Changes in skin temperature, thickness, consistency and color of the affected skin area should be compared with adjacent areas. The presence of subcutaneous edema, abnormalities of sebaceous and sweat glands should be specified. Self-inflicted trauma can greatly modify and mask the clinical picture. Skin lesions may involve some or all of the component structures of the skin- the hair, follicles, epidermal, dermal and subcutaneous tissues.

The skin lesions are primary (due to skin disease) or secondary (non-specific). Primary lesions are described as *macule* (flattened area of < 1 cm diameter with change of color and no skin thickening) ; *papule* (flat , rounded , circumscribed lesions of < 1cm diameter often raised with a necrotic centre and may be associated with the hair follicles); *nodule* (a large papule more than 1 cm in diameter); *plaque* (solid raised flat topped lesion of > 1 cm diameter) ; *tumor* (large nodular structure); *vesicle* (fluid filled lesion < 1 cm in diameter); *bullae* (vesicle of> 1 cm diameter) ; *pustule* (pus filled vesicle); *wheals* (erythematous, edematous transitory swelling); *acne* (pimple, infection of sebaceous glands), or *impetigo* (flaccid vesicle then pustule then scab up to 1 cm diameter).

Secondary skin lesions are described as *scale* (accumulation of loose, dry fragments of superficial skin layers); *crust* (dried accumulation of debris containing blood and pus); *erosion* (loss of superficial epidermal layer only with no loss of other layers); *excoriation* (erosion or deep ulcer of traumatic origin); *ulcer* (deep erosion penetrating the epidermal basement membrane); *scar* (fibrous tissue replacing damaged skin); *fissure* (split in the superficial skin layers); *keratosis* (overgrowth of dry horny keratinized epithelium); *pigment changes* (hyper or hypo) or *alopecia* (hair loss).

Discoloration (yellow, erythema or pallor) of the skin may be seen in some cases. Yellow discoloration is suggestive of jaundice. In white pigs with septicemia, skin may show red to purple discoloration. Early erythema may be seen in early stage of photosensitization.

Clinical Signs and Symptoms in skin Diseases of Different Etiologies

Some skin diseases may have specific signs that may help in diagnosis. Pruritus is an important sign seen in skin diseases. Sensation of rubbing and scratching are included with pruritus. Itching is not seen in ringworm infection where hair fibers and keratinized epithelium are involved. Pruritus is seen in sarcoptic and chorioptic mange, Aujesszky's disease, nervous acetonemia and lice infestation in cattle; scrapie and infestations of lice, mange, ked, blow fly itch mite infestations in sheep; lice and mange (sacrcoptic and chorioptic) infestations in pigs; and chorioptic mange, sweet itch, lice infestation or oxyuris equi infestation (perianal pruritus) in horses.

Parasitic etiology

Lice infestations, most common in housed bovines, is associated with signs of irritation (rubbing themselves against the walls), excessive grooming, and hair loss. Adult lice and small white egg adhering to hairs can be easily visible. Anemia is common in animals with heavy infestations by sucking lice. Lice infestation in calves is not restricted to the dorsum of the neck and back as seen in adults. Species identification of lice can be done on microscopic examination of a skin scraping.

Flies infestation during summer can cause tear scalding under the eyes, superficial skin damage, multiple skin nodules (bites of Haematobia, Stomoxys and Simulium), urticaria, or open wounds (on the lower abdomen, udder and teats). Subcutaneous nodules and cysts along the back of cattle are caused by the presence of warble fly larvae (Hypoderma family). With the presence of flies, diagnosis can be made easily.

Mange infestation in cattle is caused by chorioptic, sarcoptic, psoroptic and demodectic mites. Infestation of cattle by chorioptic, sarcoptic or psoroptic mites is associated with intense pruritus, impaired growth and reduced milk production. Alopecia and skin thickening are the result of rubbing by infested cattle. Sarcoptic mange mites may occasionally infest the tail head of cattle, but are also seen on the head, neck and shoulders. Being a burrowing mite, sarcoptic produces thickening of the damage to the skin. Psoroptic mange is associated with extensive thickening of the skin, pruritus and hair loss over the shoulders, hindquarters and perineum. Infestation of demodectic mite in cattle is often asymptomatic and non-pruritic , but the presence of the mite and alopecia can be detected. Diagnosis of mite infestation is based on the clinical signs and identification of mites in skin scrapings.

Ticks (Boophilus, Hyalomma, and Rhipicephalus) are found on the animals and are visible to naked eye examination. Anemia, and areas of granulation on the skin of lower parts of the body (possibly a local hypersensitivity reaction) are seen in cattle with heavy tick infestation. Ticks can transmit a number of diseases (babesiosis, tick-borne fever and Lyme disease).

Blow-fly strike can occur anywhere on the body. The larvae (readily visible to naked eye) penetrate the skin and the deeper body tissues and may also enter natural orifices of the body. Necrosis in adjacent skin may be seen.

Bacterial etiology

Superficial trauma, contamination, damp and dirty living conditions, or reduced dermal immunity are the predisposing factors to bacterial infections of the skin. Insertion of metal tags on the ear pinna in calves may cause local infection and oozing of pus. Staphylococcal infection is common in superficial pustular dermatitis. The infection is non-pruritic impetigo-like condition often seen on the udder or perineum. Young cattle, kept in unhygienic conditions, may develop staphylococcal folliculitis on the skin of the hind quarters.

Dermatophilus infection may occasionally be seen, on the backs of cattle kept in crowded conditions. It is manifested with thick crusty lesions overlying a granulating skin surface.

Subcutaneous abscesses are seen in cattle owing to bacterial invasion gaining access through superficial injuries. These abscesses are palpable fluctuating subcutaneous swellings.

Hygromas are seen on the knees or hocks as fluctuating, swellings. Diagnosis can be made on ultrasonographic examination and needle aspiration.

Mandible osteomyelitis in cattle is caused by the infection of *Actinomyces bovis* and is seen as lumpy jaw.

Actinobacillus lignieresii has been associated with superficial skin lesions on the face and around the muzzle and nostril.

Fungal etiology

Ringworm is a very common in young cattle. Dermatophyte infection is also seen in goats (Fig. 156). The lesion are characterized by raised grey crusty lesions associated with alopecia most often on the head or neck. Detection of fungal spores and mycelia on microscopic examination of skin scrapings or broken hairs (prepared in 10% KOH) is of diagnostic significance.

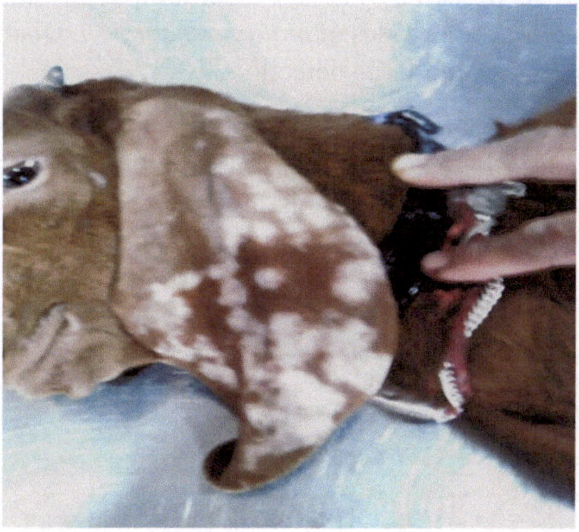

Fig. 156: Dermatophyte lesions on the ear pinna in a goat.

Viral etiology

Many viruses can cause skin disease in cattle.

Wart is the most common skin disease in cattle caused by bovine papilloma virus. Warts vary in size and distribution. Penile warts are soft and bleed easily. Diagnosis of papillomatosis or wart is based on the appearance of cauliflower like growth, their distribution pattern and finally by histopathology of the tissue.

Bovine papular stomatitis is another viral disease of zoonotic importance with lesion (horse shoe shaped small red brown rough and slightly elevated papules of about 0.5 cm diameter) distribution chiefly on the muzzle, nostrils, dental pad and buccal mucosa of young cattle. Confirmatory diagnosis requires electron microscopy of the fresh lesions for the evidence of the characteristic parapoxvirus.

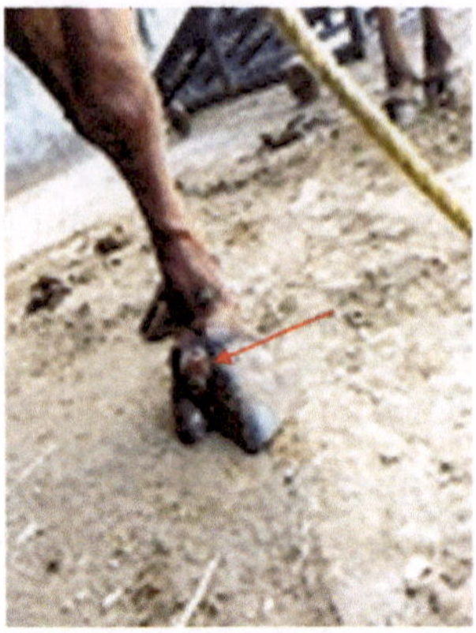

Fig. 157: FMD lesions on the foot of a cow.

Skin lesions (vesicles, bullae, ulceration) on the teat caused by *FMD virus* resemble those of bovine herpes mammillitis. FMD lesions (Fig.157) can also be seen on foot.

Skin lesions around muzzle are also seen in animals infected with *infectious bovine rhinotracheitis (IBR) virus*.

A superficial necrosis followed by severe ulceration of the oral mucosa, muzzle and coronary band is seen in cases of *malignant catarrh*. Diagnosis is based on the severe clinical signs and virus detection.

Lumpy skin disease is highly contagious viral disease affecting cattle and buffaloes. It is caused by pox virus transmitted by blood feeding insects. The disease is clinically characterized by painful, firm skin nodules on skin, nose, mouth or body (Fig.158), fever and drop in milk yield.

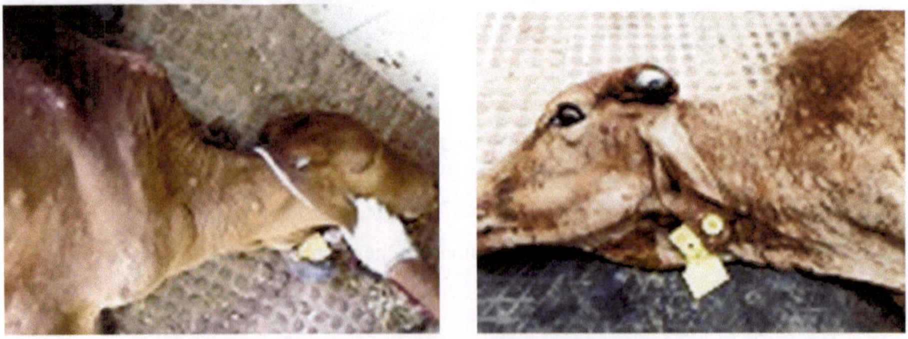

Fig. 158: Cattle showing lesions of Lumpy Skin Disease. The lesions are characterized by painful, firm skin nodules.

Neoplasms

Squamous cell carcinoma is seen in poorly pigmented areas at the mucocutaneous junctions of the body. The mucosa of the third eyelid, the periorbital skin and the vulva are common sites.

Cutaneous lymphosarcoma or may be the precursor of generalized lymphosarcoma or may be a consequence of it and are mainly located in the skin of the neck or flanks.

Fibromas and fibrosarcomas are large tumors arising from the subcutaneous tissue and covered with normal skin.

Lipomas invading the soft tissues are also observed. Confirmatory diagnosis of skin tumors is solely based on histopathology of the lesions.

Nutritional deficiencies and other toxicities

Nutritional deficiency can lead to deterioration in skin condition (dull, dry non-elastic skin with poor growth of scant and brittle hairs) and making it prone to bacterial infection. It is important to ensure that no specific condition such as chronic mucosal disease which might predispose to similar lesions is present. Copper deficiency and excess molybdenum may produce spectacle eye condition. Chronic arsenic, selenium or molybdenum toxicity has been reported with poor skin quality and in some cases changes in hair color. Excess of potassium iodide may lead to widespread dry seborrhoeic dermatitis.

Severe swelling and edema of the skin may be seen in cases of snake bites.

Management problems

Lack of bedding can be a predisposing factor to callus formation or decubitus ulcers over pressure points covered only by skin and connective tissue such as the hock. Traumatic injuries can lead to subcutaneous hematomata. Hematomas can be differentiated from the abscesses by aspiration of the contents (pus in abscess; blood in hematoma). Milk scalds are common in calves. Parts of the muzzle remain wet with milk may suffer hair loss. Photosensitization is commonly seen in individual animal fed on lush pasture and exposed to sun owing to the presence of photodynamic agents (such as present in St John's wort), or due to accumulation of phyloerythrin as a result of hepatic insufficiency. Skin lesions are restricted to non-pigmented areas on the dorsum of the body. Chronic diarrhoea may lead to the development of sore and cracked areas of skin in the perineal area.

Heredity

Heredity causes are also responsible for some of the skin problems. Affected animals are born with partial or complete alopecia. Diagnosis requires histological and genetic evaluation.

Other causes

Telogen effluvium, a non-specific and non-pruritic skin lesion characterized by hair loss on head, neck and limbs, is seen in calves recovering from severe *Escherichia coli* septicemia.

Special Procedures for the Diagnosis of Skin Disease

Confirmatory diagnosis of skin diseases requires special diagnostic procedures such as bacterial culture, skin scrapings, skin biopsy, and /or electron microscopy for viral infections.

Further Readings

Jackson, G.J. and Cockcroft, D. (2002). Clinical examination of the skin. In: Clinical Examination of the Farm Animals. Blackwell Publishing Company. Osney Mead, Oxford OX2 0EL, UK.

Varshney, J.P. (2025). Clinical Diagnostic Processes in Canine Medicine. NIPA GENX Electronic Resources and Solutions P. LTD. New Delhi-110034

18

Examination of Eyes and Ears

Eyes

Anatomy

The eye has following important structures.

- Choroid is the layer containing blood vessels that lines the back of the eye and is located between the retina and the sclera.
- Ciliary body is the structure containing muscle and is located behind the iris (that assist in focusing the lens).
- Cornea is the clear outer layer at the front of the eye that transmits and focuses (i.e., sharpness or clarity) light into the eye.
- Fovea is the depression in the inner retinal surface, the photoreceptor layer of which is entirely cones specialized for maximum visual acuity (sharp vision).
- Iris is the colored part of the eye regulating the amount of light entering the eye. In bright light, the iris closes the pupil to let in less light. In low light, the iris opens up the pupil to let in more light.
- Lens is transparent and it absorbs, focuses and directs incoming light to the retina. It changes its shape automatically to focus on objects at different distances. The lens deteriorate with age.
- Macula is an oval shaped pigmented area in the center of retina containing special light sensitive cells. The deterioration of macula is common with the advancing age.
- Optic nerve is a bundle of more than a million nerve fibers carrying visual messages from the retina to the brain. Images on retina are upside down but the brain turns images right side up.
- Pupil is the dark center opening in the middle of the iris. The pupil changes its size to adjust the amount of light available.

- Retina is the light-sensitive layer of tissue at the back of the eyeball. Images coming through the eye's lens are focused on the retina. The retina then converts these images to electric signals and sends them along the optic nerve to the brain.
- Sclera is the white outer coat of the eye surrounding the iris. It is tough, fibrous tissue extending from the cornea to the optic nerve.
- Vitreous humor or fluid is clear, gelatinous substance filling the central cavity of the eye (vitreous chamber) located between eye's lens and retina.

Examination of Eyes

Examination of eyes consists of general inspection and examination of deep eye structures. The inspection of eyes is conducted with care.

General inspection

The routine general inspection of the eye includes examination of eyes for discharge; eyelids; nictitans membrane; size and movement of the eye ball; conjunctiva and cornea. Sometimes calves are born with anophthalmia (absence of eye or eyes) or microphthalmia (small, disorganized and undeveloped eye or eyes) as congenital defects. An ocular examination should be performed in a calm and quiet, well-lit room that can be darkened as per need. The animal should initially be viewed from a distance of 2–3 feet in good light, and with minimal restraint of the head.

Eye discharge- Eyes should be examined for presence or absence of the discharge. The discharge may be watery (obstruction of lacrimal duct), serous (early inflammatory stage) or purulent (in the later stages) affecting one eye (local inflammation) or both the eyes (a systemic disease).

Eye lids - The eye lids remain close in painful eye conditions, local edema or photosensitization. Their movement becomes excessive in painful eye conditions or in nervous irritability (hypomagnesaemia, lead poisoning and encephalitis).

Nictitans membrane- Nictitans membrane becomes more apparent and is seen across the eye in tetanus, encephalitis or orbit pain.

Eyeball movement and size- Abnormal lateral movement of eyeball (nystagmus) may be observed in anoxia, or due to lesions of cerebellum/ vestibular tract. Eye ball movement in nystagmus is periodic; involuntary; horizontal, vertical or rotatory. Abnormal position of the eyeball may be due to paralysis of motor nerves innervating the orbital muscles. Unilateral eye ball protrusion is indicative of pressure from behind the orbit. Eye ball retraction may be seen in starvation or severe dehydration.

Conjunctiva- The conjunctiva is examined for its color as it is an indicator of peripheral vascular system. Pallor conjunctiva is indicative of anemia while yellow conjunctiva indicates jaundice. Dry conjunctiva is seen in high fever and acute pain. Conjunctival edema, petechial hemorrhages and engorged sclera vessels are seen in cases of gut edema of pigs and in cases of congestive heart failure.

Cornea- Cornea should be examined for its clarity, edema, ulcer or opacity. Varying degree of cloudiness of cornea is seen in keratitis. In glaucoma or hypopyon, cornea becomes more convex (due to increased pressure within the eyeball).

Deep structure examination

Deep structures of the eye are chambers (anterior and posterior), pupil, lens and retina. The examination of the deep structures of the eyes can be carried out with the help of an ophthalmoscope. Hypopyon (pus in anterior chamber) is manifested as yellow or white opacity with horizontal upper border covering the iris.

Pupil is visualized for its shape and position. Abnormality in its shape and position may be due to adhesions to cornea or other structures. Unilateral abnormal pupil dilatation may be due to orbit lesion. Bilateral excessive pupil dilatation may be due to CNS lesions affecting occulomotor nucleus, encephalopathies or functional disorders (botulism and anoxia). Miosis (constriction of pupil) is seen in cases of poisoning with organic phosphate insecticides or overdosing of parasympathomimetic drugs.

Lens opacity is readily visible in advanced stage.

Tests for Blindness

For detection of blindness, menace reflex and an obstacle test are conducted.

Other Tests

The menace response test, palpebral, corneal, pupillary light reflexes, and dazzle reflex test are performed with a strong light and magnification in both bright and dim lighting. Routine baseline tests viz. Schirmer tear test (STT), fluorescein staining, and tonometry (intraocular pressure –IOP measurement) are also recommended. An STT is performed before any drop is instilled into the eye. Culture and antibiotic sensitivity test is indicated if ocular discharge is purulent. Corneal or conjunctival cytology is performed after topical anesthesia. Tonometry is performed before pupil dilation is done. Diseases of the lens, vitreous, and ocular fundus are evaluated employing direct and indirect ophthalmoscopy Vision is evaluated by maze testing in both bright and dim lighting.

Special methods of eye examination, such as slit-lamp biomicroscopy, ultrasonography, fluorescein angiography, and electroretinography, may require sedation or local, regional, or general anesthesia, depending on the species.

Menace response test

Menace response (Fig.159) is elicited in farm animals, as with pet animals, by threatening hand motion towards eye or moving an object toward the eye. A positive response is blinking of the eye with or without an attempt to move. A normal menace response is indicative of vision and an intact facial nerve. A negative response is usually suggestive of blindness.

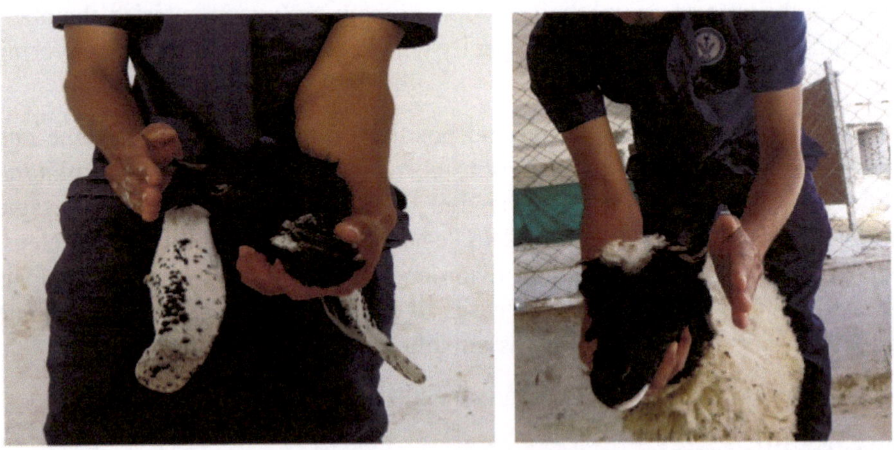

Fig. 159: Conducting menace response test in goat and sheep.

Dazzle reflex test

A dazzle reflex involves blinking or partial blinking and head aversion in response to a bright light being quickly shone on the eye. No response is suggestive of poor vision. A dazzle reflex can be elicited at a much younger age than a menace response.

Palpebral reflex

Palpebral reflex test (Fig.160) is indicated in animals showing negative menace response test or dazzle reflex test. The test is performed by gently tapping/ touching the eyelids at the medial and lateral canthi and observing a blink. A lack of blink in response to palpebral reflex suggests poor sensation or facial nerve paralysis.

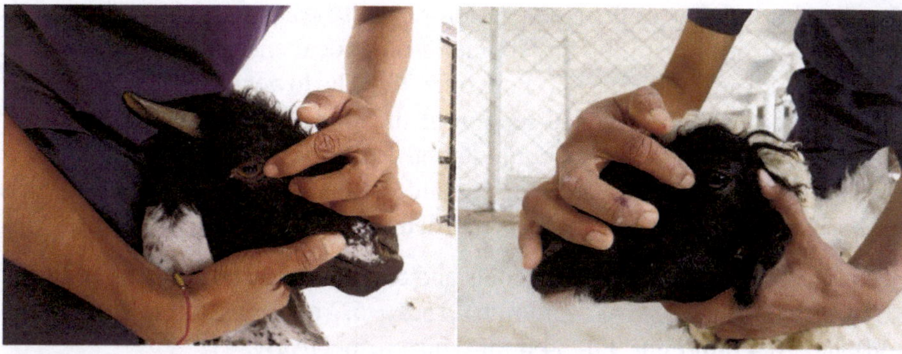

Fig. 160: Conducting palpebral reflex test in goat and sheep. The test is conducted by touching medial or lateral canthi and observing the blink response.

Pupillary light reflex (PLR) test

A bright light is shone (Fig. 161) into lateral aspect of each eye turn by turn. Constriction of the pupil of the eye (direct reflex) being stimulated by light and slightly less constriction or weak constriction of the pupil of the other unstimulated eye (indirect reflex) is the normal response. Failure of pupil to constrict in response to bright light can be due to many factors.

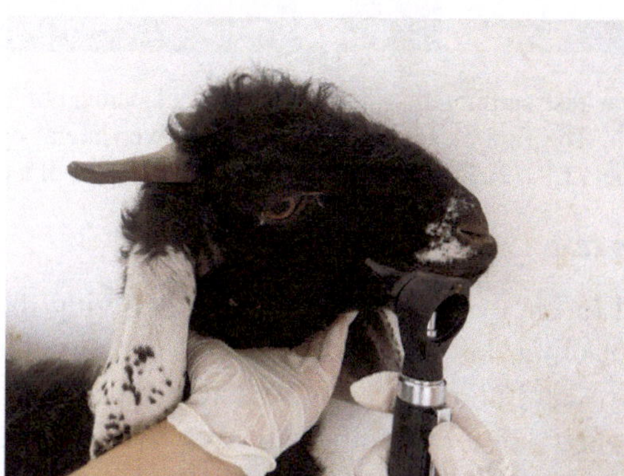

Fig. 161: Conducting a pupil light response test in a goat. The test is conducted by shining a bright light into one eye and noting the constriction of pupil of the eye being tested as well as other eye also.

Schirmer tear test (STT)

The Schirmer tear test is conducted to measure ocular tear production and is generally indicated for the diagnosis of keratoconjunctivitis sicca (KCS).

Tear test strips are available in the market. The strips are with or without notch and are calibrated with a millimeter scale (Fig.162). Strips of some manufacturers are also impregnated with a blue dye marking for an easy visualization of the result. No sedation or anesthesia is needed. The test is conducted prior to instilling any topical drops (local anesthetic or fluorescein dye). Notched area of the strip is bent over 90°. The lower eyelid is slightly and gently pulled out. The notched section of the strip is inserted over the lower lateral eyelid margin into the conjunctival fornix for 60 seconds (Fig. 163). After 60 seconds the strip is taken out and the distance travelled by the tears on the test strip in sixty seconds is recorded.

Normal values of STT for sheep, goats and calves have been reported as 18.5 ± 2.5 mm/min (range 13.5–22 mm/min) ,15.8 ± 5.7 mm/min (range 10–30 mm/ min) and 20.4 mm/min(9.0- 34.0 mm/min) respectively.

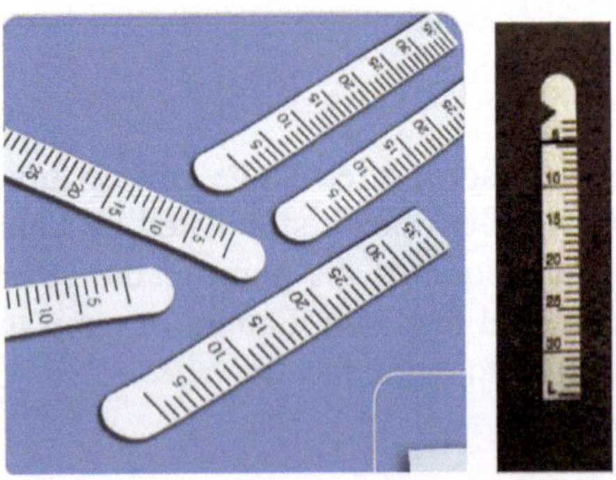

Fig. 162: Schirmer tear test calibrated strips with or with out notch. The figures are with the courtesy of Varshney, J.P. (2025).

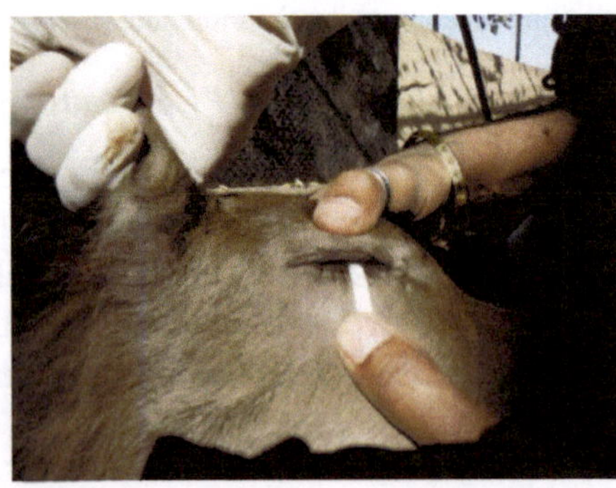

Fig. 163: Placement of Schirmer tear test strip over the lower lateral eyelid margin into the conjunctival fornix of a cow calf.

Fluorescein stain test

The test is applied to the eye to detect injuries involving the cornea. Corneal scratches, ulcers, or other defects can be diagnosed using fluorescein staining. Performing a fluorescein stain takes only a few minutes and is not painful. Fluorescein is a water soluble ophthalmic dye available as 1 to 2% sterile solution. or impregnated paper strips (Fig.164). The strips can be moistened by a drop of sterile saline or local anesthetic. One drop is applied in the eye (Fig.164). The dye is highly hydrophilic and does not remain in contact with an intact corneal epithelium. Corneal ulceration causes breach in corneal epithelium exposing the stroma. When fluorescein dye is applied in cases of corneal abrasion/ulceration, the exposed stroma absorbs and retains fluorescein dye. Examination with blue light or Wood's lamp light greatly enhances fluorescein stain uptake. This test is conducted to detect corneal injuries/ ulcers, small foreign bodies or particles in the eyes, and tear break up time. It can also be used for the diagnosis of descemetocoele as descemetocoeles do not take up stain. Appearance of the dye in the mouth or at nares confirms nasolacrimal duct patency.

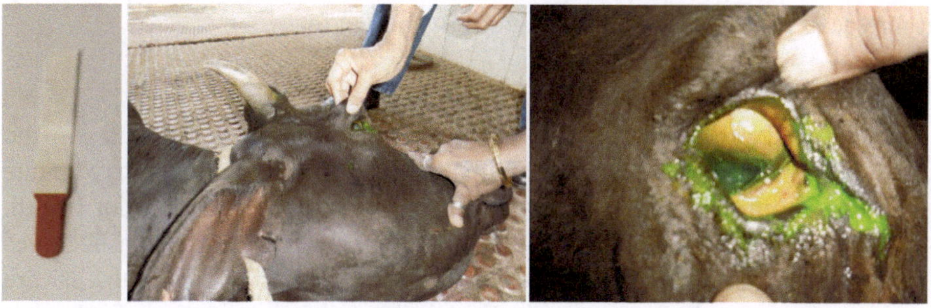

A B C

Fig. 164: Fluorescein stain/dye test. Fluorescein dye strip **(A)**, Instillation of fluorescein dye solution in the eye of a cow **(B)**. Retaining of dye in case of damaged cornea (C).

EARS

Ears are sensory organ that help in hearing and are divisible into inner, middle, and external ear. Sound waves are collected by the funnel shaped external ear, transmitted to ear drum through ear canal. The tympanic membrane is vibrated by the sound waves reaching there creating further movement of the waves due to vibration of the tympanic membrane. These waves further passes to the fluid in internal ear through auditory ossicles stimulating hair like nerve endings in the membranous wall of the canal and finally the impulses reach to the brain where these are analyzed and perceived. The external ear consists of concha, external auditory meatus and ceruminous glands (produce ear wax). The middle ear consists of tympanic cavity, ossicles and Eustachian tubes. The internal ear has membranous labyrinth and osseous labyrinth.

Common Diseases

The inflammation of the external, middle and internal ears are called as otitis externa, otitis media and otitis interna.

External parasites are the common cause of otitis externa. It is clinically characterized by scratching of ears. Another important disease of ear pinna in dogs is aural hematoma.

The inflammatory disease of the middle ear is termed as otitis media. The common cause is microbial infection (Staphylococci in all animals; *Actinomyces pyogenes* in calves, pigs and sheep; *Pseudomonas aeruginosa* in dogs and pigs). The infection of the middle ear may extend to internal ear causing otitis interna.

The inflammatory disease of internal ear is termed as otitis interna. Clinically it may lead to disturbances of equilibrium and deafness.

Examination of the Ears

i. Clinical examination -The ears are normally held horizontally to the head and drooping is abnormal. Unilateral dropping of ears may be due to ear infection and bilateral ear dropping may be due to poor health. Ear drooping may also be due to heavy ear tags. Ear pinna becomes stiff in cases of tetanus. Otitis externa and foreign bodies in the ear are not common in farm animals. Ear discharge is an indication of infection and requires examination with otoscope or small fiberoptic endoscope. Lacerations in ear are common. Aural hematomas are uncommon in cattle but more common in dogs. Deafness may be caused by damage to the auditory part of the 8th cranial (vestibule cochlear) nerve.
ii. Ear Function test-Reaction to clapping near ears or tapping a metallic object (creation of sound) should be watched. Cattle usually reacts to sounds stimuli by turning head to look in the direction of sound and ear movement.
iii. Otoscopic examination- Otoscopic examination (Fig.165) is performed using otoscope to examine the internal structures of the ear.
iv. Ear examination techniques in pet animals such as dogs and cats include history, clinical examination including ear canal palpation and visual inspection of ears, imaging (Otoscopy, radiography, CT scan, MRI, and ultrasonography) and cytological evaluation of otic contents/ discharge.
v. Cytological examination of ear canal discharge is an important diagnostic technique that provides information about aural environment and infection. It is the primary diagnostic tool in identifying bacterial or yeast overgrowth. Cytological examination should be performed prior to bacteriological culture and antibiotic sensitivity.

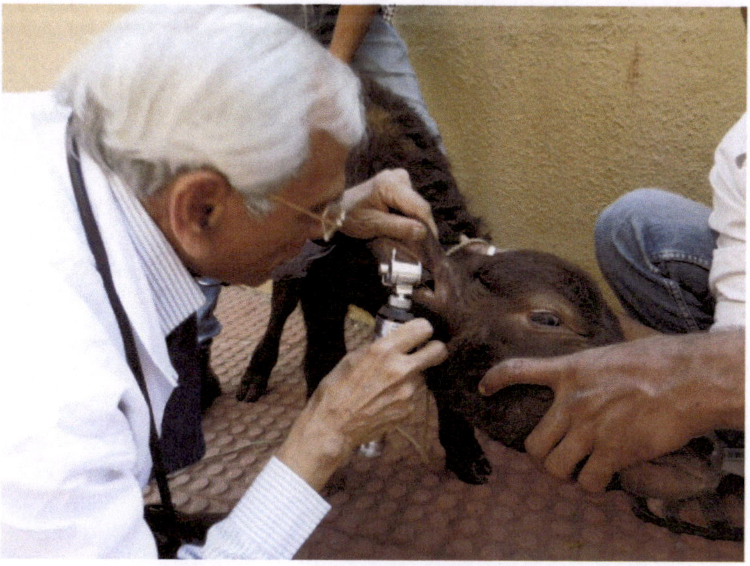

Fig. 165: Otoscopic examination of a buffalo calf.

Further Readings

Bracun, A., Ellis, A.D. and Hall, C. (2014).A retinoscopic survey of 333 horses and ponies in the U.K. Vet. Ophthalmol. 17:90-96.

Dogru, M., Ishida, K., Matsumoto, Y., Goto, E., Ishioka, M., Kojima, T., Goto, T., Saeki, M. and Tsubota, K. (2006). Strip meniscometry: a new and simple method of tear meniscus evaluation. Invest. Ophthalmol.Vis. Sci. 47: 1895–1901.

Gelatt, K.N., Ben-Shlomo, G., Gilger, B.C., Hendrix, D.V.H., Kern, T.J. and Plummer, C.E. (2021). The eye examination and diagnostic procedures. In: Veterinary Ophthalmology. Featherstone, H.J. and Heinrich, C.L. (Eds). 6th edn. John Wiley and Sons Inc. pp 601–618.

Varshney, J.P. (2025). A Fully Illustrated Handbook on Clinical Diagnostic Processes in Canine Medicine. NIPA Genx Electronic Resources and Solutions P.LTD, New Delhi-110034.

19

Cerebrospinal Fluid (CSF)

Large Animals

The collection and evaluation of cerebrospinal fluid (CSF) in farm animals with signs of diseases of nervous system provides very valuable diagnostic and prognostic information. Routinely, CSF is evaluated for total protein level, erythrocyte and leukocyte counts, and a differential cell count. Additional tests are available for specific neurologic conditions. CSF is primarily produced by the choroid plexuses in the lateral, third, and fourth ventricles through plasma ultra filtration and active transport across the blood-brain barrier. Smaller amounts of CSF are generated by the ependymal lining, pia arachnoid, meningeal blood vessels, and spinal cord's central canal. The turnover rate is about 1% per minute, meaning systemic electrolyte or acid-base changes take several minutes to affect CSF composition significantly. Located in the subarachnoid space, CSF provides cushion to the brain and spinal cord; protects from trauma; regulates intracranial pressure; maintains electrolyte and acid-base balance; facilitates transport of neurotransmitter and hormone; and aids in the removal of metabolic waste. CSF glucose is typically 60-80% of serum glucose, reflecting glucose metabolism by the nervous tissue. Sampling from the lumbosacral region is a safe and effective approach. The composition of CSF collected from lumbosacral or cisternal sites is similar. Nevertheless, sampling should generally be done near the lesion unless there is increased intracranial pressure.

Key Points for CSF Collection and Handling

- CSF is collected in a sterile tube without adding anticoagulants, even if there is blood contamination.
- Cytology should be conducted as quickly as possible (ideally within 15 minutes) because cells degrade quickly due to the low oncotic pressure in CSF.
- If cytology is to be delayed, autologous serum is mixed to create an 11% serum solution to keep bovine CSF samples viable for up to 24 hours at 4°C.

- Puncturing the conus medullaris may cause immediate pain and discomfort.
- Pelvic limb movement can dislodge the needle, risking local trauma and hemorrhage.
- In animals with a vertebral body abscess and hind limb neurological issues, obtaining CSF can be challenging.
- Repeated attempts can cause fibrosis in the epidural tissue, making it harder to obtain a sufficient sample.
- Failure to obtain CSF often results from incorrect needle placement requiring rechecking of the anatomical landmarks and realignment of the needle.

Procedure of Cerebrospinal Fluid CSF) Collection

Lumbo-sacral site

- Spinal fluid can be collected from two sites viz. lumbo-sacral and cisterna magna.
- The lumbosacral site is the preferred site for CSF collection in large animals as it does not require general anesthesia. With proper restraint and identification of anatomical landmarks, CSF can easily be collected from the lumbosacral cistern.
- The procedure can be performed in large animals either in standing or in sternal recumbency. For recumbent animals, flexing the hips and extending the pelvic limbs alongside the abdomen helps in opening the lumbosacral space and facilitating proper needle placement.
- The collection site is located at the midpoint of the lumbosacral space, between the last palpable dorsal lumbar spine (L6 in cattle, goats, and horses; L6 or L7 in sheep and pigs; L7 in New World camelids) and the first palpable sacral dorsal spine (usually S2). In poor conditioned animals, the site can be located at the midpoint of a line drawn between the caudal aspects of the tuber coxae.
- The site is clipped and aseptically prepared. Wearing gloves, 1 to 2 ml of local anaesthetic is injected subcutaneously.
- Animal is restrained and hypodermic spinal needle is inserted perpendicular to the vertebral column. A "pop" indicates entry of the needle into the epidural space. The pop is caused by the loss of resistance as the needle point penetrates the ligamentum flavum into the epidural space.

- If the needle is in right place, CSF appears in the needle hub within 2 to 3 seconds.
- 2 to 3 ml of CSF is aspirated with a syringe attached to the needle.
- If no fluid appears, needle is rechecked and realigned.

Atlanto-occipital cistern (Cisterna magna) site

- The cisterna magna is preferred for intracranial lesions because CSF flows caudally from the subarachnoid space down the spinal cord. The cisterna magna site can be used to obtain post-mortem samples for magnesium evaluation.
- This site is rarely used due to the risk of needle penetration of the brainstem.
- The site is prepared as with the lumbosacral cistern.
- The head and neck is ventriflexed in cattle to enlarge the cisterna magna, making entry easier.
- A styletted spinal needle is inserted at the intersection of the transverse line of the cranial rim of the atlas wing and the dorsal midline.
- The needle is carefully advanced directing it rostrally toward the symphysis of the lower jaw. The needle passes through the skin, ligamentum nuchae, and leptomeninges.
- The entrance to the cisterna magna is approximately 4 to 6 cm deep in adult horses, 1.5 to 2.5 cm in neonatal foals, and 5 to 7 cm in mature cattle.
- When nearing the lower end of the expected depth, the spinal needle is advanced 1 to 2 mm at a time.
- On puncturing the leptomeninges, the animal may slightly move its head. At this point, the needle is advanced only 1 to 2 mm more and the stylette is removed.
- CSF starts flowing freely when the needle is correctly placed in the cisterna magna. CSF will flow freely.
- The CSF pressure can be measured using a manometer attached to the spinal needle. CSF analysis is more diagnostically valuable than hematology for nervous system diseases. Measurement of CSF pressure is indicated in animals with signs of cerebral disease. Pressure can be affected by movements like tenesmus.
- CSF pressure is 5 to 15 cm H_2O in cattle, and 28± 4 cm H_2O in horses.

Interpretations of CSF Analysis

- CSF pressure increases in conditions like polioencephalomalacia (PEM), bacterial meningitis, and hypovitaminosis- A.
- Cytological examination is often done using Cytospin to preserve cell structure due to low CSF cell counts.
- Bacterial infections of central nervous system (CNS) increases CSF protein and leukocyte counts, sometimes up to 2000 cells/µL with >70% neutrophils. Neutrophilic pleocytosis strongly suggests CNS inflammation. Pleocytosis refers to an increase in the CSF leukocyte count above 5 cells/µL. Pleocytosis is categorized as mild (6-49 cells/µL), moderate (50-200 cells/µL), or marked (>200 cells/µL). Normal CSF primarily contains lymphocytes and monocytes, with no erythrocytes.
- Turbid CSF generally indicates high cell counts (>500 cells/µL) and increased protein levels.
- Dipstick (practical tool for assessing CSF protein concentration on farms) range from trace (<25 mg/dL) to 3+ (470–590 mg/dL), with a reading of 2+ or higher indicating increased CSF protein concentration in dogs. However, similar studies have not been conducted in large animals.
- CSF analysis for cytokine concentrations and gene expression in nucleated cells can assist in diagnosing specific nervous system diseases.
- The presence of eosinophil in CSF is rare and typically suggests parasite migration or fungal encephalitis.
- In bacterial meningitis, CSF glucose may decrease, and lactate may increase, but these signs are often unreliable.
- Bacteria can sometimes be cultured from CSF.
- CSF creatine kinase (CK) and lactate dehydrogenase (LDH) activities have limited clinical value. High CSF CK activity (>19.5 U/L) in sheep with listeriosis strongly suggests a poor prognosis, although CK activity is unreliable in horses due to potential contamination with epidural fat. The clinical utility of CSF LDH activity in large animals remains unclear.
- The Pandy test is an on-farm method for assessing CSF protein concentration using a 10% solution of carbolic acid crystals in water (Pandy's solution). The test relies on the precipitation of proteins (globulin and albumin) by phenol, where a drop of CSF added to 1 mL of Pandy's solution causes turbidity at the interface if protein levels are elevated. Turbidity is ranked from 1+ (faint) to 4+ (strong), with normal

CSF showing faint turbidity. A positive control (4+) can be created using serum or plasma. Safety precautions should be observed when handling phenol, including wearing gloves and protective eyewear and properly disposing of reagents.

Dogs and Cats

Cerebrospinal fluid of dogs and cats is clear, colorless and transparent with few cells and protein. It plays an important role in providing physical support; pressure modulation; transport of metabolites, nutrients, and neurotransmitters; and maintenance of ionic balance (Di Terlizzi and Platt, 2009). Evaluation of cerebrospinal fluid is of diagnostic value in dogs and cats with central nervous system signs. As compared to blood, cerebrospinal fluid analysis provides more accurate reflection of ongoing pathology in brain, meninges and spinal cord.

Indications

- Cerebrospinal fluid tapping is indicated in dogs and cats with seizures, incoordination, circling, nervous signs, and pain in neck or back, CNS disorders, encephalopathy and myelopathy.
- The pets suspected with an infectious or inflammatory disease, tumor, degenerative disease of the brain and or the spinal cord, are good candidate for examination of cerebrospinal fluid.

Contraindications

- Increased intracranial pressure as seen in space occupying lesion, cerebral trauma, hydrocephalus, and inflammatory diseases of central nervous system.
- Underlying coagulopathy.
- Atlanto-axial subluxation.
- Cervical trauma.

Risk Factors

- Anesthesia risk.
- Trauma to brain stem or spinal cord from the spinal needle.
- Invasion of infection if aseptic procedures are not maintained.

Material Needed

- Spinal needle (7- 9 cm, 22-20 gauge).
- Cotton swab.

- B.P. blade with handle.
- Sterilized test tubes, test tube with EDTA, and plain test tubes.
- Sterilized test tubes, test tube with EDTA, plain tubes
- Razor
- Rectified spirit
- Sterile surgical gloves

Sites for CSF Tapping

- Cerebrospinal fluid can be collected from cistern magna (atlanto-occipital or cerebellomedullary cistern puncture) or lumbar puncture. In dogs and cats cistern magna site is preferred.

Preparation of the Site

- The site for CSF tapping is prepared aseptically by shaving, cleaning and applying antiseptic on the skin.

Restrain and Control

- Spinal tapping at cistern magna site is generally recommended under general anesthesia to reduce the risk of animal movement. Short acting barbiturate provides satisfactory anesthesia for dogs.

Positioning

- The dog is placed in lateral recumbency with head and cervical vertebrae at the edge of the table. Skull is flexed to create a 90^0 angle with the cervical spine. The nose is slightly elevated to keep the long axis of the muzzle parallel to the table.

Tapping Procedure

- Tapping procedure is done at the aseptically prepared site.
- For cisternal tapping, a spinal needle with stylette (keeping it perpendicular to dorsal laminae of vertebral column) is inserted slowly at the level of atlanto-occipital space (mid line between occipital crest and the cranial edge of the wings of the atlas) in to the spinal column penetrating the duramater and arachnoid membranes to subarachnoid space.
- No resistance is felt as soon as the needle enters the subarachnoid space.
- Stylette is removed. A sterile syringe is attached to the spinal needle to draw out CSF or fluid can also be allowed to drip directly into the collection tube.

- Once CSF collection is completed, the needle is gently removed.
- Lumbar tapping is comparatively more difficult to perform in dogs and cats and is associated with more chances of blood contamination. Tapping at lumbar site is done in lateral recumbency with the pelvic limbs fully flexed. The appropriate intervertebral space is L5-L6 in dogs and L6-L7 in cats (Oliver and Lorenz, 1997).
- The needle (20 G) is inserted caudal to L6 perpendicular to the dorsal laminae of the vertebrae, along the cranial border of the caudal dorsal spinous process until it touches the bone of ventral aspect of the spinal canal.
- Insertion of the needle in the canal may cause a slight twitch of the tail or leg.
- The stylette is taken out and the fluid is collected. The flow of the fluid is slow and quantity of the fluid, that can be retrieved, is also less as compared to cisternal tapping.

Fluid Collection

- CSF can be collected in EDTA tubes or plain sterile tubes. CSF collected in EDTA tubes is used for hematological evaluation. The volume of CSF collection should not be more than 1.0 ml/5 kg body weight in dogs and cats (Carmichael, 1998).

CSF Analysis

- CSF should be processed for examination within 30-60 minutes of collection. It is subjected to physical, chemical, cytological and microbial examinations. Physical examination includes color, turbidity and coagulation. Indices of chemical examination are proteins, glucose, chloride, sodium, cholesterol, enzymes (lactic dehydrogenase, ALT and AST). Cytological examination of CSF includes total nucleated cells and differential cell count. CSF samples having high cell count and increased protein should also be subjected to cultural examination for isolation and identification of the organisms. Normal CSF contains total white cell up to 5 cells/μL and total protein as 10-25 mg/dl (much lower than blood).

References and Further Readings

Carmichael, N. (1998). Nervoussystem. In: Manual of Small Animal Clinical Pathology. Davidson, M., Else, R. and Lumsden, J. (Eds.). British Small Anim. Vet. Assoc. Cheltenham, UK, pp. 235–240.

Di Terlizzi, R. and Platt, S.R. (2009). The function, composition and analysis of cerebrospinal fluid in companion animals: Part II – Analysis. Vet J. 180: 15-32.

Oliver, J.E. and Lorenz, M.D.(1997). In: Handbook of Veterinary Neurology.Oliner, J.E. and Lorenz, M.D. (Eds.), Saunders, Philadelphia, PA. Nelson, R. W., and Couto, C. G. (2019). Small Animal Internal Medicine-E-Book: Elsevier Health Sciences.

Varshney,J.P.(2025). A Fully Illustrated Handbook on Clinical Diagnostic Processes in Canine Medicine. NIPA GENX Electronic Resources And Solutions P. LTD. New Delhi - 110034.

20

Special Examination of Musculoskeletal System

Musculoskeletal system consists of muscles, cartilage, bones, joints ligaments and tendons. Primary function of the musculoskeletal system is to support the body, helps in mobility, and protects vital organs of the body. Diseases of these organs have clinical manifestations such as lameness, failure to support weight, inefficient joint movement and deformity. Examination of musculoskeletal system includes general as well as special examination. General examination consists of examination of posture in sitting/recumbency, standing position and while rising; examination of gait; and any other abnormality.

General Examination

The animal is observed for conformation (outline and shape), symmetry and posture in standing position from the cranial, caudal and lateral positions. Asymmetrical shape of the body or limbs may be due to swelling that should be examined by palpation, percussion or imaging (radiography or ultrasonography). The limbs and joints are compared with their contralateral limbs or joints for size and shape.

Special Examination Techniques

Special examination techniques for musculoskeletal system are examination of the history (nutritional), examination of the environment (housing and floor), analysis of gait and conformation, close examination, radiography, muscle biopsy, arthroscopy, arthrocentesis and serum biochemistry including enzymology.

History

Nutrition plays a significant role in the diseases of musculoskeletal system. Most of the clinical cases of osteodystrophies and myopathies are of nutritional origin, therefore information about feeding feeds and fodder is very vital for further analysis of the feed.

Examination of the environment of the animal

The housing, floor and its quality should invariably be inspected when many cases of lameness are occurring in housed cattle and swine.

Analysis of gait and conformation

Gait is the movements of the limbs exhibited in terms of rate, range, force and direction of the movement. Abnormalities in rate, range, force and/or direction of the limb movements make the gait abnormal. Limb movement is affected in true cerebellar ataxia. Range and force of the limb movement becomes excessive (high stepping gait) in sheep suffering from louping ill. The range of limb movement is diminished in arthritis or laminitis. Walking in circles may be seen in cases with head rotation or head deviation; listeriosis; or acetonemia and pregnancy toxemia. Compulsive walking is seen in encephalomyelitis and liver insufficiency. Inspection of gait can assist in localizing the site of lameness and provide information whether lameness is in fore or hind limbs. Most gait evaluation assess changes relative to normal gait. But stoic nature of the cattle (painful conditions of hoof or limb may cause little or no obvious change in behavior) can sometimes make visual inspection of gait difficult to judge. Gait is generally assessed by subjective observational methods. In research studies observational methods of gait assessment such as kinetic and kinematic techniques have also been evaluated. Numerical rating scores (NRS), and to a lesser extent, visual analogue scales (VAS) are used for subjective observation of the gait. NRS rates individual cow on a five-point scale, for the presence or absence of certain behaviors and postures related to gait. VAS, developed for use by human patients to rate pain (Scott and Huskisson, 1976), have been adapted for lameness assessment in animals. The objective methods of gait assessment use force plates or force-recording shoes to obtain force data. Both devices (kinetic and kinematic) can measure horizontal and vertical forces exerted by hooves in contact with the ground and have been used successfully to assess forces involved in walking of dairy cows (Scott, 1988; Van der Tol *et al.*, 2002).

Examination of conformation can provide information about factors contributing to lameness.

Close examination

To localize the lesion, a detailed physical examination should be carried out. The limbs can be subjected to passive movements to detect pain, dislocation or fracture. Muscles need to be palpated to detect pain, enlargement or atrophy.

Radiography

Radiography is often considered the first step for the diagnosis of diseases of bones, joints and soft tissues of the limbs. Positioning of the animal for taking radiograph is very important for obtaining a clear image for an accurate diagnosis. The restrain of the animal is equally important for getting clear and sharp radiographic images. In most of the cases no sedation or anesthesia is needed. Nevertheless, highly uncooperative animals may require sedation. Radiographs are usually taken in lateral (right or left), the dorsal and sternal recumbancy position.

Ultrasonography

Ultrasonography (USG) is a very useful tool in evaluating soft tissue structures like tendons, abscesses and muscle lesions. It also identifies abnormalities and inflammation in and around joints. It is cost effective, and there is no exposure to radiation as is in case of X-ray.

Arthroscopy and Arthrocentesis

Arthroscopy involves taking a radiograph of a joint after injecting a radiopaque dye into a joint space to outline the structures such as ligaments inside the joint. This helps in viewing a torn ligaments and fragmented cartilage in the joint, inflammation of the synovium/synovitis, etc.

A small diameter fiberoptic devise or scope (arthroscope) is inserted into a joint space under general, local, or spinal, (or a combination) anesthesia to visualize the joint. Degenerative joint disease, ruptures of joint capsules, etc. may be diagnosed with arthroscope. A piece of tissue/ biopsy of joint cartilage or the joint capsule may be taken during the procedure for analysis. Surgery can also be done, if necessary, to correct the condition. Recovery time after arthroscopic surgery is much faster than after traditional surgery.

Arthrocentesis (Fig.166) is tapping of the joint or withdrawing fluid from the joint capsule. This is usually indicated in cases of septic arthritis to ascertain the pathogenic etiological agent and their sensitivity to antibiotics. It is also done to know the presence of uric acid crystals for diagnosis of gout and calcium pyrophosphate dihydrate crystals for diagnosis of calcium pyrophosphate arthritis (pseudo gout).

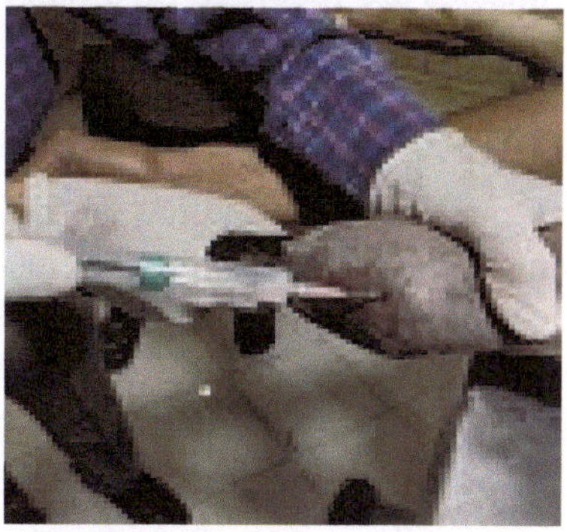

Fig. 166: Arthrocentesis in a cow calf.

Normal joint fluid contains about 2500-3000 WBCs/ microliter. Mostly mononuclear cells (about 90-95) and a few RBCs. In infectious or inflammatory condition, the joint contains more than 5000 WBCs mostly neutrophils. For arthrocentesis (Fig.67), the joint site is aseptically prepared, a local anaesthetic is infiltrated, a sterile 1.5 inch needle (19 G) is inserted into the joint capsule and the joint fluid is slowly withdrawn. A sterile 9 cm lumbar spinal needle of size 18 G (1.20mm) with a stylet may be used to avoid blockage of the lumen by soft tissue. A sample of synovial fluid may also be obtained under the guidance of ultrasonography.

The collected sample is placed in two tubes- 1) containing EDTA for cytology and protein content analysis and 2) into a plain sterile tube for bacteriological culture. An elevated total leukocyte count with a neutrophilia in synovial fluid is suggestive of septic arthritis. The protein content is also elevated. For bacteriological examination, a thin smear can be prepared on a glass slide and stained with Giemsa and examined under microscope. Other findings like presence of uric acid crystals confirms a diagnosis of gout and calcium pyrophosphate dihydrate crystals confirm a diagnosis of calcium pyrophosphate arthritis (pseudo gout).

Muscle biopsy and histopathology

Muscle biopsies are useful to diagnose and identify white muscle disease and other affections of muscles caused by neurogenic disuse e.g. as in case of femoral nerve paralysis where there is atrophy of the hind limb muscles.

Electromyography

This test records electrical impulses in the muscles on a graph paper. It helps in diagnosing cases of lameness due to nerve affections. The electrodes are placed in the muscle tissue and electrical impulse generated during the working of muscles are recorded on a piece of graph paper. It is helpful for the diagnosis of peripheral nerve disorders.

Nerve conduction test

It provides information about the functioning of the nerve supplying the muscle. Nerve conduction test is of value in the diagnosis of peripheral neuropathy, carpal tunnel syndrome and ulnar nerve palsy. Nerve conduction test along with electromyography are indicated for the diagnosis of myositis, muscle dystrophy, neuropathy, myasthenia gravis.

Thermography

A thermal camera with infrared radiations is used to detect temperature difference and capture and create the images emitted from the tissue. The device can detect small temperature differences and thus, is useful in identifying the presence of inflammation.

Scanning or scintigraphy of bone

Bone scanning or bone scintigraphy involves use of a radioactive substance (e.g. technetium-99m–labeled pyrophosphate). The radioactive substance is administered intravenously and a bone-scanning device detects the affected area and creates an image of the bone on a computer screen. It is used to diagnose a fracture, particularly if other tests, like x-rays and computed tomography (CT) or magnetic resonance imaging (MRI), do not reveal the fracture. It is useful particularly in conditions of the hip joints in larger animals where radiography is a difficult task.

Serum biochemistry and enzymology

Estimations of serum calcium, phosphorus, alkaline phosphatase, creatine phosphokinase (CPK), serum glutamic oxaloacetic transaminase (SGOT or AST) are of assistance in cases where involvement of bone and muscle is speculated. Muscle enzymes (CPK, SGOT or AST) are sensitive indicators of muscle damage.

References and Further Readings

Scott, G.B.(1988). Studies of the gait of Friesian heifer cattle. Vet. Record 123, 245–248.

Scott, J.and Huskisson, E.C. (1976). Graphic representation of pain. Pain 2, 175–184.

Van der Tol, P.P.J., Metz, J.H.M., Noordhuizen-Stassen, E.N., Back, W., Braam, C.R. and Weijs, W.A. (2002). The pressure distribution under the bovine claw during square standing on a flat substrate. J. Dairy Sci. 85: 1476–1481.

21

Lameness in Dairy Animals

Lameness in dairy animals is a serious health problem all over the world causing substantial economic losses to dairy industry owing to reduced milk production, reduced reproductive performance, cost of treatment including medicines, increased risk of culling, death, and the development of other diseases. Lameness can be defined as a state of being unable to walk properly because of pain in the limbs or foot. The incidence of lameness in the dairy herd is variable (1 to 50%) depending on the farm. In USA its incidence varies from 13 to 55% (Hoffman *et al.*, 2014; Cook *et al.*, 2016).

Lameness may be due to diseases of limb and/or hoof of infectious and/or non-infectious origin. Lameness usually affects hind limbs because of greater load of the body on hind limbs. A lame animal adjusts gait and posture depending on the origin of the pain (foot or upper limb). Clinical signs of lameness include changes in locomotion (e.g. decreased speed, decreased stride, length and tracking, increased abduction, increased asymmetry), posture (e.g., arched-back, cow-hocked, hind-limb protraction or retraction, cross-legged, knuckling of the fetlock), weight distribution (e.g., weight shifting between limbs, uneven weight bearing) and changes in behavior (e.g., lying, standing, feeding, socializing, ruminating).

Etiology

- Lameness is a multifactorial disease.
- Predisposing causes include inadequate building size, excessive stocking density, unsuitable walking surfaces (slippery or slatted floors), sharp turns at the entrance or exit of the cowshed, or an unsuitable resting surface.
- Dietary factors includes high energy fodder, diet deficient in fiber, excessive protein, deficiencies of minerals and vitamins, sudden change in diet in post partum period.
- Concomitant diseases such as sub-acute ruminal acidosis may lead to laminitis and white line ulcer leading to lameness.

- Genetic factors such as inherited hoof structure, some breeds (Holstein Frisian cows) are also prone to lameness in dairy animals.
- Infectious agents such as *Fusobacterium* spp., *Campylobacter* spp., *Staphylococcus* spp., *Streptococcus* spp., *Treponemma* spp., *E. coli,* or *Mycoplasma* may cause digital dermatitis, paronychia, interdigital dermatitis, heel horn erosion, arthritis leading to lameness.

Clinical Signs

- Limb and hoof diseases are manifested by lameness.
- Clinical signs of lameness include changes in locomotion (e.g. decreased speed, decreased stride, length and tracking, increased abduction, increased asymmetry), posture (e.g. arched-back, cow-hocked, hind-limb protraction or retraction, cross-legged, knuckling of the fetlock), weight distribution (e.g. weight shifting between limbs, uneven weight bearing) and changes in behavior (e.g. lying, standing, feeding, socializing, ruminating).
- Lameness may be graded as no lameness, minor lameness, moderate lameness, and severe lameness depending on the severity of clinical manifestations (posture at rest, gait changes and position of the head).

Diagnosis of lameness

- Detection of lameness is typically based on visual observation of the cow's locomotion.
- Cows need to be assessed during walk with consistent flow for at least four complete strides on a flat surface without any obstacles, turns, slopes, and steps.
- Number and type of lameness indicators may be used to score the lameness (back arch, head bob, tracking up, asymmetric steps, weight bearing, stride length, speed, leg abduction and/or adduction, joint stiffness/flexion). Locomotion scoring is a key component of lameness prevention programs and welfare quality assurance programs. There are many challenges in detecting lameness such as masking of the pain by stoic nature of the cows, lag between the time experiencing pain inside hoof and the time when lesion becomes visible on the sole.
- Development of automated lameness detection system (kinetic methods) is taking care of limitations of visual methods of lameness detection.
- As lameness is a sign, not a disease and its cause at cow level should be ascertained and located.

- The history and environment of the animal should always be taken into account in conjunction with the findings of general clinical and detailed physical evaluation.

Steps in a lameness examination

- *Evaluation of locomotion*- The posture of the animal (e.g. back, shoulders, pelvis), movements (at rising and lying down), examining each limb and digit, and gait is to be assessed. Evaluation of gait can assist in determining whether lameness is in the front or hind limbs. Detailed foot examination should invariably be done before upper limb.
- *Physical examination of limb*- After identifying the affected limb, the animal should be properly restrained for detailed examination of the limb and foot for swelling, lesions, or foreign material. Hoof testers can be used to evaluate pain reaction in the hooves. Functional hoof trimming followed by therapeutic trimming should be adopted when foot is detected as a seat of problem. Careful palpation of joints, muscles, tendons, and bones, as well as limb manipulation (flexion, extension, abduction, and adduction) should be undertaken in cows having no foot lesion and no pain response.
- *Nerve blocks*- When examination of locomotion and limbs fail to ascertain lameness, the use of ancillary diagnostics is recommended. Digital nerve blocks or intravenous regional anesthesia (2% lidocaine, 20–30 mL), administered distal to a tourniquet above the dewclaw, can assist in differentiating lameness of foot from upper-leg.
- *Other diagnostic tools*- Imaging (radiography, ultrasonography), arthroscopy, arthrocentesis, and synovial fluid analyses, are of diagnostic value in complicated cases. Arthrocentesis with synovial fluid analyses facilitates diagnosis of lameness associated with joint diseases. Radiography can detect fractures, luxation, septic arthritis, and osteochondrosis. Ultrasonography is valuable in examining soft tissues of the musculoskeletal system, and can facilitate the diagnosis of septic arthritis, tendonitis, osteochondral fragments, and muscle injuries.

References and Further Readings

Cook, N.B., Hess, J.P., Foy, M.R., Bennett, T.B. and Brotzman, R.L.(2016). Management characteristics, lameness, and body injuries of dairy cattle housed in high- performance dairy herds in Wisconsin. J. Dairy Sci. 99:5879–5891. doi: 10.3168/jds.2016-10956. [DOI] [PubMed] [Google Scholar]

Hoffman, A.C., Wenz, J.R., Vanegas, J.and Moore, D.A.(2014). Estimated prevalence of lameness in 53 Northwest US dairy herds. Bov. Pract. 48:89–94. doi: 10.21423/bovine-vol48no2p89-94. [DOI] [Google Scholar]

Urban-Chmiel, R., Mudron, P., Abramowicz, B. and Stachura, R. (2024). Lameness in cattle-Etiopathogenesis, prevention and treatment. Animals (Basel) 14(12):1836. doi:10.3390/ani14121836.

22

Bone Marrow and Lymph Node Aspirate

Bone Marrow

Bone marrow is a soft, gelatinous tissue that fills the medullary cavities of bones. It plays a crucial role in the production of blood cells. It makes up about 3% of body mass in rats, 2% in dogs, and 5% in humans. Bone marrow develops in the embryo during the second trimester of pregnancy. This tissue is vital because it contains stem cells that can differentiate into various cell types. Bone marrow is of red and yellow types. In young mammals, the marrow is red and active (with minimal fat). With aging, accumulation of adipose tissue, and long bones transition turn bone marrow to yellow, fatty, inactive. In adult mammals, active hematopoiesis continues in flat bones, such as the sternum, ribs, pelvis, vertebrae, and skull, as well as the epiphyses of long bones. Erythropoiesis in these regions contributes significantly to red blood cell production, with the axial skeleton accounting for about 35% and the appendicular skeleton for 65%. In birds, hematopoiesis is mainly restricted to bone marrow, while in reptiles, it occurs in the bone marrow, spleen, liver, and thymus.

Indications for Bone Marrow Examination

- To study the distribution of intermediate forms in erythrocyte development.
- Abnormal complete blood count (CBC) of unknown etiology.
- Unexplained non-regenerative anemia, neutropenia, thrombocytopenia, or severe leukocytosis.
- Detection of immature blast cells or atypical mature cells on a blood smear examination.
- For easy detection of *Leishmania*, *Cytauxzoon*, and *Histoplasma* in suspected cases in dogs and cats.
- To detect occult neoplasia (lymphoma, leukemia, multiple myeloma, and histolytic sarcoma).

- Subjective evaluation of iron stores in dogs.
- Confirming diagnosis of iron deficiency anemia.
- To rule out megakaryocytic hypoplasia in dogs.
- To isolate mesenchymal stem cells (MSCs) for regenerative therapies.
- For the diagnosis of bone marrow disorders (Bone marrow or erythroid aplasia-hypoplasia, thrombocytopenia of unknown cause).
- In cases of bracken fern poisoning in cattle, radiation injury and phenylbutazone toxicity causing bone marrow suppression.
- For demonstrating a regenerative response, especially in horses.
- Investigating anemia (chronic blood loss: Bone marrow- decreased myeloid: erythroid ratio). Aplastic anemia has very high myeloid: erythroid ratio as a result of lack of red cell series.
- Investigating bone-marrow neoplasia (myeloproliferative disease, lymphosarcoma).

Contraindications

- There are very few contraindications to bone marrow aspiration, as the risk of bleeding from this procedure is very low.
- A severe coagulopathy involving deficiency or impairment of coagulation factors is a concern and marrow aspiration should be delayed until it is controlled.
- The risk of causing a bone fracture is low but it may occur if an inappropriate sized needle is used for a patient or if the bone is already compromised by any disease.

Collection of Bone Marrow

- Bone marrow collection or sampling is done by aspiration and core biopsy techniques. Aspiration is done for cell morphology and differential count of myeloid and erythroid cells. While bone marrow biopsy is done for the assessment of the overall cellularity and structure of the marrow. Sometimes there is discordance between the results of cytology and histology with respect to MER (myeloid to erythroid ratio). Even bone marrow collection sites (iliac crest or humeri) can also influence MER findings in dogs. M: E ratios is 1:1 to 2:1 in healthy dogs and cats; and 0.75 to 2.53 in cows.
- In bovines, bone marrow samples can be obtained from the sternebrae (preferred site is the 3rd or 4th sternebra), proximal aspects of the ribs, or

tuber coxae. The sternebrae are the unpaired segmental series of bones forming the body of sternum. The sternebrae are united by intervening cartilages of fused by synostoses but remain separate in some species of domestic animals. The xiphoid process is the most caudal sternebrae. This site can be located by counting the ribs starting with the last one; however, the location on the sternum coincides with an imaginary line joining the olecranon. The iliac crest is unsatisfactory because of bovine pelvic anatomy. The adjustable depth control of the Turkel trephine needle helps prevent penetration of the pleural cavity, but does not eliminate the risk. The preferred site in adult animals, and in calves, is the cranial sternum. Collection of bone marrow in cattle is done in standing position. The site (sternum) is shaved and aseptically prepared. The animal is sedated and ensured that it remain in standing position. A small area over the sternum, is infiltrated with local anesthetic 2.0% lidocaine (anesthetizing all tissue layers, including the periosteum). A small stab incision is made in the skin, and a Jamshidi needle (with the stylet in place) is inserted until it reaches the bone. The needle is rotated slowly to penetrate the bone, advancing until it is firmly seated. A 13 to 15 gauge, 5 to 7 cm needle with a stylet is used for marrow aspiration, while an 8, 11, or 13 gauge, 100 to 150 mm biopsy needle is employed for core biopsies. Intravenous xylazine provides adequate sedation, and the needle should be advanced about 1 to 1.5 cm, with a noticeable reduction in resistance when entering the marrow cavity. In calves, the proximal humerus is an excellent site for marrow collection, and the method for bone marrow aspiration and biopsy is similar to that used in dogs and cats. A 15 gauge × 5 cm Jamshidi or Illinois sternal biopsy needle can be used for aspiration, and an 11 gauge × 10 cm Jamshidi needle is appropriate for core biopsies.

- In adult horses, bone marrow aspirates are collected by inserting a needle about 2 to 3 cm into the sternebra. After removing the stylet, a 5 to 10 mL syringe is attached to aspirate the marrow. The collected sample is either spread onto a clean glass slide to air-dry or placed in a Petri dish with 0.5 to 1.0 mL of 2% EDTA. For core biopsies, the biopsy needle is inserted 2 cm into the cortex of the sternebra. The stylet is removed, and the needle is rotated and advanced 2 to 3 cm before being rapidly withdrawn, which may require considerable force in adult horses. The bone marrow sample, which is pink to red, is rolled onto a clean, dry glass slide for cytological examination and then placed in 10% neutral buffered formalin for histological processing.

- In sheep, the preferred site for bone marrow collection is the sternum, specifically at the 3rd or 4th sternebra. Sheep are usually restrained in a sitting position, known as the "sheep sit" or "rump position," with an assistant holding each foreleg. The collection area is aseptically prepared and anesthetized, using 2% lidocaine, following similar procedures as described under bovine. For bone marrow aspiration, a 15-gauge × 5 cm Jamshidi or Illinois sternal biopsy needle is commonly used. To ensure precise penetration, the needle guard is generally removed. The depth of insertion from the skin surface is typically 2.5 to 3.0 cm, with bone penetration occurring at about 0.5 cm. For core biopsies, an 11-gauge × 10 cm Jamshidi needle is appropriate.
- In dogs and cats, common sites for bone marrow collection are the anterior greater tubercle of the proximal humerus and the iliac crest. For small iliac crests, as found in toy breeds and cats, the femoral trochanteric fossa serves as an alternative site. The sternum is generally avoided due to the risk of penetrating the thoracic cavity and potentially injuring intrathoracic organs. It is considered only as a last resort or during necropsy.
  (i) Materials and equipment -For bone marrow aspiration, Rosenthal and Illinois sternal/iliac needles are used, while the Jamshidi biopsy needle is used mainly for biopsies, though it can also be used for aspiration. In dogs, a 15-gauge Illinois sternal/iliac needle is commonly used, while an 18-gauge needle is preferred for cats and small dogs. A 1-inch (2.5 cm) needle suits many animals, but a 1.5-inch (4 cm) needle is necessary for large breed dogs. Jamshidi biopsy needles are available in 13-gauge and the larger 8-gauge size, the latter being reserved for larger dogs. All needles should be sterile and consist of a hollow stainless-steel needle and a solid stylet. The Illinois sternal/iliac needle has a removable plastic cap and needle guard, and while it can be re-sterilized, it is best used once as its tip dulls quickly. The Jamshidi needle does not have a needle guard, and while it can also be re-sterilized, repeated use can dull the tip, making future use more challenging and potentially more painful for the patient. Battery-powered drills are available for bone marrow aspiration and biopsy in humans, cats, and dogs, as well as for intraosseous (IO) catheter placement. Additional materials for bone marrow aspiration include microscope slides, a 12-mL syringe, sterile anticoagulant (4% EDTA), a Petri dish or weighing tray, and forceps or a pipette. For core biopsies, a formalin jar, syringe, anticoagulant, and Petri dish or weighing tray are needed.

(ii) Sedation and local anesthesia -Since bone marrow aspiration and core biopsies are painful, dogs usually require heavy sedation or general anesthesia. The dog is positioned in lateral recumbency, and the site is shaved, aseptically prepared, and anesthetized locally. Anesthetic application starts at the periosteum, then moves to the subcutaneous tissues, and finally to the skin, using 2% lidocaine.

A small stab incision made with an 11-scalpel blade can help reduce skin contamination during the procedure. To maintain sterility, surgical gloves should be worn, and all instruments and supplies should be kept in a sterile field, with a small surgical drape used as needed. If suitable, administer a non-steroidal anti-inflammatory drug or gabapentin after the procedure.

(iii) Positioning -For bone marrow aspiration, the clinician typically chooses the humerus on the same side as their dominant hand. For a right-handed clinician, this means standing ventrally to the patient (with the patient's head pointing right) and stabilizing the right forelimb. The lateral aspect of the greater tubercle serves as the ideal aspiration site due to minimal muscle and fat coverage. The needle is inserted through a stab incision and aligned along the bone's long axis. Firm pressure and a twisting motion help advance the needle through the cortex. For the proximal femur, the patient is positioned in lateral recumbency. The trochanteric fossa, found just medial to the greater trochanter, is targeted, and the limb is rotated medially to avoid the sciatic nerve. The needle is aligned parallel to the femur to enter the marrow cavity. For iliac crest aspiration, the patient is positioned in sternal or lateral recumbency, and the widest part of the iliac crest is palpated. The needle is inserted into the crest, aiming slightly caudal.

(iv) Procedure -A Klima-Rosegger, Rosenthal, or Illinois sternal needle (13– 19 gauge, 2.5 cm) with an interlocking stylet is commonly used. The needle is inserted either perpendicularly to the humerus or at a 45° angle, aligned with the long axis of the ilium, using slow rotating motions. Once the stylet is removed, a syringe containing 5–10% EDTA anticoagulant is attached, and bone marrow is quickly aspirated using negative pressure, collecting about 1 mL depending on the animal's size. The choice of anticoagulant depends on the diagnostic protocol; heparin is preferred for flow cytometry but is unsuitable for cytological examination due to potential staining and morphological artifacts. The sample is transferred to a Petri dish where bone marrow particles are isolated, spread on slides, and prepared for diagnostic evaluation, with six to eight quality smears needed for optimal diagnosis. After collection, pressure is applied to the site to control bleeding.

Core biopsy specimens should be collected at least 1 cm away from the previous needle site or from a different location to accurately reflect bone marrow status, avoiding areas near the aspiration site to prevent hypo cellularity in the core sample. An 11–13 gauge Jamshidi needle, with a cylindrical shape, tapered distal end, and beveled cutting tip, is used for this procedure. Once the needle is seated in the cortical bone, the stylet is removed, and the needle is advanced with rotating motions. Repositioning or rotating the needle helps break off the sample, allowing it to remain within the needle core. After withdrawing the needle, the stylet is reinserted in reverse to gently push the core sample out, which is then placed in a tissue cassette. Rolling the core sample on a glass slide before placing it in formalin provides additional cytological material. The core sample is fixed in neutral buffered 10% formalin for at least 2 hours and should ideally measure 1 cm with about 10 trabeculae visible under a microscope. Before histologic processing, the sample must be decalcified. For transport, 70% ethanol is used to reduce toxicity and limit exposure to formalin fumes.

- In dromedary camels, bone marrow collection is performed over the $5^{th}$to 8^{th}dorsal rib. This area is aseptically prepared and anesthetized in a manner similar to cattle. For llamas, the preferred collection site is the sternum, specifically at the 3rd or 4^{th}sternebra. Camelids are notable for having a higher proportion of eosinophil in their bone marrow compared to other species.

  Procedure-For sedating adult llamas and camels, intravenous xylazine is used, and animals are kept in standing position, in a chute, or in left lateral recumbency. After aseptic preparation of the site, 2% lidocaine is administered as a local anaesthetic. An 11–16-gauge bone marrow biopsy needle is employed to collect samples, which are typically aspirated into EDTA.

  If bone marrow smears are prepared immediately, anticoagulant may be avoided. General evaluation of bone marrow samples is conducted using Wright or Wright-Giemsa stains, while Perls iron stain is specifically used to assess iron stores. The average myeloid-to-erythroid (M: E) ratio in healthy camelids is around 1.2 in dromedary camels, 0.9 to 2.9 in llamas, and 0.5 to 1.0 in alpacas.

Lymph Node

Lymph node aspiration in animals is performed to diagnose the underlying cause (inflammatory, infectious, immunological or neoplastic) of lymphadenopathy. When multiple lymph nodes are enlarged, it is recommended to collect samples from more than one site. While advanced imaging technologies can evaluate and "score" enlarged lymph nodes, microscopic examination remains the gold standard for accurate diagnosis.

The size of lymph nodes varies considerably in normal animals, even within individuals of the same species. Lymph nodes are typically larger in young animals than that of in adults, Normal lymph nodes are firm to the touch; smaller nodes feel smooth on palpation, whereas larger nodes, such as the sub maxillary lymph node in horses, often have palpable lobulation. The skin over a superficially located lymph node is always movable, and the node itself is somewhat mobile in relation to the surrounding tissues.

Structure and Function

- Lymph nodes in mammals are typically rounded or bean-shaped structures that function as both structural and functional components of the lymphatic system. These nodes are enclosed in a collagenous capsule, with trabeculae extending inward and linking to lymphatic vessels. Lymph nodes are divided into two primary regions: the cortex and the medulla. The medulla contains lymphatic tissue arranged into medullary cords, separated by medullary sinuses, and supported by a reticular tissue framework. It is in the medulla that immune response products, including activated cells, effector lymphocytes, and inflammatory mediators, are generated. The cortex, where immune responses are initiated, consists of two layers: the outer cortex and the inner cortex. The cortex, which houses follicles just beneath the capsule, extends toward the medulla through interfollicular regions called the paracortex. In the outer cortex, B lymphocytes are organized into nodules or follicles, which may be primary (without a germinal centre) or secondary (with a germinal centre). Deeper within the node, the paracortex is rich in T helper lymphocytes.
- Lymph nodes play a crucial role in phagocytosis and the initiation of immune responses. Migrating dendritic cells reach the lymph nodes via lymph, where, together with resident dendritic cells in the cortex, they present antigens to activate B lymphocytes. This activation occurs at the interface of the cortex and medulla, leading primary follicles to transform into secondary follicles with germinal centers, where lymphocytes

proliferate. Activated B lymphocytes differentiate into memory cells and plasma cells. The memory B cells then migrate to other lymphatic tissues, while most plasma cells move to the bone marrow to produce antibodies. When infection or other processes occur in a specific area of the body, the activity in the nearby lymph nodes increases, leading to their swelling. This swelling can sometimes be painful. The most common cause of swollen lymph nodes is infection, with nodes near the infected area swelling first.

Clinically Important Lymph Nodes

The clinically important lymph nodes and their associated conditions in different species of the animals are given in the table 10.

Table 10: Clinically important lymph nodes in Animals

Sr. no	Species	Name of Lymph nodes	Associated Condition
1.	Dog	Femoral LN	Lymphoma, Mountain spotted fever, Bartonellosis, Leishmaniasis, Leukemia, Systemic mycoses
		Internal LN	Neorickettsiosis (Salmon poisoning)
		Mandibular LN	Cryptococcosis
		Cervical and Submandibular LN	Bubonic plague
		Axillary and superficial inguinal lymph nodes	Mammary gland tumor
		Tonsillar and Cervical lymph node enlargement	Canine Distemper
		Perihilar lymph nodes (enlarged)	Coccidioidomycosis causing lymphangitis
		Hilar (tracheobronchial)	Lymphoma or fungal pneumonia (histoplasmosis, blastomycosis, coccidioidomycosis)
		Subcutaneous, Fascial, Mediastinal, Hepatic	Lymphangioma
		Iliac lymph node	Anal Sac Adenocarcinoma
		Lymphadenopathy	Leishmaniosis, Bartonellosis
		Generalized lymphadenopathy	Trypanosomiasis
2.	Cattle	Pre-scapular	Theileriosis
		Inguinal	Brucellosis
		Mesenteric	TB, Salmonellosis, Johnes disease
		Superficial	Malignant catarrhal fever (MCF), BVD-MD
		Peripheral LN	Bovine ephemeral fever

Sr. no	Species	Name of Lymph nodes	Associated Condition
		Prescapular and precrural	Bovine farcy
		Enlargement of several or all lymph nodes	Bovine lymphosarcoma
		Supramammary	Mastitis
		Prescapular, femoral, and Supramammary nodes	Bovine Lymphoma
		Sub maxillary and Parotid	Actinobacillosis
		Superficial	Trypanosomiasis
3.	Horse	Submandibular, Retropharyngeal, Mesenteric	Strangles
		Submandibular and Sub maxillary	Glanders
		Retropharyngeal	Guttural pouch empyema
		Bronchial	Rhodococcosis (Granulomatous lymphadenitis
		Submandibular LN	Equine viral rhino pneumonitis
		Lymph nodes at hocks	Epizootic lymphangitis (equine blastomycosis)
4.	Sheep/ Goat	Retropharyngeal, Mediastinal, Mesenteric, Mandibular, Parotid, Prefemoral, or Prescapular(Abscessation)	Caseous Lymphadenitis
		Ileum and ileocecal	Johne's disease
		Tracheobronchial and Mediastinal	Ovine Progressive Pneumonia
		Sub maxillary or Parotid	Malignant edema
		Supramammary	Mastitis, Lymphosarcoma
5.	Pig	Cervical	Tuberculosis, Brucellosis
		Mandibular and Retropharyngeal	Streptococcal lymphadenitis of swine (jowl abscesses, cervical abscesses
		Bronchial and Mediastinal	Enzootic pneumonia (EP)
		Mediastinal and Inguinal	Porcine circo virus, Nipah virus
		Bronchial, Submandibular and Mesenteric	Porcine Reproductive and Respiratory Syndrome Viruses (Porcine Arteriviruses)
6.	Camel	Submandibular and Renal	Caseous Lymphadenitis

Lymph Node Aspiration

- The large animal is first properly restrained, and the area over the enlarged lymph node (submandibular, prescapular, or inguinal nodes) is clipped and disinfected. Local anesthesia may be administered to minimize discomfort. A fine needle (22- to 25-gauge) is inserted into the lymph node at a shallow angle and gentle suction is applied using a syringe to collect a sample. The needle is then removed, and the aspirate is transferred on to a glass slide for cytology or placed in a fixative for histopathological examination. In case of multiple enlarged nodes, it is advisable to collect samples from more than one site. After aspiration, the animal is monitored for signs of bleeding or swelling. If the aspiration is inconclusive, a core biopsy may be needed for a more accurate diagnosis.

- Lymph node aspiration in small animals can be done using a 22-gauge needle and 5 or 10 ml syringe. The lymph node is stabilized, and either suction or "tattooing" aspiration is performed to minimize blood contamination. The sample is expelled onto a slide, gently spread, and air-dried. Slides are initially examined at low magnification for cell count and blood contamination, then further analyzed at high power or oil immersion if necessary.

Lymph Node Aspirate Examination and Interpretation

- A large number of immature lymphocytes in an aspirate can indicate lymphoproliferative disorders, such as lymphoma or lymphosarcoma. The presence of neutrophilic granulocytes at various stages of maturation in a lymph node confirms neutrophilic leukemia, often associated with hematological malignancies. Lymphocytes containing macroshizonts are indicative of theileriosis, a protozoal infection caused by *Theileria* species, affecting cattle, horses, and ruminants. Additionally, an increased number of eosinophilic granulocytes suggest a hypersensitivity reaction or parasitic infections, such as nematodes or protozoa. Macrophages with ingested particles may signal an ongoing infection or inflammatory process, such as tuberculosis or histoplasmosis. A rise in plasma cells can point to chronic inflammation, immune response, or plasma cell neoplasms. The presence of mast cells in significant numbers may suggest mast cell tumors, particularly in dogs, or other neoplastic processes. Lastly, the identification of atypical cells with irregular nuclei and increased mitotic figures raises suspicion for metastatic tumors, including carcinomas, melanomas, or sarcomas.

References and Further Readings

Constable, P. D., Hinchcliff, K. W., Done, S. H., & Gruenberg, W. (2016). Veterinary Medicine: A Textbook of the Diseases of Cattle, Horses, Sheep, Pigs and Goats – Two-Volume Set. Saunders Limited.

Ettinger, S. J., Feldman, E. C., and Côté, E. (2017). Textbook of Veterinary Internal Medicine: Diseases of the Dog and the Cat. Vol. 1. Nelson, R. W. and Couto, C. G. (2019). Small Animal Internal medicine. Elsevier.

Pugh, D. G., Baird, N., Passler, T., and Edmondson, M. (2020). Sheep, Goat, and Cervid Medicine. Elsevier.

Smith, B. P. (2015). Large Animal Internal Medicine. Mosby. Sorenmo, K. (2024) , March 6). Mammary tumors in dogs. MSD Veterinary Manual. https://www.msdvetmanual.com/reproductive-system/mammary-tumors-in- dogs/mammary-tumors-in-dogs.

Zimmerman, J. J., Karriker, L. A., Ramirez, A., Schwartz, K. J., Stevenson, G. W., and Zhang, J. (2019). Diseases of Swine. John Wiley and Sons.

23

Udder Examination and Mastitis Testing

Udder Examination

The udder is responsible for milk production. It is composed of four mammary glands each with its own teat (Fig. 167). Swelling of udder or teats is an indication of udder disease warranting clinical exploration and palpation. Visual examination the udder (before and after milking) and palpation is of great significance in detecting abnormalities. The udder is visually inspected from back and sides for symmetry and also palpated thoroughly. Palpation is a simple, quick and easy routine diagnostic tool for mastitis detection. Teats and supra mammary lymph nodes should also be palpated. These examinations aid in detecting fibrosis, inflammation or atrophy. Diffuse fibrosis gives the quarter a firm feeling. Areas of local fibrosis may vary in size and shape. Hot, painful and diffuse swelling with marked abnormality of mammary secretion is seen in acute mammary inflammation (acute mastitis). Atrophy of the gland is the terminal stage of chronic mastitis. With the development of gangrene, the area becomes cold and bluish. Five point palpation scores (Petrovski *et al.*, 2011; Hogan *et al.*, 1995; Scaletti and Harmon, 2012) and three point score for udder firmness have been developed. Udder firmness can be determined by palpation and by using a dynamometer (Burtulat *et al.*, 2012). Palpation is conducted by pressing the fingertips of all fingers of one hand into the udder tissue.

Nowadays ultrasonography is being increasingly used to visualize udder parenchyma and blood flow. Udder parenchyma of healthy cows is uniformly echogenic and milk appears anechoic. The echogenicity of the milk in mastitis is increased due to more cellular contents.

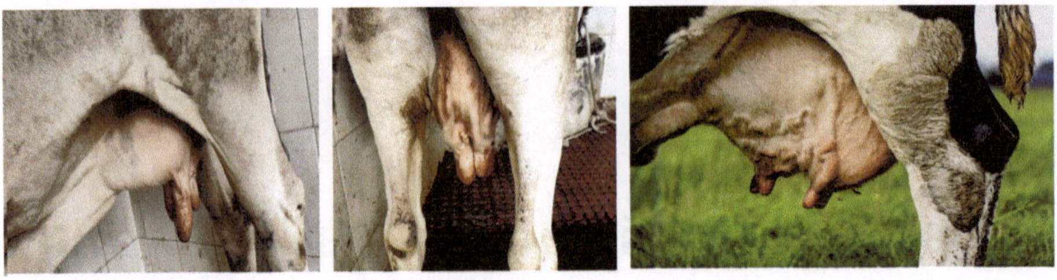

Fig. 167: Udder of healthy cows from different angles.

Mastitis Testing

Somatic cell count (SCC) and bacteriological culture (BC) of milk are the most frequently used diagnostic test for mastitis. Measurement of N-acetyl-b-D-glucosaminidase (NAG-ase), lactate dehydrogenase (LDH), electrical conductivity (EC), and molecular methods such as polymerase chain reaction (PCR) technology are the other advanced diagnostic tests advocated for mastitis diagnosis.

Changes in Milk

Mastitis is associated with abnormalities in milk (physical and chemical). A strip cup is used to detect milk discoloration, clots, flakes and pus and is a valuable diagnostic tool for detecting mastitis in the field. The milk may be more watery in chronic mastitis. Blood staining in milk may be seen in cases of blood in milk. Clots and flakes with discoloration are indicative of severe inflammation. Flakes at the end of milking have been observed in some cases of mammary tuberculosis.

Indirect Tests on Milk

Indirect tests are based on determining the quantity of DNA and therefore on the number of leukocytes. There are numerous mastitis diagnostic tests available, and they differ markedly in sensitivity, specificity, simplicity, rapidity, and cost. California mastitis test (CMT) was the first indirect test developed. It is conducted by mixing 3.0% sodium lauryl sulphate with milk in a white container and graded as negative, trace, 1,2 or 3 on the basis of amount of gel formed (Fig. 168). There is a close relationship between CMT reaction and somatic cell count.

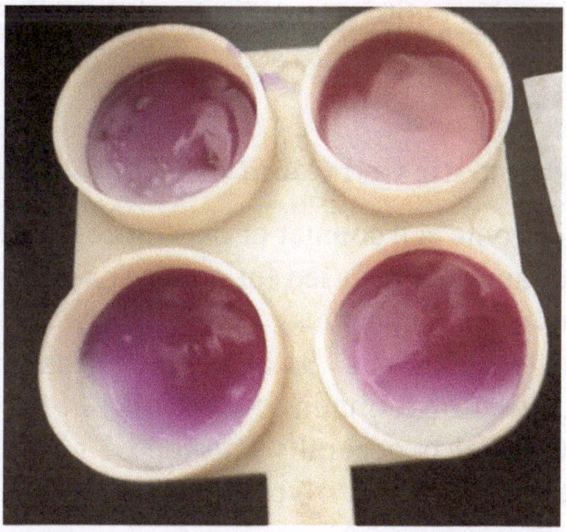

Fig. 168: CMT test for mastitis.Figure with the courtesy of Dr. D.K. Gupta, Principal Scientist, Department of Veterinary Medicine, GADVASU, Ludhiana

There is another indirect test called NAGase based on the amount of the cell associated enzyme (N-acetyl-β-D-glucosaminidase). Its higher concentration in the milk are indicative of high cell count. Milk NAGase levels are high at the beginning and the end of lactation in the similar manner as cell count.

The level of enzyme lactate dehydrogenase also increases in the milk in mastitis. The enzyme level depends on the number of lymphocytes and granulocytes in the milk.

A direct capture ELISA measuring elevated polymorphopnuclear granulocyte antigens provides an accurate estimate of cell counts in the milk and therefore valuable in the diagnosis of mastitis.

Indirect Chemical Tests on milk

Electrical conductivity of the milk depends on the sodium and chloride concentration of the milk. With the increase in milk sodium and chloride ions, electrical conductivity of the milk also increases. In mastitic milk there is an increase in sodium and chloride ions and therefore there is an increase in electrical conductivity.

Estimation of serum albumin concentration in the milk is another indirect chemical test that can be used for the diagnosis of mastitis. Its higher concentration in the milk is suggestive of epithelial damage. A radial immunodiffusion test is commercially available to estimate serum albumin.

Antitrypsin test, measures trypsin inhibitor capacity of the milk, is another indirect chemical test. Trypsin- inhibitor capacity of the milk is higher at the beginning of lactation due antitrypsin activity of the colostrum. But after the first one month, this activity is due to serum antitrypsin and therefore can be used for the diagnosis of the mastitis.

Milk Cell Count or Somatic Cell Count (SCC)

It can be done by electronic cell counter or on stained milk smear on slide. Cell count is done on the same sample used for culture examination. In sored milk, there is break down of the leukocytes/ therefore, smears should be prepared, fixed and stained within 1-2 hours. Breakdown or lysis of the leukocytes can be prevented by addition of formalin to milk (40% formaldehyde solution). Cell count < 2,50,000 /ml is generally considered normal. A bulk milk cell count > 3,00,000 cells /ml is suggestive of mastitis.

SCC is an indicator of intramammary infections. SCC can be measured by electronic digitalized machine or conventional microscopic method. In conventional microscopic method, a microscopic slide is marked with 1cm square area (1 x 1 cm). 0.01 ml milk is transferred to the marked area with the help of a platinum loop. Milk is gently spread with the loop to cover whole 1cm^2 area on the slide. The slide is completely dried and then stained with Broadhurst stain, Loeffler's modified methylene blue stain or Modified Newman Lampert stain. Cells are counted in 50 microscopic oil immersion fields. The count of the cells is multiplied by a working factor of 10,000 (the area of one oil immersion microscopic field is 0.02 mm^2. There will be 5000 oil immersion field in 1 cm^2. having 0.01 ml milk. In this way 1 ml milk will need 5,00,000 mm^2. Counting is being done in 50 oil immersion fields. Therefore working factor will be 5,00,000 ÷50 =10,000) to get cell count per ml of milk. Direct SCC measurement, of course, is much more accurate. The most important factor influencing SCC is intramammary bacterial infection. Milk samples for SCC measurement should be taken immediately before milking, after removing three squirts of milk. The SCC in uninfected quarters is approximately 70,000 cells/ml, optimal sensitivity and specificity of SCC at a threshold of 2,00,000 cells/ml as indicator of presence of IMI are estimated at 73% and 86%, respectively. Cell count < 2,50,000 /ml is considered insignificant for mastitis in cows.

Differential Cell count in milk

High proportion of neutrophils (90%) is indicative of acute inflammation. Neutrophil count <40% may indicate small chronic lesion.

Bacteriological Culturing

Bacteriological culturing from individual or bulk milk is the gold standard for the diagnosis of mastitis. Culture and antibiotic sensitivity is essential for mastitis. Laboratory culturing is time consuming and costly. Before taking milk sample for culture, technique of teat cleaning is very important. Teats should be washed and dried and teat tip should be cleaned with a swab soaked with 70% alcohol. First few streams of milk are discarded and next few streams should be taken for culture. When tuberculosis is suspected last few streams of the milk should be collected for culture examination. The collected milk samples for culture examination should be refrigerated or frozen if delay is expected in laboratory examination. The sample may be incubated on blood agar or selected media.

References

Bertulat, S., Fischer-Tenhagen,C., Werner, A. and Heuwieser, W. (2012).Technical note: Validating a dynamometer for noninvasive measuring of udder firmness in dairy cows.J. Dairy Sci., 95 :. 6550-6556.

Hogan, J.S., Weiss, W.P., Smith, K.L.,Todhunter,D.A., Schoenberger,P.S. and Sordillo, L.M.(1995). Effects of an Escherichia coli J5 vaccine on mild clinical coliform mastitisJ. Dairy Sci., 78 : 285-290.

Petrovski, K.R., Caicedo-Caldas, A., Williamson, N.B., Lopez-Villalobos, N., Grinberg, A., Parkinson, T.J. and Tucker, I.G. (2011). Efficacy of a novel internal dry period teat sealant containing 0.5% chlorhexidine against experimental challenge with Streptococcus uberis in dairy cattle.J. Dairy Sci., 94 : 3366- 3375

Scaletti,R.W. and Harmon,R.J. (2012). Effect of dietary copper source on response to coliform mastitis in dairy cows.J. Dairy Sci., 95 : 654-662.

24

Restraint, Physical Examination and Medication of Zoo Animals

Restraint

Various techniques and equipment are employed to capture and restrain both zoo and free-ranging wild animals for physical examinations and medical treatment. The primary methods of restraint are physical and chemical restraint. Assessing the physical condition of the animal and determining the necessity of capture are essential, as all restraint methods can impact an animal's behavior, physiology, and response to medication. Certain animals, such as black bucks, chinkaras, and deer, are particularly sensitive to restraint and prone to thermoregulatory issues, capture myopathy, and shock. In these species, even a routine capture can lead to injury or death.

Terminology

The following terminology is generally used during restraints of wild animals.

Anesthesia: Causes unconscious with loss of awareness and pain relief.

Analgesia: A state in which pain and sensation are effectively blocked.

Tranquilization/Sedation: It reduces anxiety without causing unconsciousness and animal remains responsive to stimuli.

Induction Time: The interval between drug administration and when the animal is fully immobilized.

Thermoregulatory Effect: Refers to an increase or decrease in the core body temperature due to the drug.

Antagonists: Drugs that counteract the effects of the primary anesthetic or tranquilizer; ideally administered intravenously, though intramuscular injections are also effective but slower to act.

Muscle Relaxant: It targets muscle relaxation without inducing mental sedation.

Reversal: Drugs used to counter the tranquilizing or sedative effects.

Local Anesthesia: It numbs a specific area, blocking sensation only in the injected region.

Protocols for wild animal capture and release

- Most large zoos and state wildlife agencies provide in-house training for their personnel on wildlife capture and immobilization methods. Veterinarians, especially zoo veterinarians, should be thoroughly knowledgeable about the various drugs, equipment, and physiology and anatomy of wild animals. Given that conditions in zoos and the field settings can vary greatly, operators must be able to recognize how these factors might impact a capture or immobilization event. Incompetent handling can lead to death of the animal or even harm the handler.
- Equipment must always be ready, in good working condition, and up-to-date, with proper inventory management to ensure drugs are not outdated, as expired drugs may be less effective. The drug box should be clean and organized, and the handler should take personal responsibility for ensuring that all equipments are ready for use. After a capture, equipment should be promptly replaced and readied for the next call.
- Before releasing a wild or sedated animal back into the zoo or a natural habitat, it is crucial to inspect the surrounding environment for safety, both for the animal and for the personnel involved. After all the resources and effort spent on capturing the animal, it is important to release it in an area that maximizes its chances of survival.

Considerations for wild animal capture and restraint

When responding to a wild animal situation, initial contact helps determine the appropriate response and the equipment needed. Upon arrival, following factors are to be considered.

- Capture Necessity- Not all situations require capturing an animal. Many issues, such as nuisance animals, can be resolved through non-capture methods like education, scare tactics, or habitat changes. The need for capture is evaluated based on whether the animal has caused damage, poses a threat to human safety, and if capturing the animal is safe for both the animal and yourself. Public expectations often lean toward capture, but capturing should be a last resort.
- Physical vs Chemical Restraint: Physical restraint should be used unless it poses a danger to human or animal safety. Non-lethal methods such as education, deterrents, traps, or darting are preferred where appropriate.

- Terrain and Animal Condition: The environment and the animal's condition affects approach. If the terrain could conceal a downed animal, reconsider chemical immobilization, as the animal may escape. If the animal is aggressive or in a vulnerable state (e.g., during rutting or with young ones), strategy is planned carefully. Excited animals may require higher doses of drugs, while debilitated animals are typically easier to immobilize.
- Equipment and Assistance: It is to be ensured that the right equipment and necessary assistance are available before engaging in the operation. However, there should be flexibility to adapt if circumstances change on site.
- Human Safety: The safety of zoo staff and others involved in the operation is to be kept as top priority. Visitors should not be allowed near the operation area, and it's important to have assistance for crowd control, ideally from local authorities. Never dart an animal unless it is certain that the public is at a safe distance. All darts should be accounted. Only trained personnel should assist with the operation.

Selection of restraint method

- The restraint procedure must be under the control of experienced zoo veterinarian. It is crucial to note that in many instances, individuals such as laypersons, zookeepers, or forest officers with limited experience may be involved. These personnel may lack the necessary knowledge of pharmacology, drug dosages, or the physiological needs of zoo and wild animals, which can lead to complications. Proper training and understanding of these factors are essential for a safe and effective restraint operation.

Restraint tools

Restraint tools are categorized in to seven categories.

Psychological tools: These helps manage animals by understanding their biological traits. The use of voice is key, as animals can detect emotions such as fear or confidence, which influences their behaviour. A handler must also understand the animal's anatomy and physiology to anticipate its movements and reactions.

Diminishing senses: Reducing an animal's sensory input, such as vision, is an effective technique. For example, placing a parakeet in a dark room before handling or using blindfolds can reduce stress and allow for calm manipulation, though sedation is needed for painful procedures. Animals that are sedated and handled in sunlight should be blindfolded to protect their retina from damage caused by direct sunlight, as their eyes may not adjust quickly enough to the light.

Confinement: The degree of confinement depends on the species. Wild animals experience stress when confined in small areas, such as in transfer cages or squeeze cages. These cages are crucial but must be designed specifically for the species' anatomical and physiological needs.

Tools for strength or reach extension: These help increase the handler's physical capacity to control the animal.

Physical barriers: Used to protect handlers or allow closer inspection of the animal.

Physical force: Applied to subdue an animal when necessary.

Chemical agents: Sedatives, tranquilizers, and anaesthetics are used to immobilize or sedate animals for procedures.

Physical restraint

- Persons involved in restraining zoo or wild animals should have a good understanding of the animal's behavioral traits and psychological characteristics to ensure their well-being. He should also be familiar with the tools and techniques of both manual and chemical restraint, including the use of voice, and be proficient in handling specialized restraint devices.
- The tools required for effective restraint can vary significantly depending on the species being handled. Some tools may be useful for one species but may not be suitable for other. With experience it becomes easy to select which restraint method is to be applied and when. Misuse of technique can be ineffective and potentially harmful to the animal and the handler.
- *Physical restraints with extended arms* – It involves various tools to help capture and restrain animals. Ropes are useful for extending the handler's reach, while snares are commonly used, but must be applied carefully to avoid suffocation or unnecessary harm. Commercial snares often include swivels for more humane handling, and quick-release snares allow animals to twist without suffocation. Home made snares can be made from metal or plastic pipes, with flexible rope or cable. Nets are another vital restraint tool, available in various sizes for different species. The net mesh should be appropriate for the animal's size, as

larger mesh can cause strangulation if the animal gets its head through. The material of the net also matters, with nylon, cotton, and manila each offering different durability, though carnivores may chew through nets, potentially causing escape.

- *Physical barriers* - These are used to protect both the handler and the animal, allowing for closer proximity without causing stress to the animal. Shields, such as simple plywood or plastic sheets with handles, help facilitate close interaction with animals and can be used between transfer cages. Plastic shields are particularly useful for handling non-venomous reptiles, small mammals, and birds. This technique is best suited for small to medium-sized primates, carnivores, and rodents that cannot easily push away the shield. Blankets can also be used to shield animals from handlers or protect handlers from injury, such as from antlers or horns. A small antelope, for example, can be captured by coaxing it into a blanket, which then fully encloses the animal. Small mattresses can serve the same purpose. Animals tend to recognize opaque plastic sheets as barriers, making them useful for directing animals into crates or chutes, especially for herding hoofed animals.
- *Physical force* - It is often necessary for manipulative procedures but with care to protect the hands. A deep understanding of the animal's behavior and anatomy is essential for applying the appropriate amount of force. Over-applied pressure can lead to suffocation or injury, such as broken ribs or limbs. Gloves are commonly used for protection, with materials ranging from thin cotton for small animals to thick leather for large primates and birds of prey. However, gloves reduce tactile sensitivity, making it harder for the handler to gauge the force they are applying. While gloves can offer some protection, it is not fool proof, as carnivorous animals may bite through it, and gloves cannot prevent injury from crushing force, such as that from a tiger's jaws or a macaw's beak.

Physical restraint should be prioritized over the use of drugs whenever the animal can be safely handled without causing harm or stress. Generally, the least amount of restraint required is the most effective approach. Understanding species-specific restraints is crucial for planning the operation, as outlined below in table 11.

Table 11: Restraint equipment for different animals

Animals	**Equipments Required**
Small carnivores	Nets, pole syringes, blow dart equipment, crates, squeeze cages
Hoofed stock:	Projectile guns and darts, crates blow dart
Reptiles	Nets, bags, plastic tubes, snake tong, snake hook
Amphibians, birds and fish	Nets, gloves

- *Large mammal restraint*- Effective restraint of large mammals in zoos requires careful planning, the right methods, appropriate equipment, and trained personnel. It involves using proper techniques, scheduling, medication, transportation, and emergency supplies like ECG and oxygen. The timing of handling free-living ungulates is also important, with some species being easier to manage in winter. Pregnant animals should not be restrained late in gestation, and young animals may be captured when separated from their mothers.

  Restraint tools like squeeze cages, nets, and muzzles are used based on the animal's species and size. Wild ungulates may require barriers like burlap to recognize corral enclosures. For large animals, body squeezes, curved chutes, and escape routes can improve handling efficiency. Immature bears can be controlled with nets, while larger and large felines require squeeze cages. Proper care must be taken to prevent overheating when muzzles are used on canids, and any extended procedures for large felines should be done with careful restraint to prevent injury.

- *Small mammals restraint* - Small mammals can often effectively be restrained by gripping them firmly across the shoulders with a gloved hand, ensuring they are secure and unable to twist or bite. For smaller species, the loose skin along the neck and back can be grasped. For larger or more active animals, conical cloth bags with zippered openings, net bags, and wire cones are helpful tools, especially when weighing or marking the animal.

- *Bird restraint* -When restraining birds by hand, it is important to secure both the wings and legs to prevent injury. Cloth bags, sacks, or hoods used for restraint should be handled carefully to avoid overheating or feather damage, as excessive feather loss can lead to hypothermia after release. Birds can be inspected and weighed while suspended upside down, using a leather thong around their legs. However, capturing or marking certain bird species may alter their behavior and increase the risk of death.

- *Restraint of amphibians and non-hazardous reptiles-* Most amphibians and reptiles are small and slow-moving, making them easy to capture and restrain by hand or with a net. However, excessive force can cause injury. Restraining lizards by the tail may result in tail autonomy (self-amputation). While this is not usually a severe injury, it can affect the specimen's usability.
- *Restraint of hazardous reptiles* - Venomous snakes, crocodilians, and large turtles are potentially dangerous and require specialized restraint methods, depending on the species and the purpose. To ensure safety when handling these reptiles, the following guidelines should be followed:

  Handling time should be minimized and direct contact with the animal should be avoided as far as possible.

  Anaesthesia and/or physical restraint should be used whenever possible.

  Another experienced person should also be available for assistance and emergency response.

  Prior to handling, consult with knowledgeable workers and review the relevant procedures, as handling techniques for dangerous species are often shared informally rather than being widely published.

Chemical Restraint

- *Authorization letter-*The use of drugs for the safe restraint of wild and zoo animals has become increasingly essential, enabling procedures that help manage their health and well-being with minimal stress. Written authorization from the appropriate authority (such as Forest officials or the jurisdiction's collector) should be obtained before capturing any wild animal. Additionally, they should be well-informed about the current list of threatened and endangered species and adhere strictly to all regulations governing the capture and handling of these animals. For these drugs to be effective and safe, certain properties are highly desirable.
- *Characteristics of an ideal chemical restraint-* An ideal chemical restraint agent should have high therapeutic index (higher margin of safety to minimize the risk of overdosing), muscle compatibility (the drug should be gentle on muscle tissue, causing no long-term damage), short induction period (currently available drugs often require 10 to 20 minutes to take effect. An ideal agent should provide faster immobilization), antidote availability (to reduce the risk of respiratory arrest and complications), solution stability at room temperature for extended period and potent low dose.

- *Delivery modes* -Delivery modes of chemical agents for immobilization and restraint poses unique challenges, as the method used must ensure effective absorption. Techniques vary widely, influenced by factors like species, animal size, operator skill, confinement level, and distance from the animal, and available equipment.

  (i) Oral sedation of wild animals has proven largely ineffective. Many species cannot tolerate oral sedatives, and these drugs are often either poorly absorbed or destroyed within the digestive system, resulting in minimal effects.

  (ii) The intravenous route is only feasible when the animal is secured in a squeeze cage, making the intramuscular route the preferred method for administration.

  (iii) Intramuscular injections can be administered swiftly using a hand-held syringe, ideally with a large-gauge needle to allow quick liquid delivery. It is essential to secure the needle tightly on the syringe hub to prevent detachment from pressure buildup. A Luer-Lok connection is recommended to maintain secure needle placement during rapid injections.

  (iv) Modern chemical restraint techniques require tools that can project syringes over distances and release the drug on impact. Jack Crockford and collaborators developed the first effective tools, and today, Palmer Cap-Chur equipment is standard in the U.S., with other manufacturers emerging globally.

  Palmer projectors (guns) are short range (pistol - a modified pellet gun powered by compressed carbon dioxide (CO_2), with a range of up to 15 meters), long range (riffles –CO_2-powered, with a range of 35 meters), and extra long range (powered by percussion caps, allowing for range adjustment based on target distance, with a maximum range of 80 meters).

  Blow guns are increasingly popular in zoos and captivity settings because of silent projection and gentle impact making them suitable for small animals. These are easy to aim, have no mechanical parts requiring maintenance, and are straight forward to use. However, blow guns are limited in range to approximately 15 meters and can be cumbersome in tight spaces.

  Various homemade and commercial stick syringes are available to help operators for administering drugs to dangerous animals at close range.

These require immediate injection upon needle insertion. The operator must apply quick pressure to ensure full delivery of the drug, maintaining contact until the animal either moves away or the injection is complete.

(v) While using chemical immobilization, animal species (different species react differently to restraint agents owing to unique physiological and behavioural characteristics), physiological status (age, sex, general health, and lactation status), animal's physical condition (hydration status, body temperature, fitness), and emotional status (alarmed state, fear, excitement, or rage, releases high levels of catecholamine and cortisol altering the intended effects of the restraint drug) should be kept in mind.

(vi) Failure in immobilization of a wild may be due to equipment malfunctioning (mechanical issues), operator's error (lack of skill in handling equipment), and uncontrollable conditions (weather, wind velocity, temperature changes etc).

- *Chemical agents for restraint of wildlife* -Several chemical agents, detailed below, are commonly used for immobilizing wild animals. their effects, and precautions.

(i) Etorphine Hydrochloride -An opium alkaloid, derived from thebaine, etorphine hydrochloride is especially useful for immobilizing large mammals like elephants, rhinos, and hippos. Anesthesia typically begins within 10 to 20 minutes after intramuscular injection. The commercial preparation Immobilon combines 2.45 mg/ml of etorphine with 10 mg/ml of acepromazine maleate. Its antidote Diprenorphine, administered at double the etorphine dose, reverses the effects of etorphine within 4 to 10 minutes.

(ii) Fentanyl and Droperidol- Fentanyl, a morphine derivative, combined with droperidol, a tranquilizer, is effective for short procedures across a variety of species, especially carnivores, non-human primates, and smaller mammals. This combination may slightly lower blood pressure. Effects appear within 10 to 15 minutes and last around 40 minutes. Naloxone hydrochloride (0.006 mg/kg) via intravenous or intramuscular injection is the antidote for fentanyl and droperidol that provides immediate reversal.

(iii) Ketamine Hydrochloride -A phencyclidine hydrochloride derivative, ketamine maintains pharyngeal and laryngeal reflexes, reducing the risk of aspiration. It crosses the placental barrier, affecting fetuses without inducing abortion. Ketamine is effective for wild carnivores, reptiles, and birds but generally unsuitable for most large herbivores. It can be administered orally or by injection, with a variable dosage of 2 to 50 mg/kg. Effects begin within 5 to 10 minutes, with smooth recovery. Some domestic and wild felids may experience tonic-clonic convulsions, which can be managed by administering diazepam (0.25 to 0.5 mg/kg). Yohimbine (0.125 mg/kg) is the antidote for ketamine.

(iv) Tiletamine hydrochloride and Zolazepam hydrochloride- This combination utilizes tiletamine, a dissociative agent like ketamine, with zolazepam, a tranquilizer. Together, they provide desirable sedation while minimizing side effects. Effects occur within 5 to 12 minutes after intramuscular injection. No antidote is available.

(v) Xylazine- A non-narcotic sedative, it induces a sleep-like state in animals. It is widely used alone or combined with other agents for various species, administered intramuscularly or intravenously. Immobilization is achieved within 15 to 30 minutes for intramuscular injections or 3 to 5 minutes for intravenous injections, lasting 1 to 2 hours. Antidote: Dopram can speed up recovery, and yohimbine (0.125 mg/kg) may also be used.

(vi) Acepromazine maleate: Typically used in combination with other agents like etorphine or ketamine. It is rarely used alone.

(vii) Diazepam- It serves as an effective anticonvulsant and is also used as a pre-anesthetic to calm agitated animals. Dosage: 0.5 to 1.5 mg/kg, with effects appearing within 1 to 2 minutes after intravenous injection.

Physical Examination

Zoo animals should be carefully observed daily for their physical state, consumption of food and water, urination, defecation, alertness and activities. Their dens are also required to be examined carefully. Periodic screening for parasitic, bacterial and viral infections assists in an early detection of infection. Collection and examination of fecal, blood and feed samples plays a significant role in disease diagnosis and health management of zoo animals.

Visual health check

A visual health check of all animals should be carried out daily for prompt identification of the abnormalities in health and in behaviour. If daily inspection is negatively affecting the animals (during breeding cycle), it should be deferred. Animals should be weighed at least monthly as it can assist in early identification of health problems. A bright and alert animal with good physical condition, absence of pain, and normal behaviour reacting to stimuli is in a state of good health.

Animal showing lethargy, excessive sleepiness or depression, lameness or abrasions, vomiting, unusual levels of aggression, reduced social interaction, hiding for extensive periods of time, over-grooming, reduced appetite, and abnormal repetitive behaviors is in the state unhealthiness. Animal behavior should be closely watched as change in behavior may be a reflection of pain. A diseased or injured wild animal often hides obvious signs of being ill making it important to identify even the smallest changes in movement, body and skin condition, dietary intake or behavior. Normal eyes are clear, free from secretion and reactive to environment. Normal ear and nose are free from secretion and clear of debris. Deviation from these in eyes, ear and nose may indicate illness. Mouth, teeth and beaks should be inspected for sores, broken teeth or an overgrown injured beak. Hooves, claws or horns should be checked whether these are overgrown, damaged, cracked or impeding movement. Animals should be watched when moving to detect change in gait or lameness. Body coat/skin/feather/ scales should be inspected for the presence or absence of parasites, smoothness, uniformity, shine, scales, scratches, growth or injury. Hind quarters and anal area is looked whether it is clear from parasites and clean of feces. Feces are inspected for its color, consistency, smell, mucus, blood etc. Normal feces are well formed. General body condition is assessed as normal or abnormal.

Many animals live longer in captivity than their wild counterparts. In captivity, the provision of good nutrition, a safe environment and medical care extend their longevity. Caring for elderly animals can be challenging as they will experience physical and mental changes as they age. These changes will require extra consideration in terms of husbandry to ensure a good standard of welfare is experienced throughout their entire life. Dental problems such as tooth fractures, gum disease and missing teeth; joint degeneration; decreased muscle mass and increased fat are common health problems of aging animals. In addition, mental changes such as a gradual deterioration of cognitive and sensory function as well as changes to sleep cycles, appetite and social behavior may occur too.

Individual sick animal is examined in a manner similar to other animals using clinical (examination of vitals such as temperature, respiration , pulse, hydration status, general and specific organ examination using palpation, percussion, auscultation etc), specialized diagnostic techniques (radiography, sonography, electrocardiography, echocardiography, ophthalmoscopy, endoscopy etc.), and laboratory support. Dead animals are subjected to necropsy examination and histopathological investigation as per requirement.

Body weights of wild animals: It is very important to know body weight of zoo animals for restraining by medication. Their approximate body weight is given in the table 12.

Table 12: Approximate Body Weight of Wild Animals

S.No.	Animal	Male(Kgs)*	Female(Kgs)*
1.	Tiger	180-230	145
2.	Lion	100-120	90-100
3.	Leopard	68	50
4.	Snow leopard	50	40
5.	Elephant	2500-5000	2000-4500
6.	Rhinoceros	1000-2200	900-2000
7.	Sambhar	150-250	100-200
8.	Barasingha	125-200	100-150
9.	Nilgai	150-200	100-200

(* Reference Jani RG, Basic of Wildlife Health Management Book; 2010)

Medications for wildlife species

Following are the some commonly used drugs for restraint for various wild species:

- **Carnivores**

(i) Canidae: Ketamine hydrochloride or Xylazine are commonly used. Acepromazine maleate can be administered to reduce the risk of convulsions from ketamine.

(ii) Ursidae: For immobilizing small bears, intramuscular ketamine hydrochloride is the preferred option.

(iii) Procyonidae and Mustelidae (e.g., raccoons, minks, otters, skunks, beavers, badgers): Ketamine hydrochloride and Phencyclidine hydrochloride are often used.

(iv) Viverridae (e.g., mongooses, civets): Ketamine hydrochloride and Xylazine are effective choices.

(v) Hyaenidae: Ketamine hydrochloride serves as an effective sedative.

(vi) Felidae: Ketamine combined with Xylazine is recommended, providing sufficient analgesia for surgeries such as castration, laparotomy, and tooth extraction in lions, tigers, leopards, and pumas. Insufficient analgesia may lead to cardiac arrhythmias.

- **Primates**

  (i) Ketamine hydrochloride is highly effective as an immobilizing agent across all primate classes.

- **Elephants**

  (i) Etorphine hydrochloride is valuable for elephant immobilization, with a dose of 0.0022 mg/kg for Asian elephants and 0.0017 mg/kg for African elephants.

- **Rhinoceros**

  (i) Etorphine hydrochloride is the preferred drug for immobilizing rhinoceroses, with a sedative dose ranging from 0.5 to 1 mg (total dose).

- **Cervidae**

  (i) Etorphine, xylazine, and the combination of tiletamine and zolazepam are effective for different cervid species.

- **Birds**

  (i) Psittaciformes and Passeriformes (e.g., parrots, macaws, cockatoos, and perching birds): Ketamine hydrochloride at a dose of 0.05 to 0.2 mg/kg. A standard dose of 25-50 mg ketamine hydrochloride is commonly used as an anesthetic and immobilizer for birds.

- **Reptiles**

  (i) Ketamine hydrochloride is the anesthetic of choice for snakes, lizards, and turtles. Small snakes and lizards can be deeply sedated with a dose of 44 mg/kg.

Common issues in the capture and handling of wildlife

- *Injury*: Injuries to both handlers and animals are a frequent risk during the handling of wild animals. In some cases, these injuries can even be fatal. However, scientific handling techniques can significantly reduce the risk.
- *Stress*: Wild animals are much more vulnerable to stress and injury than domesticated species, especially during capture, handling, restraint, and transport. Even minor procedures, like blood collection or routine examination, can be highly distressing. Many wildlife species pose risks to handlers, not only due to potential injuries but also from zoonotic diseases, making human safety a critical consideration.

- *Capture myopathy*: It is a serious condition that arises from various stressors, such as fear, chasing, capture, and restraint, which can deplete the body's energy reserves needed for escape. Capture myopathy presents with symptoms like uncoordinated movements, paralysis, muscle breakdown (myoglobinuria), and acute muscle damage. Clinical signs are associated with elevated serum enzymes (AST, CK, and LDH) and blood urea nitrogen (BUN) levels. Early symptoms may involve increased heart and breathing rates, elevated body temperature (hyperthermia), and, in severe cases, death. Other initial signs include lack of environmental awareness, weakness, stiffness, and paralysis in the muscles used for movement.

Guidelines for zoo design, animal welfare, and veterinary facilities

Zoo Enclosure Design and Safety

- All animal enclosures should be designed to ensure the complete safety of animals, caretakers, and visitors. Barriers and clear warning signs must be in place to keep visitors at a safe distance from animals.
- Enclosures should meet the full biological needs of each species. They must be spacious enough to allow free movement and exercise, prevent domination within groups, and avoid provocation for public viewing. Measures should also be taken to reduce stress from neighboring enclosures.
- The conditions within enclosures should simulate each species' natural habitat as closely as possible. This includes planting shade-providing trees and adding shelters that blend with the enclosure's environment. Where feasible, enclosures should use moats as barriers.
- Enclosures for endangered species (as listed in Appendix I of these guidelines) must include feeding and retreat cubicles of specified dimensions. Each cubicle should have facilities for resting, feeding, drinking, and exercising to meet species-specific needs, along with proper lighting and ventilation for comfort.
- Effective drainage and waste removal systems must be provided in all cubicles and enclosures.
- Plans for new enclosures for endangered species must be developed in consultation with the Central Zoo Authority.

Hygiene, Feeding and Maintenance

- Zoos must provide wholesome, unadulterated food in sufficient quantities for each animal, tailored to individual needs to prevent under nutrition.
- A proper waste management system for both solid and liquid waste must be in place to maintain zoo cleanliness.
- Left over food, animal waste, and other refuse should be regularly cleared from enclosures and disposed of hygienically.
- Round-the-clock access to potable drinking water must be available in every enclosure.
- Disinfectants should be applied periodically in all enclosures, following the guidance of the zoo's authorized veterinary officer.

Animal Care, Health and Treatment

- Only trained and experienced staff should handle wild and exotic animals to minimize discomfort, stress, or physical harm.
- Daily health check of all animals is mandatory, with immediate reporting of any illness, injury, or unusual stress to the veterinary officer for prompt treatment.
- Regular health screenings, parasite checks, and preventive treatments, including vaccinations, should be conducted according to schedules set by authorized veterinary officer.
- Staff responsible for animal care should undergo medical examinations at least biannually to ensure they are free of infections that could affect zoo animals.
- Zoos must keep detailed health records and treatment histories for all endangered species, as identified by the Central Zoo Authority.

Veterinary Facilities

- Large and medium-sized zoos must have well-equipped veterinary facilities, including a hospital with diagnostic equipment, a variety of medicines, and separate isolation and quarantine wards for newly arrived or ill animals to reduce infection risk.
- Veterinary hospitals should be equipped with tools to safely handle and restrain sick animals, such as tranquilizing devices and syringe projectors. A reference library on animal health and care should also be maintained.

- Small and mini zoos without a full veterinary hospital must have at least a treatment room where routine check-ups and immediate care can be provided.
- Each zoo must have a dedicated post-mortem room for detailed examinations of deceased animals, with records maintained for at least six years.
- Zoos should have a grave yard to safely dispose of animal remains. Large and medium-sized zoos are additionally required to have incinerators for the disposal of carcasses and other waste materials to maintain cleanliness.

25

Restrain, Physical Examination, and Medication of Laboratory and Exotic Pets

Proper restraint and handling of laboratory and exotic pets reduces stress for both animal and examiner. Handling-related stress introduces a variable into experiments, so its minimizing is essential. Improper handling can lead to significant injuries for both animals and humans.

Laboratory Animals

General Care

- A healthy and regularly behaving animal is in a state of well-being. Maintaining the state of wellbeing requires effective health management and appropriate exercise.
- New animals on arrival should be quarantined, carefully examined, and placed in clean cages. Shipping containers should not enter the main facility and must be discarded. Proper identification and documentation of each animal, including details like origin and condition, is essential. Any sick animals should be euthanized immediately.
- Animals should be housed in separate rooms by species when possible. If animals from different sources need to be combined, efforts should be made to group those that are compatible and share similar environmental needs.
- Animals are identified by their cage or group. Detailed records should be maintained, including the animal's arrival, sex, age, weight, breed, colour, and any physical anomalies if any. Species identification cards should be placed on animal room's door.
- Laboratory animals must be provided with clean and nutritious food appropriate to their species. Bulk food should be stored in contamination-free conditions, in cool, dry, and well-ventilated rooms. Feed containers should be regularly cleaned and disinfected.

- Slightly acidified water is commonly provided to laboratory animals, unless otherwise specified by experimental protocols. Water bottles should be transparent, clean, and regularly inspected. Regular cleaning and sterilization are necessary to prevent bacterial contamination, such as *E. coli* or *Pseudomonas* spp. Freshly filled, sterile water bottles are recommended for all animals.

Restraint

- While using a mechanical restrainer, it is ascertained that the animal is comfortable and not left unattended. The restrainer should be warm, comfortable and fit closely to prevent movement but making the animal to rest comfortably and breathe normally.
- Chemical restraint should be considered for any procedure that is prolonged or may cause pain.

Handling of Laboratory Animals

While handling lab animals, especially rodents the following points are to be considered.

- Approach slowly- Rats have poor vision, so sudden hand movements or grabbing motions should be avoided. Instead, place hand gently over the cage and allow the rat to sniff it first.
- Stay calm - Avoid fear or nervousness as this can alarm the animal.
- Observe carefully- Before opening the cage, watch the animals closely. Nervous or young rats may attempt to escape quickly.
- Handle gently- Avoid squeezing the thorax or holding the scruff around the throat, as this can restrict their breathing and cause them to struggle.
- Tail handling- Do not pick up larger rats or mice by the tip of their tail, as the skin can tear if they struggle.
- Hold smaller rodents properly- For smaller rats and mice, lift them by the base of the tail, never further back. Lifting from the tail's base prevents them from turning and possibly biting due to fright.
- Familiarization- Ensure that animals in an experimental group are handled regularly before the experiment begins to reduce stress.

Handling methods

Mice

- Juvenile and adult mice can be picked up by gentle holding the base or middle third of their tail with the fingers. After catching the mouse, it can be restrained by placing it on a wire cage lid and grasping the loose skin behind its neck and ears with thumb and forefinger. The tail can be held against the palm using the fourth and fifth fingers. Care should be taken not to pull the skin around the neck too tightly, as this could restrict the mouse's breathing. This method is often used for quick examinations or to administer injections. Pregnant or obese mice should be handled gently, with one hand supporting their feet.

- It has been observed that picking up mice by the tail triggers aversion and elevates anxiety. The stress can be reduced by using a tunnel or cupped hands instead. Mice handled with these methods perform better in behavioral tests compared to traditional tail handling. They also quickly become accustomed to tunnel handling and can then be restrained by the scruff or tail base for procedures or health evaluations.

Rats

- Rats respond well to gentle handling and tend to become more relaxed with increased handling. They can be picked up by the base of the tail, with precautions similar to those used for mice especially for brief tasks like cage cleaning. However, a safer approach for both the rat and the handler is to grasp the rat's entire body.
- The rat is gently lifted by the base of the tail and then one hand is placed over its back, using the thumb and forefinger to hold its forelegs gently but firmly near its head. For additional control, support the tail or hindquarters with the other hand. This grip prevents the rat from biting, even if it attempts to. Care should be taken not to squeeze the thorax too tightly, as this may restrict the rat's breathing.

Hamsters

- Hamsters have sufficient loose skin around their neck, shoulders, and back. For restraining hamsters, their loose skin is grasped between thumb and fingertips. It's crucial to support the hamster's body with the other hand for proper restraint.

Guinea pigs

- Guinea pigs are generally easier to restrain as compared to other laboratory rodents due to their limited agility and calm temperament. But they are easily frightened and will vocalize and squirm to avoid restraint. Scuffing a guinea pig is not encouraged because of their short necks with minimal loose skin to grasp securely. Guinea pigs can be restrained by placing one hand over their shoulder area, positioning the thumb and forefinger just behind the front legs and placing the other hand under their hind quarters to support their weight. For additional control, the fingers of the hand supporting the hindquarters can gently hold and restrain the hind legs.

Rabbits

- Rabbits are timid by nature but are excitable. Occasionally they may resist handling that could endanger both themselves and the handler. Restraint methods (manual or mechanical) depend on the procedure being performed. Gentle but firm manual restraint often calms the rabbit. To remove a rabbit from its cage, scruff of its neck is held with one hand while supporting its hindquarters and back with the other to prevent kicking and reducing the risk of injury to both the handler and rabbit. Restraining by the ears should be avoided, as it can damage ear cartilage. Mechanical restrainers made of plastic or stainless steel are advisable when ear access is needed (for injections, bleeding, tattooing, or treatments).When transferring a rabbit to an examination table, light restraint, like gentle petting, is often sufficient to keep calm rabbits in place. When more restrain is required, rabbit's head is tucked between body and arm, supporting its body like a football. Improper restraint can cause the rabbit to kick with its hind legs, risking back fractures. Rabbits are highly prone to lumbar spinal luxation, which can lead to paralysis.

Chinchillas

- Most pet chinchillas are accustomed to handling, making them relatively easy to capture and restrain. To catch a chinchilla, place one hand under its thorax or abdomen while gently holding the base of its tail with the other hand. Once secured, the chinchilla can be placed on the examination table. Chinchillas can experience "fur slip"—a shedding of fur patches—if they become frightened or are handled too roughly. Due to this, scruffing is generally not recommended. Though uncommon, "fur slip" can occur unexpectedly during an examination, so it is recommended to inform the client beforehand.

Ferrets

- Ferrets can be restrained in much the same way as cats. For most of the examination, gentle petting and treats, like a high-calorie paste are often sufficient. If additional restraint is needed, the ferret can be held in a similar manner to a cat. Common venipuncture sites include the cephalic, lateral saphenous, jugular, and cranial vena cava. Peripheral vessels are small, so obtaining enough blood for a full panel can be challenging.

Cats

- Cats possess sharp, curved, retractable claws, along with sharp teeth. They tend to strike with their paws, often targeting the handler's hands or eyes. Therefore, protective measures are essential to prevent injuries from bites or claws during handling. The neck skin of a cat is typically loose, making it a suitable area to grasp for control during examination. Once the neck skin is securely held, then the legs can be firmly restrained for further examination. For especially difficult cats, the front and hind legs may occasionally be taped together in pairs.

Dogs

- A slip lead is strongly recommended when handling dogs, and proper support should be provided while carrying them. For injections and minor procedures, dogs can be restrained either in a lateral recumbent position or seated. During venipuncture, the handler may restrain the dog on a table, placing one arm around its neck, while the other hand can either secure the body or help occlude the vein for the person administering the injection. Shy or fearful dogs may benefit from additional time spent helping them feel more comfortable, with slow movements and a calm tone to avoid startling them. For a highly uncooperative dog, a muzzle may be necessary. A commercial muzzle can be used, or one can be improvised with gauze.

Nonhuman primates

- Nonhuman primates, regardless of size, can pose a risk. Chemical immobilization, typically with ketamine, is commonly used. Injections can be administered to a contained animal using a squeeze cage. Physical restraint of an alert primate should only be performed by trained, experienced staff. If frequent handling is required, primates may be trained to accept a pole and collar. Tether systems are advisable when regular drug administration or frequent blood collection is necessary.

Medication

- Laboratory animals receive substances through various routes depending on whether local or systemic effect is desired. Parenteral methods, which bypass the digestive tract, usually provide higher bioavailability. Regulatory guidelines may also influence the route of administration to ensure it aligns with humane administration for specific studies, like nonclinical safety tests.

- Substances can be given orally, by gastric gavage, intravenously, subcutaneously, intramuscularly, or by other routes (directly into the lungs or brain). Many methods require restraint, sedation, or anesthesia, so route selection should aim to minimize invasiveness and stress to the animals. Each route has specific advantages and impacts pharmacokinetics. The timing of administration, in line with animal's circadian rhythms, may also influence the effectiveness or toxicity of the substance.
- *Enteral administration*- In laboratory animal research, substances are commonly administered orally, either directly into the mouth, mixed with food, or through orogastric or nasogastric gavage. Oral administration is convenient, cost-effective, and safe, with some animals able to be trained to voluntarily consume the substance. However, this method may not be reliable for all animals, as individual preferences, flavor issues, and behavior changes over time can affect consistency, especially in long-term studies. For safety testing, oral dosing mimics human administration, but it is important to ensure animals swallow the full dose by placing tablets or capsules far back in the mouth.

  Gavage, which is used to deliver substances directly into the stomach or esophagus, ensures accurate dosing. Nasogastric tube, typically soft pediatric feeding tube, is commonly used for enteral nutrition in rabbits and for dosing in nonhuman primates. When administering substances orally, restraint is often required, which can cause stress, especially in species like rodents and nonhuman primates. Stress can be reduced by habituation and positive reinforcement. For oral gavage, smallest volume (5 mL/kg) is recommended to minimize stress. Large volumes should be delivered slowly.

  Oral administration is associated with slower onset compared to parenteral methods, first-pass liver metabolism, potential lack of absorption due to the substance's properties, and difficulty in unconscious or sick animals. Oral gavage also requires skilled technique to ensure that it is performed quickly and safely. It requires training to reduce adverse effects.
- *Intravenous (IV) administration*- Intravenous (IV) administration is the most efficient method for delivering substances to laboratory animals, as it bypasses the absorption process and directly introduces substances into the blood stream. The choice to use IV administration depends on factors such as the pharmacokinetics of the substance, maximum tolerated dose, and dosing intensity, which influences the timing and consistency of blood levels. Aseptic techniques are essential for intravenous injections.

Intra-arterial injections should be avoided due to the risk of serious complications.

For repeated IV treatments, refinement techniques can enhance animal comfort, such as using smaller needles, butterfly needles for single injection, indwelling catheters for comfort, topical anesthetics to reduce pain, and external pump packs to allow greater movement.

- *Intraosseous administration (IO)-* It is an alternative to IV for certain cases, especially in emergency situations, though it requires advanced training and can be invasive. This technique carries risks, including infection and injury. All substances administered IV or IO (intraosseous) must be sterile, free of particulates, and non-irritating to prevent complications such as vasculitis or hemolysis. While efficient, IV administration requires proper training and careful technique to minimize risks, including blood loss, hematoma formation, and fluid overload in chronic infusions. Regular maintenance and cleaning are necessary for implanted catheters and vascular access ports to prevent infection and ensure proper function.
- *Administration to skin and muscle-* Substances can be administered through the skin (epicutaneous administration) for topical effects, with absorption depending on factors like surface area, concentration, skin integrity, and hydration. Hair should be clipped to maximize contact with the substance, and the area should be cleaned before application. Caution is necessary to avoid irritation or caustic reactions, as well as to minimize systemic toxicity, especially if the animal can groom the area.
- *Transdermal delivery-* It involves applying substances via a patch for absorption into the systemic circulation, providing a constant blood level of the substance. This method avoids repeated animal restraint and injections, and materials can be removed if adverse effects occur. However, care is needed to avoid over dosage, as human transdermal patches may contain high doses unsuitable for animals. Animals should be monitored for toxicity, and patches should be protected from ingestion or removal.
- *Subcutaneous administration* – It is an easy and efficient method for parenteral substance delivery, providing slower absorption and a sustained effect. This route is suitable for fluids, oily materials, or implantable pumps. It is especially useful for small or dehydrated animals, as excess fluid is excreted through the kidneys. While subcutaneous injections are simple, training is required to avoid incorrect injections, such as inadvertent intravenous administration. Proper needle size and

aseptic techniques are important to minimize discomfort and avoid complications like abscess formation.

- *Intramuscular injections-* Intramuscular injections are typically used in large animals, offer rapid absorption due to the rich blood supply in muscle tissue but are less commonly used in smaller animals due to their reduced muscle mass. This route requires more skill and can cause issues such as nerve damage, muscle necrosis, or inflammation with repeated injections. Like subcutaneous injections, intramuscular injections require attention to technique to avoid complications.
- *Epidural and intrathecal (subarachnoid) administration-*Epidural and intrathecal (subarachnoid) techniques are used to deliver substances directly to the spinal cord or cerebrospinal tissues, bypassing the blood-brain barrier for rapid effects. These methods are commonly employed to induce spinal anesthesia or administer contrast media for imaging. Animals must be heavily sedated or under general anesthesia, and the skin at the injection site must be prepared aseptically. The injection site and volume of substance vary between species, with intrathecal and epidural routes differing in their spread of the substance.

  Confirmation of intrathecal placement is done by visualizing cerebrospinal fluid after needle insertion. If fluid is seen during an epidural injection, the needle should be repositioned to avoid incorrect delivery. Epidural and intrathecal techniques require significant technical skill and knowledge of spinal anatomy and anesthesia. These procedures should be performed by experienced personnel. Potential side effects of epidural administration in small animals include delayed hair regrowth, itching, urinary retention, nausea, vomiting, and severe respiratory depression.
- *Intraperitoneal (IP)-*Intraperitoneal (IP) administration involves injecting substances into the peritoneal cavity, commonly used in small rodents but rarely in larger mammals or humans. It is an alternative when intravenous access is difficult and can safely administer large volumes of fluids or serve as a site for osmotic mini pump implants. The absorption of substances via this route is slower than intravenous delivery, with substances often undergoing hepatic metabolism before entering systemic circulation, similar to oral administration.

  In mammals, IP injections are performed with the animal in manual restraint, with the abdomen angled downward to move internal organs away from the needle site. In rodents, injections are made in the lower right abdomen to avoid hitting the bladder or cecum. Incorrectly angling the needle may cause subcutaneous rather than intraperitoneal injections.

Substances used for IP injections must be sterile, isotonic, and non-irritating to avoid complications like peritonitis or ileus. Irritating substances should be diluted to minimize adverse effects. Though technically simple, the procedure requires proper training to ensure accuracy and avoid incorrect injections into the bladder or other organs.

- *Intranasal, intratracheal, and inhalational routes*-Intranasal, intratracheal, and inhalational routes are used for administering substances to animals, typically requiring sedation or anesthesia to minimize discomfort. Intranasal administration involves small volumes to avoid suffocation and is best for local or systemic delivery, such as vaccines or decongestants. The nasal mucosa, with its rich blood supply, allows for rapid absorption and systemic effects, bypassing the hepatic first-pass metabolism. Lipophilic substances are absorbed faster than larger or polar molecules. Intratracheal instillation involves injecting small volumes directly into the trachea, which leads to localized and uneven distribution in the lungs. This method requires careful control of volumes to prevent suffocation, and competent intubation skills are necessary.

  Inhalational delivery uses vapors or nebulized aerosols and can efficiently target the lungs, with particles smaller than 3 μm reaching the alveoli. Inhalation requires specialized equipment and careful handling, as particles are deposited by gravitational sedimentation, impaction, or diffusion. Larger particles deposit in the upper airways, while smaller particles reach deeper lung areas. This method requires non-irritating substances to avoid respiratory issues like edema, bronchospasm, or fibrosis, and animals should be conditioned to restraint devices for effective administration.

Blood Collection Sites

Blood collection sites in animals are given in the table 13.

Table 13: Blood Collection Sites in Laboratory Animals

Species	Site of collection
Mouse	Cardiac (terminal only), orbital sinus (anesthetized only), tail vein (incisional method not permitted), saphenous vein, facial vein.
Rat	Cardiac (terminal only), orbital sinus (anesthetized only), tail vein (incisional method not permitted), saphenous vein, facial vein, subclavian vein.

Species	Site of collection
Hamster	Femoral vein
Guinea Pig	Cardiac (anesthetized only), anterior vena cava
Rabbit	Cardiac (anesthetized only), marginal ear vein
Ferrets	Cranial vena cava
Chinchillas	Cephalic, lateral saphenous, and jugular veins
Dog, Cat & Nonhuman Primate	Cephalic, saphenous, femoral and jugular veins
Ruminants	Jugular vein
Chicken	Brachial wing vein, right jugular vein

Injection Sites for Laboratory Animals

Injection sites and guide lines for injectioning in laboratory animals are given in the table 14.

Table 14: Injection sites and guidelines for injectioning in laboratory animals

Species	Route	Injection Site	Volume	Needle Gauge
Mouse	Intravenous	Lateral tail vein	0.2 ml	25 gauge
	Intraperitoneal	Peritoneal cavity	2-3 ml	21 gauge
	Intramuscular	Quadriceps/posterior thigh	0.05 ml	23 gauge
	Subcutaneous	Scruff	2-3 ml	20 gauge
Rat	Intravenous	Lateral tail vein	0.5 ml	23 gauge
	Intraperitoneal	Peritoneal cavity	5-10 ml	21 gauge
	Intramuscular	Quadriceps/posterior thigh	0.3 ml	21 gauge
	Subcutaneous	Scruff/back	5-10 ml	20 gauge
Hamster	Intravenous	Femoral or jugular vein	0.3 ml	25 gauge
	Intraperitoneal	Peritoneal cavity	3-4 ml	21 gauge
	Intramuscular	Quadriceps/posterior thigh	0.1 ml	21 gauge
	Subcutaneous	Scruff	3-4 ml	20 gauge
Guinea Pig	Intravenous	Ear saphenous vein	0.5 ml	23 gauge
	Intraperitoneal	Peritoneal cavity	10-15 ml	21 gauge
	Intramuscular	Quadriceps/posterior thigh	0.3 ml	21 gauge
	Subcutaneous	Scruff/back	5-10 ml	20 gauge
Rabbit	Intravenous	Marginal ear vein	1-5 ml (slowly)	21 gauge
	Intraperitoneal	Peritoneal cavity	50-100 ml	20 gauge
	Intramuscular	Quadriceps/posterior thigh, lumbar muscles	0.5 ml	20 gauge
	Subcutaneous	Scruff/flank	30-50 ml	20 gauge

Physiological Data of Common Laboratory Animals (CPCSEA guidelines)

Physiological data pertaining to common laboratory animals is given in the table 15.

Table 15: Physiological data of laboratory animals

Parameters	**Rat**	**Mouse**	**Guinea pig**	**Rabbit**	**Monkey**
Daily intake of feed (gm)	10 - 15	5 - 10	30 - 50	100 - 200	100 - 300
Daily water intake (ml/100 gm body weight)	15	5 - 10	15	20	30 - 40
Urinary output (ml/100 gm body weight)	5 - 8	3 - 4	4 - 9	7 - 8	5 - 6
Daily faecal output (gm)	9 - 13	6 - 9	15 - 18	20 - 30	100 - 150
Pulse rate (no's/min)	300	600	150	155	90 - 100
Respiratory rate (no's/min)	85 - 113	163	82 - 90	38 - 60	39 - 60
Rectal temp (°F)	99 - 100	96 - 100	100 - 102	102 - 103	100 - 102
Room temperature (°F)	65 - 75	68 - 74	65 - 75	62 - 68	68 - 72
Relative humidity (%)	45 - 55	45 - 55	45 - 55	45 - 55	45 - 55

Life Span of Common Laboratory Animals

Life span of rat, mouse, guinea pigs, rabbit and monkey is 2-3, 1-2, 3-5, 5-6 and 15-30 years respectively.

Haematological Data of Laboratory Animals (CPCSEA guidelines)

Haematological indices of healthy rats, mice, guinea pigs, rabbits and monkeys are given in the table 16.

Table 16: Haematological parameters of laboratory animals

Parameters	Rat	Mouse	Guinea pig	Rabbit	Monkey
Total blood volume (ml/kg body weight)	58	78	75	70	75
Blood volume (ml/100g body weight)	20	30	35	35	40
Clotting time (seconds)	20	14	-	60-300	90
RBC life span (days)	45-68	20-30	-	45-70	-
RBC count (cells/mm^3)	7.2-9.6	7.7-12.5	4.5-7.0	4.5-7.0	4.5-6.5
Haemoglobin (g/dl)	14.8	14.8	12.4	13.6	12.5
Blood plasma pH	7.3	7.3	7.3	7.3	7.3

Physical Examination

Rabbits

A routine physical examination includes examination of eyes, nares, ears and teeth; auscultation of chest (heart and lungs) and abdomen (gut sounds); palpation of abdomen; visual inspection of skin, fur, perineum and foot pads; and general inspection for body condition and gait. Weight, temperature, respiration rate and pulse rates should always be recorded. Rabbits brought to clinics should be made comfortable and restrained by wrapping a bath towel around their body and put them on a table with rubber mat. Temperature is taken by inserting clinical thermometer in rectum as is done in other animals. Respiration rate is determined by counting the rise of the chest wall in a minute or by putting the hand in front of the nose and counting exhalation per minute. Pulse is taken by putting fingers over femoral artery in the upper inner thigh and counting beats in one minute. Normal rectal temperature (101.0 to 104.0^0 F), respiration rate (30 to 60 per minute), pulse (120 to 330 per minute) and heart rate (130 to 350 bpm) varies in healthy rabbits depending on many factors.

Rodents

Proper history is mandatory as many diseases are due to poor husbandry and lack of knowledge. Heart rate and respiration should be observed in natural condition without restraint. Rodents with severe depression or dyspnea may die of stress of physical examination. Coprophagy is normal behavior in rodents. Weight of the ailing rodents should be recorded. After proper restrain, the mouth is examined for ptylism, swelling, condition of the incisors (normal or over grown) or any discharge and its nature. Cheek pouches of hamsters are examined for swelling, discharge or impaction of material. Ophthalmic examination is conducted in case of eye problems. Dry white tears in guinea pigs are suggestive of vitamin C deficiency. Ears are examined for abnormalities if any. Discolored ears are sign of illness as bluish discoloration indicates cyanosis and bright red septicemia. The lymph nodes are palpated as their enlargement is indicative of infection or neoplasia. Heart may be auscultated in large rodents. Abdomen is palpated gently for abnormalities, if any. The ano-genital region is also inspected. Males have two external opening (anus and urethra) while females have three (anus, vagina and urethra). Skin scrapping or cellophane tape examination is conducted for the diagnosis of ectoparasites and dermatophytes. Faecal examination is essential for the diagnosis of endo parasitic infestation. Radiographic examination (40 kvp and 3-10 MAS exposure) is a valuable diagnostic tool. Haemato-biochemical investigations are of great diagnostic assistance in rodents as in other animals.

Exotic Pets

Birds

Budgerigar, canary, cocktail, cockatoo, finches, linnet, love bird, mynah, parrot and whydahs are common birds being kept as pet in India. Their brief introduction is given below.

Budgerigar- Budgerigar is a colorful bird of Australian origin. Its plumage is bright yellow and green, with a blue cheek and black scalloping on its wing feathers. Its tail is slender and dark blue. They are roughly 18-20 cm (from top to tail) and weigh 30-40 gm. Their diet consists of 6.5 to 8.5 grams of food per day, with grit added to aid digestion. They need plenty of water and enjoy eating greens. Budgerigars can live for 10-15 years and require two nesting boxes, as the female may start laying eggs again before the first brood is fully grown.About two weeks after mating, the female will lay one egg per day. She will incubate the eggs for 17 days, and the chicks will hatch.

Canary - Canaries are small birds measuring between 7.6 to 17.7 cm. They are popular for their singing, with males being the primary vocalists. A pair of canaries can live comfortably in a cage measuring 0.6 x 0.5 x 0.5 m. Regular cage cleaning is essential. Their diet includes canary food, vegetables, and occasionally hard boiled eggs mixed with arrowroot or biscuit. They live 10-12 years and need their beak trimmed if it grows too long. Canaries require a nesting box with soft, non-absorbent material for breeding.They breed quite successfully in captivity and require a nesting box placed in a corner of the cage, filled with soft, non-absorbent material. After a few days, the female will lay her eggs in the box and begin incubating them. It takes about 14 days for the eggs to hatch. During this time, the mother should be provided with a slightly moist diet, consisting of chopped boiled egg and canary seed, along with a small amount of milk.

Cocktail - Cocktails are colourful members of the parrot family, known for their friendly and social nature. They should be fed small seeds, fruits, and fresh greens.

Cockatoo - Cockatoos have a variety of striking colors. They are known for crest or tuft of feathers on the top of the head and can mimic words and sounds. They should not be kept in small cages and are best suited for homes away from busy areas, as their loud calls can be disruptive. They eat seed mixtures, nuts, and fresh fruits.

Finches -There are various types of finches, such as Waxbills, Linnets, Chaffinches, Bullfinches, and Zebra. They feed on insects, particularly when nesting, laying eggs, or raising young.

Linnet- Linnets are small, attractive birds known for their pleasant song. The male is colourful, while the female is brown. They eat seeds and small insects and can breed in captivity with proper care.

Lovebirds-They are part of the parrot family. Lovebirds are hardy and cheerful. They breed easily and should be kept in pairs.

Mynah -Mynahs are medium-sized birds measuring 20-25 cm. They require a large cage (91 x 61 cm) and should have a diet of meat, vegetables, fruits, green leaves, raw milk, and honey. They need plenty of water and are prone to aspergillosis. Mynahs do not breed in captivity and have a lifespan of 20 years.

Parrots - Parrots, weigh from 750g to 7 kg, are known for their mimicking abilities, especially African Greys. They are herbivores, enjoying fruits like apples and oranges. Parrots can live up to 50 years and can breed in captivity.

Whydahs- Whydahs are known for their long tails, often three to four times the length of their body. They are parasitic birds that tend to lay their eggs in the nests of other birds. Their diet includes canary seeds, vegetables, small insects, and grit. Whydahs should not be kept in cages.

Medications

- Medicines should be administered correctly in right dosage and frequency. Incorrect dosing, such as giving antibiotics or antifungals for too short a period, can lead to drug resistance.
- Restrain of the birds is essential for drug administration. Bird should be covered with a towel, its head is held securely by grasping the jaw to prevent movement and the bird is kept against the body ensuring the wings are safely tucked in. The neck of the bird is extended gently and the medicine is administered slowly through the corner of the beak ensuring that the bird swallows the medicine.

  Alternatively, if proper restraint is difficult, the medicine may be administered by mixing it with their favourite food, like peanut butter or nut butter, spread on a cracker, cookie, or cornbread. The medicine may also be applied.
- After medication, bird should be closely monitor. Some birds may stop eating or reduce their food intake if they detect medication. It may simply be due to the taste of the medication rather than a sign of illness. If the bird's behavior remains normal, the reduced appetite likely stems from dislike of the medicated food.

Route of parenteral drug administration

Drugs can also be administered parenteral by intramuscular, intravenous, intraosseous and subcutaneous route. Their site is given in the table 17.

Table 17: Injection sites in birds

Route of Administration	Injection Sites
Intramuscular	Pectoral muscles (for most birds), Thigh muscles (for ratites)
Intravenous	Right jugular vein, Brachial vein, Medial metatarsal vein
Intraosseous	Useful for very small or hypotensive birds
Subcutaneous	Inguinal region (not recommended due to slow absorption)

Physical examination

The birds should be initially examined in its natural environment in cage without manual restrain for its perching or standing on one or both legs, alertness, closed/open eyes, fluffed or sleek feathers, wing droop, tail bobbing up and down and breathing pattern (respiratory rate is varying from 30 to 60 per minute in small birds; and from 15 to 30 per minute in comparatively large sized birds). Respiration rate increases in birds affected with hyperthermia, stress, or obesity. The crop is palpated for the presence of food or fluid. The bird should be weighed for monitoring health, body condition and dose calculation. The ears, eyes, nares and oral cavity is examined for exudate, mass or swelling if any. Examine the condition of feathers and skin, symmetry of beak and nails. The integument of feet is examined for intactness. Determine the body condition by palpating the pectoral muscles. A keel scoring system (1 to 5) is generally used. Score 1 indicate very thin bird, 5 indicate obese bird and 3 as an appropriate score. In restrained birds heart rate is rapid. It may be >250 beats/min in large restrained parrots. Arrhythmias cannot be characterized by auscultation because of the rapid heart rate. The cloaca is examined for its tight closure, moist skin, and cleanness of feathers around the vent.

Reptiles (Iguanas and Geckos)

Restrain

Reptiles are generally less prone to stress than birds, so restraint can often be performed with less risk, even in debilitated animals. However, it's important to consider factors that could make restraint potentially harmful to both the animal and the handler. These considerations are

Respiratory distress - Excessive handling could worsen the condition if the animal is struggling to breathe.

Fragility- Some species, such as day geckos, are delicate and can drop their tails when handled. Others, like green iguanas, may suffer from conditions like metabolic bone disease, making them susceptible to fractures.

Aggression - Certain species, including snapping turtles, tokay geckos, and rock pythons, are naturally aggressive and may require more cautious restraint.

Need for restrain: Restraint may be necessary for administering medication or performing a physical exam. Additionally, many reptiles naturally harbor *Salmonella* bacteria in their gut, so personal hygiene is essential to prevent the spread of zoonotic diseases when handling them.

Techniques and equipments involved in restraining

When handling lizards, especially aggressive or large species, several important considerations must be kept in mind for both the handler's safety and the reptile's well-being. Aggressive iguanas and similar species may require careful restraint using tools like thick towels to control their tails and claws, and for very large or aggressive lizards, gauntlets should be worn to protect against bites. It's essential not to apply excessive force during restraint, as this can cause injury, particularly in reptiles with skeletal issues such as metabolic bone disease. Additionally, since lizards lack a diaphragm, over-restraining can put excessive pressure on their lungs.

Fragile species like day geckos should be handled in clear plastic containers, while those with delicate skin, such as other geckos, require soft cloths and latex-free gloves. Small lizards can be cupped gently in the hand, and their heads controlled to prevent bites. It's crucial to avoid grabbing lizards by their tails, as many species will shed their tails in stress, and some, like green iguanas, will not regrow them once they are older.

For restraint, it's best to secure the neck and pectoral region with one hand and support the body near the pelvis with the other. Care must be taken not to damage delicate features, such as dorsal spines or fragile skin. Lizards' claws, teeth, and sometimes tails, can pose risks, especially in species like iguanas, which may lash out.

The vasovagal reflex, which involves gently applying pressure to the eyeballs, can induce a trance-like state, allowing non-invasive procedures like radiography. However, this state is easily disrupted by loud noises or physical stimulation. Overall, careful handling and restraint are crucial to avoid injury to both the lizard and the handler.

Key points

- Netting can be used but may not be reliable for larger or more aggressive iguanas.
- Hand-grabbing works well for younger or calmer iguanas, with the head secured by thumb and forefinger.
- Larger iguanas require two-handed restraint: one for the head, one for the pelvic region.
- Restraining the tail under the arm and controlling the rear legs is an effective method.
- Using two handlers can make restraint easier for large iguanas.
- Blind folding an iguana can reduce stress and protect the handler.
- Tame iguanas can be handled without restraint, but caution is needed as they can quickly become agitated.

Chelonians (Tortoises and Turtles)

Chelonians, including turtles and tortoises, are generally harmless but can be surprisingly strong. The restraint techniques vary depending on the species, size, and temperament of the animal.

Restrain

Tortoises (Mediterranean species) can be held with both hands, one on either side of the shell, behind the front legs. To keep them still for examination, they can be placed on a cylinder or stack of tins, with their legs raised off the table, balancing on the underside of the shell (plastron).

Aggressive Species (Snapping turtles, Alligator Snapping turtles) can be restrained by holding the shell on both sides, behind and above the rear legs, to avoid bites. Chemical restraint may be necessary for examining the head region in these species.

For restraining soft-shelled and aquatic species of chelonian, use soft cloths and latex-free gloves to prevent damage to the shell. These species often require gentlle restraint to protect their delicate shells. Many chelonians are curious and may walk around during an exam, allowing the handler to simply grasp the head while securing the body. Head can be restrained by positioning thumb on one side of the neck and fingers on the other side, just behind the base of the skull. Restraint of strong or healthy chelonians may require a significant amount of force, and two people may be necessary for more active or large chelonians.

To extend a chelonian's head, food can be offered, or they can be placed in shallow, warm water. Once the head is extended, the same techniques for head restraint can be used. If necessary, a blunt ear curette or spay hook can be gently used to pull the head out, but care must be taken to avoid damaging the beak. Box turtles, with their hinged shells, are particularly challenging to restrain, as they can retract completely into their shells. To extend their head, the cranial portion of the carapace and the plastron can be gently propped open with a padded object, or one of the forelimbs can be extended, as this prevents the turtle from closing the shell on its own leg.

Restraint techniques can cause significant stress to the animal, and if initial attempts fail, chemical restraint may be necessary, particularly for large or strong chelonians, such as box turtles and large tortoises. It is crucial to use gentle but firm restraint to control the head and body, especially with healthy chelonians, and always to take care with biting species, like snapping turtles. Soft materials and latex-free gloves should be used to avoid shell damage, and caution must be taken to prevent injury, particularly when dealing with species known for aggressive behaviour.

Physical examination

Physical examination of reptiles is similar to those performed on mammals, with the focus on thoroughness and efficiency. Before starting, it's important to have all necessary equipment ready to minimize stress for both the animal and attending staff. A visual pre-capture examination can provide insight into the reptile's behavior and alertness. The physical exam should begin at the head, moving down to the tail, ensuring no areas are overlooked. The eyes should be bright, clean, and free of discharge, and any abnormalities such as redness or opacity should be noted. The ears should also be examined for cleanliness and any signs of infection.

The oral cavity should be checked for moisture, colour, and lesions. The mouth can be safely opened using soft materials, such as a spatula, and any signs of irritation, fractures, or plaque should be noted. After examining the head, the thoracic and coelomic cavities should be palpated, along with the extremities and tail. Abnormalities such as swelling, masses, or trauma should be recorded. In species like snakes, the heart, gallbladder, ribcage, and any prey items or faeces can be palpated, while in lizards, it may be difficult to palpate internal organs, although larger species may have palpable kidneys. Turtles and tortoises are more challenging due to their shell, but the inguinal area can provide access to palpate the coelomic cavity.

Shell or skin quality, along with any petechiae or ecchymosis, should also be assessed. These signs may indicate systemic infection or septicaemia. A Doppler is essential for measuring heart and respiratory rates in reptiles, as stethoscopes are not effective for most species. The Doppler probe should be placed at specific areas depending on the species—on the thoracic cavity for lizards and the thoracic inlet or carotid artery for chelonians. The heart and respiratory rates can be obtained by counting beats and breaths per minute.

Chelonians often show some non-specific symptoms with a variety of diseases. Though, clinical examination plays a significant role in disease diagnosis, bony carapace and plastron restricts its utility and diagnosing a disease becomes a challenging task even for an experienced veterinarian. Most of the time the clinical signs exhibited by chelonians are subtle and are seldom pathognomonic. Diagnostic imaging has been recommended as a routine procedure for clinical evaluation of ailing chelonians. The Chelonians are subjected to radiography without any sedation. If feasible their head and limbs are extended. They are placed in dorso-ventral and lateral position to obtain radiographs at 40 mAS and 100 kVP. Radiographic examination has been advocated as an important tool for the diagnosis of the diseases of respiratory, gastrointestinal and musculo-skeletal systems; and dystocia in chelonians. In most of the cases dorso-ventral radiographs without any sedation serve the purpose. In clinically normal chelonians, heart, ovaries, liver, pancreas, urinary bladder and kidneys are generally not discernible radiographically.

Physiological peculiarities of reptiles- Reptiles exhibit a wide range of normal physiological values, which can vary greatly depending on factors such as species, age, ambient temperature, and health. Heart rates typically range from 10 to over 80 beats per minute, while respiratory rates can vary from just a few to over 20 breaths per minute. Body weight is influenced by age, sex, nutritional status, and species, with reptiles ranging from just a few grams to several kilograms. Body condition scoring (BCS) for reptiles is similar to that used in mammals, with a scale from 1 (emaciated) to 9 (grossly obese). Key landmarks for scoring include the palpability of ribs, the girth of the tail and pelvic girdles.

- Glottis and breathing mechanism: Reptiles have a glottis, similar to birds, located at the base of the tongue. It opens briefly during inhalation and exhalation but remains closed at rest.
- Trachea variations: The trachea varies across reptilian species. Chelonians and crocodiles have complete cartilaginous rings, while snakes and lizards have incomplete C-shaped rings.
- Lung Structure: Snakes and lizards have simple, elastic lungs, with many snakes having only one functional right lung. Chelonian lungs are more complex, positioned dorsally under the shell.
- Absence of Diaphragm: No reptiles have a diaphragm. Most reptiles, excluding chelonians, use intercostal muscles to move the ribcage for breathing.
- Breathing Adaptations: Chelonians use their limbs, neck, and head to bring air in and out of the lungs due to their shell. Some reptiles, like chelonians and green iguanas, can survive for extended periods in oxygen-poor environments, making inhalant anaesthesia difficult.
- Respiratory Control: Respiration in reptiles is driven by low oxygen (pO_2) levels rather than high carbon dioxide (pCO_2). Many reptiles can bypass lung circulation through right-to-left heart shunting during breath-holding. Chelonians rely mainly on stretch receptors. Hypoxia increases breathing frequency, and hypercapnia (high CO_2) increases tidal volume.

Vein puncture site in reptiles

- When performing venipuncture on lizards, the most common site is the caudal tail vein (ventral coccygeal vein). While other sites like the cephalic, jugular, and ventral abdominal veins can be used, they are less preferred due to challenges such as small veins, difficulty in restraining the animal, or potential complications like lymphatic fluid contamination.

  The caudal tail vein can be accessed using either a lateral or ventral approach with a 1" or 1½" 27–20 gauge needle attached to a 1ml or 3ml syringe, depending on the lizard's size. Restraining the tail gently is key for a successful blood draw. The lateral approach involves inserting the needle between scales at a 90° angle, with slight negative pressure applied to the syringe. If the needle touches the vertebrae, it should be slowly backed off and redirected into the vessel. The ventral approach is similar, with the needle inserted along the midline. It is crucial to maintain the lizard in sternal recumbency to avoid struggles during the procedure.

- In chelonians, blood can be drawn from the brachial plexus, subcarapacial venous sinus, jugular vein, or femoral vein (fresh water adult turtle) depending on the species and size of the turtle or tortoise.

  For jugular venipuncture, the turtle is placed in lateral recumbency, and the head and neck are extended away from the shell. The vein is located in the same plane as the eye and tympanum, and a 27–20 gauge needle attached to a 1–3 cc syringe is typically used.

  The subcarapacial venous sinus is an alternate site when jugular access is difficult. The subcarapacial sinus is a point of convergence of numerous vessels. It is located on the animal's midline, where the eighth cervical vertebra joins the carapace . A 1" to 1½" needle is inserted at a 60-degree angle, and slight negative pressure is applied until blood is obtained or bone is encountered, in which case the needle is redirected.

  For the brachial plexus, the front limb is pulled away from the body, and a 22–20 gauge needle is inserted at a 90-degree angle to the skin near the radio -humeral joint. Additionally, the tail vein can be accessed using the same technique as in lizards.

  Blood samples can also be collected from femoral vein in adult fresh water soft shell turtles. The hind limbs are cleaned with distilled water-soaked cotton and after drying swabbed with alcohol. The femoral vein is located in the mid ventral part of the femur above the knee joint.

A small muscle depression is the correct site for venipuncture. Before inserting the needle in the skin, 50-100 μl of anticoagulant is taken into the syringe. The plunger of the syringe is withdrawn gently (up to the 0.1 to 0.3 ml mark) to create a vacuum for easy withdrawal of blood. The needle is inserted 2-6 mm deep depending on the musculature and age of the animal. The correct insertion of needle requires a 30°-40° angle downward, between the thigh and the needle. If blood does not appear in the syringe immediately, the syringe is gently rotated and the angle is slowly decreased until blood enters into the syringe.

Medication

- The oral route for drug administration is less preferred in reptiles owing to large variation in drug bioavailability. In systemically sick reptile, parenteral administration (i.e. subcutaneous or intramuscular injections) is the most preferred route of medication. Nevertheless, certain drugs, such as antifungals (e.g. terbinafine, itraconazole) can only be given by the oral route. Esophageal feeding tube in chelonians and lizards can be placed if long-term oral drug administration is required. Though, intramuscular (IM) drug administration is considered superior to the subcutaneous route, the subcutaneous route is the preferred method in practice as it is less painful, allows for delivery for larger volumes and can be performed with less restraint. Intravenous (IV) drug administration is rarely indicated in reptiles, excepting administration of certain anesthesia induction agents (e.g. propofol, alfaxalone, ketamine). Intraosseous catheters are a suitable alternative to IV catheters in lizards, but are painful. Fluids can be administered by intracoelomic route but less preferred in routine practice. Other routes of administration include nebulization of antimicrobials for skin and respiratory tract infections, intranasal administration of anesthetics and their reversal agents, and the topical administration of antiparasitics.
- Intramuscular injection is given into pectoral muscle mass from cranial site in chelonians (turtle, tortoise). Syringe is filled with required dose of medicine. Holding it at 45^0 angle, needle is inserted gently through the skin into pectoral muscle and the drug in injected. If the needle is not in muscle mass a swelling develops on injecting the drug indicating that the injection is not in the muscle. Subcutaneous injection is given to deliver the drug just below the skin. The best place for subcutaneous injection in chelonians is skin folds around front legs. Another route of drug administration is intracoelomic injection. Intracoelomic injection in chelonians is given from cranial side between bridge and thoracic limb.

Needle is directed parallel to the bridge and caudally towards coelomic cavity. Intravenous injection in turtle/tortoise can be given in right jugular vein. Sedation or anesthesia may be required for jugular catheterization. It is very difficult in small size chelonians. Oral administration of drugs is a common approach that can be followed by the owners themselves at schedule time without any specialized technical input. The drug is filled in the syringe or in a dropper and is delivered into the mouth . Though it is easy but many times it is not possible due to retraction of head into shell and shutting of mouth.

References and Further Readings

Jani, R. (2021). Basics of Wildlife Health Care and Management. 1st ed.. CRC Press.

Lemm, J. (2004). Iguana Restraint and Handling. Iguana, 11(3), 172-175.

Varshney, J. P. (2016). Diagnosis and management of anasaca in a turtle. Intas Polivet. 17: 209-210

Varshney, J. P. (2017). Electrocardiographic studies in conscious turtles. J. Anim. Research. 7: 741-744.

Varshney, J.P. and Monapara, H.D. (2019). Radiographic diagnoses in chelonians. Intas Polivet.20:201-206.

Varshney, J.P. (2021). Handbook of Exotic Pet Medicine. Brillion Publishing. Karolbagh, New Delhi, India.

Part III
Skin, Endocrine, Metabolic and Deficiency Diseases

26

Clinical Examination for Skin Disease

The skin is the largest organ in the body. It not only defends the body but also assists in the maintenance of homeostasis including water conservation and in body temperature conservation through insulation and in heat loss through perspiration. Pain and temperature extremes are recognized through sensory nerves of the skin. The skin also provides protection against physical injuries to some level, supports hair growth and offers some defence against microbial invasion. Skin diseases are multifactorial in etiology. But the response of skin to varying etiological stimulus is limited leading to same clinical manifestations in different skin diseases. Therefore, a comprehensive clinical examination is necessary for accurate diagnosis of the disease.

Clinical Examination

The clinical examination of skin diseases involves a systemic approach including history, physical examination and diagnostic testing.

History

Obtaining an accurate and detailed history regarding the owner's chief complaint is very much necessary as some of the diseases are age, breed, sex, and coat colour specific. Information regarding earlier diagnosis, treatment and adverse drug reactions is also very valuable. The information about the skin lesions such as the date and age at onset, original locations, initial appearance, progression and/or regression of lesions, factors affecting the course, and any previous treatments should also be noted. The history of the herd and a knowledge of the geographical region is valuable as in copper deficient areas, changes in coat colour may be observed. Previous history of skin disease on the farm and its management details provide background information for building diagnosis. Information regarding recent contact with other cattle may provide good thread for further investigation. Recent changes in diet and management are also important in ascertaining a diagnosis. History of frequent rubbing is suggestive of pruritus. Diets poor in mineral and vitamins leads to poor skin health. Abrasive surfaces or sharp corners in cubicles, parlour or floor can also lead to skin injury.

Overcrowding and insufficient feeding facilities are also contributory factors in causing superficial skin damage.

Physical, visual and manual (by touching)examinations

A complete information pertaining to rectal body temperature, heart rate, respiratory rate, body condition score, attitude and presence of any abnormalities involving any systemic diseases should be noted. Observations should be made whether the lesions appear localized, multifocal (affecting multiple distinct areas like the head, feet, and ventral abdomen), or generalized with or without symmetry. Symmetric lesions often indicate internal factors such as allergies, endocrine, metabolic, or immune-mediated diseases. In contrast, asymmetric lesions typically arise from external factors like infections, specific ectoparasites, or neoplasia.

Visual inspection-The whole body should be visually inspected from a distance and closely for areas of abnormal skin or hairs. Healthy animals may have lick marks on the body. Lick marks on the body may be excessive due to pruritus in cases of heavy louse infestation. Repeated rubbing of the body due to severe itching can lead to hair loss and skin thickening. The presence of any swelling, discharge or abscess should be noted on close inspection. In warm weather or in pyrexic animals sweat marking may be noted. Close skin examination may also reveal gangrenous changes, if any, owing to loss of blood circulation.

Manual appraisal - Manual appraisal of skin by touching should be done to detect invisible lesions beneath the matted hairs and lymph node enlargement. Examination by touching the skin (manual examination) may also provide information about the presence of subcutaneous edema, skin thickness, skin infection or skin turgor. Pathological skin thickening is observed in sarcoptic mange. Callus formation is also another form of skin thickening in areas specially joints (elbow, hock) subjected to repeated trauma.

Dermatological examination

The texture, elasticity, and thickness of the skin should be noted. Otoscopic examination (Fig.169) and evaluation of degree of aural erythema should be included in examination. Aural erythema is one of the first sign to develop in allergic reactions.

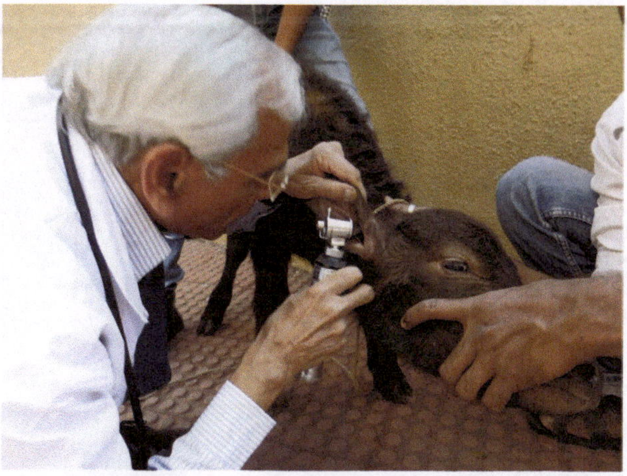

Fig. 169: Otoscopic examination of a buffalo calf

It is important that a clinician must be able to differentiate primary and secondary lesions. A primary lesion is the initial eruption that develops spontaneously as a direct reflection of underlying disease. Secondary lesions evolve from primary lesions or are artifacts induced by the patient or by external factors such as trauma and medications. Secondary lesions are more chronic and give clues about the presence of previous primary lesions. Therefore, identification and characterization of both primary and secondary lesions are important.

Primary lesions include macule or patch, papule (Fig. 170A) or plaque, pustule (Fig.170 B), vesicle or bulla, wheal (Fig. 170C), nodule and cyst. Lesions that may be primary or secondary include alopecia, scale, crust, follicular casts, comedones and pigmentary abnormalities. Secondary lesions are epidermal collarets, scar, excoriation, erosion or ulcer, fissure, lichenification (Fig. 170 D) and callus (Fig. 170 E). The configuration of skin lesions is also helpful in establishing a differential diagnosis.

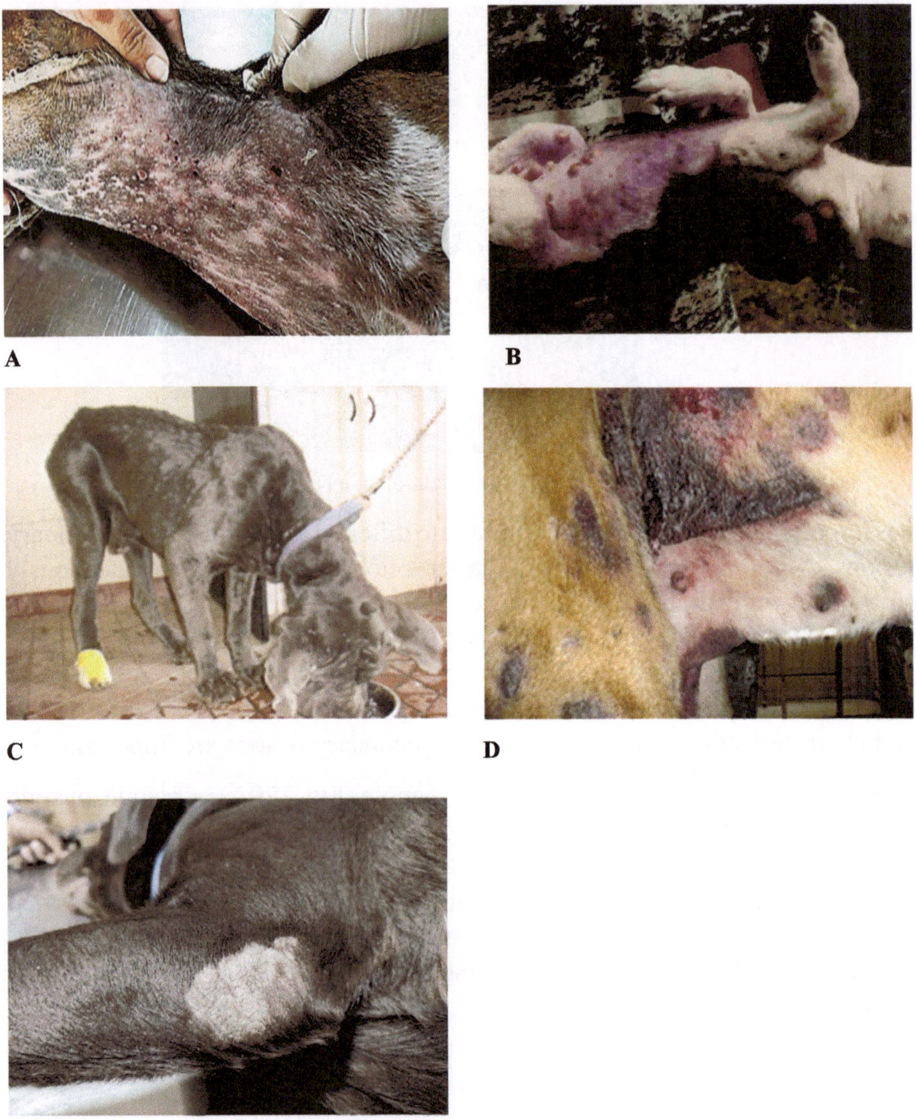

A B C D E

Fig. 170: Different types of skin lesions. Papules (A), pustules (B), urticarial wheels (C), Lichenification (D) and callus formation at elbow (E). Images B,C and D are with the courtesy of Dr. J.P.Varshney (2025).

Distribution of skin lesions- Photosensitization in farm animals causes lesions in lightly pigmented area on the dorsal parts of the body. In calves ringworm lesions are commonly distributed on the head and neck. Dermatophyte lesions may also be seen on other areas also.

Type of lesions- Temperature, thickness, colour and consistency of the the skin lesions should be recorded. Edematous (subcutaneous edema) lesions are suggestive of cardiac failure or hypoproteinemia. More than one condition can be present at the same time. Skin lesions may involve some or all skin components (hair, follicles, epidermal, dermal and subcutaneous tissues). The lesions may be primary or secondary. Primary lesions are the result of the skin disease. While secondary skin lesions are non-specific owing to further development of primary lesion or self inflicted damage.

Primary lesions are macules (flattened area of changed colour without thickening and of < 1cm), papule (flat circumscribed, rounded, raised of < 1cm with necrotic centre), nodule (papule of > 1cm diameter), plaque (solid raised flat top of > 1 cm diameter), vesicle (fluid filled blister of < 1 cm diameter), bullae (vesicles of > 1cm diameter), pustule (pus filled vesicle) and tumour (large nodular lesions).

Secondary lesions are scale (loose dry fragment of superficial skin), crust (dry accumulation of debris, blood and pus), erosion (loss of superficial epidermal layers with intact inner layers), excoriation (erosion or deep ulcer of traumatic origin), ulcer (deep erosion penetrating epidermal basement membrane), scar (fibrous tissue that has replaced damaged skin), fissure (breach in superficial layers of skin), keratosis (overgrowth of dry horny keratinized epithelium), alopecia (no hairs in the area) and pigment changes (hypo or hyper pigmentation).

Skin lesions associated with some bovine diseases- Skin lesions observed in some important bovine diseases are enumerated below:

(i) Pediculosis is associated with signs of skin irritation (rubbing against objects), excessive grooming, hair loss and adhering of small white eggs on the hairs. Adult lice are visible with naked eyes. Heavy lice infestation causes anemia.

(ii) Flies during summer may cause tear scalding under eyes. Biting flies (Haematobia, Stomoxys and Simulium) may cause multiple skin nodules, urticarial or open wounds.

(iii) Mange in cattle is caused by chorioptic, sarcoptic, psoroptic and demodectic mange mites. Chorioptic mange causes crusty lesions around tail head. Intense rubbing lead to secondary changes such as alopecia and skin thickening. Sarcoptic mange may infest the tail head, head, neck and shoulders. It is a burrowing mite and causes skin thickening, intense pruritus and secondary skin damage. Psoroptic mite is the surface living casual mite and is associated with pruritus, extensive skin thickening, and hair loss. Demodectic mange is not common in cattle. It is mostly asymptomatic. It may cause alopecia with out pruritus.

(iv) Heavy tick infestation may lead to anemia and skin lesion (granulation and hypersensitivity reaction) in the lower parts of the body. Ticks are responsible for babesiosis, theleriosis and Lyme disease.

(v) Blow fly strike can occur on any part of the body. Its larvae penetrate the skin and deeper body tissues. Necrosis may be seen in adjacent areas.

(vi) Pus is seen in lesions infected with bacteria. Superficial pustular dermatitis is caused by Staphyloccocal infection. Occasionally Dermatophilus infection is seen on the backs of the cattle kept overcrowded and in unhygienic conditions. Its lesions are thick crusty. Subcutaneous abscess, around tail head and other parts of the body surface, are mostly associated with *Arcanobacterium pyogenes* infection. Lumpy jaw is caused by *Actinomyces bovis* and is clinically characterized by osteomyelitis of the mandible.

(vii) Dermatophyte infection (ringworm) is common in young ones and clinically characterized by focal alopecia, raised grey crusty lesions on head and neck. Diagnosis can be confirmed by skin scrapping examination, Wood's Lamp examination (Trichophyton infection) and cultural examination.

(viii) Warts in cattle are caused by bovine papillomatosis virus. It is characterized by cauliflower like skin lesions. Foot-and-mouth disease causes lesions (vesicles, bullae, ulcer) on the teat, muzzle, tongue, coronary band. Infectious bovine rhinotracheitis, mucosal disease, infectious vulvovaginitis/balanoposthitis and malignant catarrh are the other viral infection with possible skin lesions.

(ix) Tumors such as squamous cell carcinoma (the mucocutaneous junctions of the body, the mucosa of the third eyelid, the periorbital skin and the vulva), cutaneous lymphosarcoma (skin of the neck and flank), fibromas and fibrosarcomas (subcutaneous tissues) and lipomas (soft tissues of the head) are seen on the skin.

(x) Gross nutritional deficiency may cause dull, dry non-elastic skin with poor growth of scant and brittle hairs. Vitamins A, C and E deficiencies have also been associated with skin diseases. Subtle coat color changes (specially around eyes) are seen in copper deficiency and excess molybdenum.

(xi) Chronic arsenic, selenium or molybdenum toxicity is associated with poor skin quality,change in hair color. Excess of potassium iodide may cause dry seborrhoeic dermatitis.

(xii) Snake bites may also be associated with severe swelling and oedema of the skin.

Diagnostic Testing

Surface sampling

Skin scraping- It (Fig. 171) is one of the most frequently used tests for the differential diagnosis of microscopic ectoparasites. Skin scraping is most commonly used to rule out the diagnosis of demodectic mange. It is also commonly used to establish the diagnosis of sarcoptic mange, *Cheyletiella* infestation, and some other ectoparasitic diseases. The equipment needed to perform a skin scraping is mineral oil, a scalpel blade, microscope slides, coverslips and a microscope.

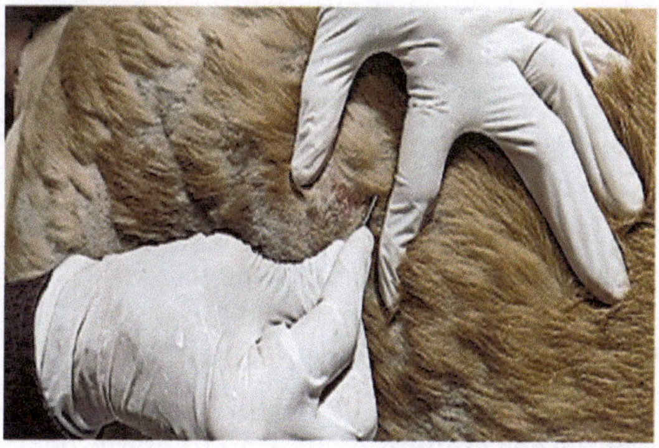

Fig. 171: Taking skin scrapings from a lesion in dog. Figure is with the courtesy of Dr. J.P.varshney (2025).

For the diagnosis of demodex multiple deep scrapings from new lesions should be obtained. The affected skin should be squeezed between the thumb and forefinger to extrude mites from hair follicles. If hair is present, it is helpful to clip the area prior to scraping. Applying a drop of mineral oil to the skin site being scraped or to the scalpel blade will facilitate adherence of material to the blade. The blade should be held at an angle between 45 and 90 degrees to the skin surface, and moderate pressure is used to scrape the skin in the direction of hair growth. The obtained material is then scooped up and placed on a microscope slide. This procedure should be repeated to scrape the skin more deeply until capillary bleeding is produced. It is important that true capillary bleeding is obtained, and not blood from laceration. Skin scrapings may be negative in dogs with chronic demodectic pododermatitis. Diagnosis is made by demonstrating multiple adult mites, finding adult mites (Fig.172) from multiple sites, or finding immature forms of mites (ova, larvae, and nymphs) of demodex. When dogs are fractious or sites are difficult to scrape, hairs from the affected area can be plucked, placed in mineral oil on a slide, and examined.

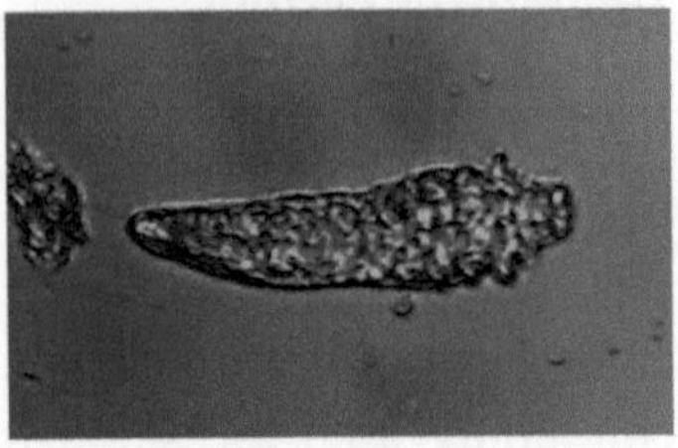

Fig. 172: Adult *Demodex canis* in skin scrapings of a dog.

Canine sarcoptic mites, *Sarcoptes scabiei* (var. *canis*) and feline scabies mite Fig.173), *Notoedres cati* reside within the superficial epidermis. Multiple superficial scrapings are indicated, with emphasis on the pinna margins and elbows in dogs and head, face, or ears in areas with crusts and scales in cats. Skin that has not been excoriated, preferably skin with red, raised papules and yellowish crusts on top, should be scrapped. Extensive amounts of material should be accumulated in the scrapings and spread on microscope slides. A second microscope glass slide may be used instead of a normal coverslip to compress the thick crusts. The clinician examines each field until a mite is found or all material has been examined; one mite is diagnostic. Dark brown round or oval faecal pellets or ova, if found, are also diagnostic. In difficult cases, it may be useful to accumulate an even larger amount of hair and keratin debris from scrapings and place the material in a warm solution of 10% potassium hydroxide (KOH) for 20 minutes to digest keratin. The mixture is then stirred and centrifuged. Mites are thus concentrated and can often be picked off the surface film and identified with a microscope.

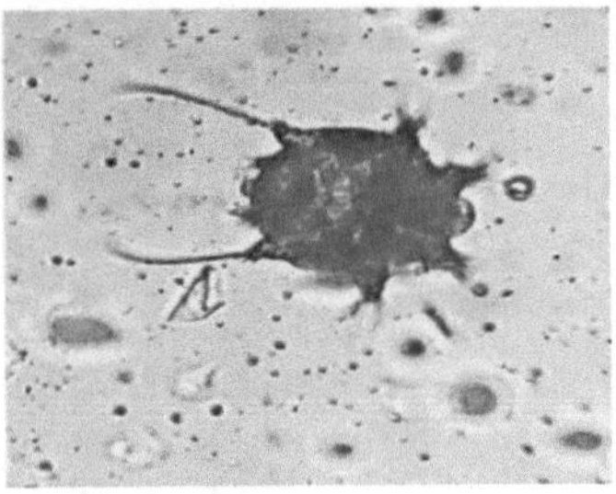

Fig. 173: Sarcoptic mites in skin scraping of a cat.

Cheyletiella mites are relatively large compared to scabies mites. Superficial scrapings, acetate tape impressions or examine the material collected through use of a flea comb can be used for diagnosis. *Otodectes cynotis* mites are usually located in the external ear canal of dogs and cats, but may also be found on the skin, especially around the head, neck, rump, and tail. They may be found by superficial scraping or acetate tape methods.

Acetate tape impression test- Transparent, pressure-sensitive acetate tape (Fig. 174) or packing tape is applied to the surface of the hair and the surrounding skin where the hair is parted or in shaved areas. This method is used to collect superficial scales and debris when cheyletiellosis or poultry mites are suspected. The tape is subsequently firmly pressed onto a microscope slide and examined.

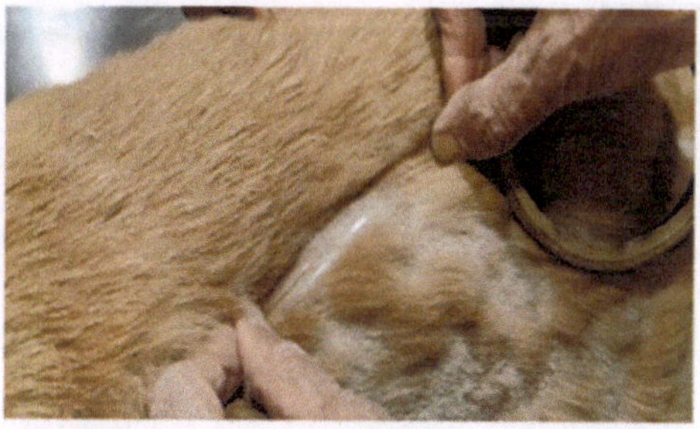

Fig. 174. Taking acetate tape impression smear. Figure with the Courtesy of Dr. J.P. Varshney (2025).

Flea comb and floatation test- In this technique, large areas of the body are combed, and the collected scale and debris are put in a faecal flotation solution. Material and scale that fall on the table or on a sheet of paper placed under the patient during combing should also be added to the flotation solution. A coverslip is applied to the surface of the flotation solution, which is allowed to stand for 10 minutes, and then transferred to a microscope slide and examined.

Skin debris examination- The material collected by briskly rubbing the pet's skin and hair coat while the animal is standing on the table or over a piece of paper and examined directly or mixed with a fecal flotation solution. Direct examination may find otherwise undetected flea dirt, flea eggs, or (rarely) mites. Suspect flea dirt may be placed on moistened white paper or cotton and will dissolve to an orange-red stain.

Trichography

Plucking hairs from the skin and examining them under the microscope is referred to as trichography (Fig.175) and is helpful for diagnosing self-inflicted alopecia, dermatophytosis, color dilution alopecia, nutritional or congenital hair dysplasias, endocrine alopecia, and pigmentary disturbances of hair growth. It is also beneficial in some cases of demodicosis or *Malassezia* dermatitis. Trichography involves gently grasping a small cluster of hairs using either the fingertips or rubber-covered hemostats, and then completely removing them from the skin in the direction they naturally grow to reduce potential damage. These hairs are then placed in the same orientation on a microscope slide with mineral oil, and observed using the low-power objective of a microscope.

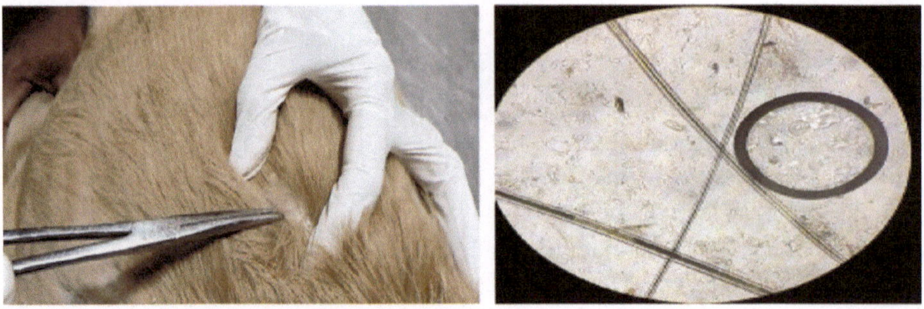

A B

Fig. 175: Plucking of hairs from the periphery of the lesion for trichography (A).A normal hair shaft of uniform diameter in seen (B). Figures are with the courtesy of Dr. J.P.Varshney(2025).

A normal hair shaft is uniform in diameter and tapers gently to the tip. Hairs will have either an anagen or telogen root (bulb). Anagen bulbs are rounded, smooth, shiny, glistening, and often pigmented and soft, so the root may bend. Telogen bulbs are club- or spear-shaped, rough-surfaced, nonpigmented, and generally straight. Catagen hairs are rarely seen and have bulbs intermediate in appearance between anagen and telogen bulbs. Normal adult animals have an admixture of anagen and telogen hairs, the ratio of which varies with season, management factors, and a variety of other influences. No normal animal should have all of its hairs in telogen, suggestive of telogen effluvium or follicular arrest. Inappropriate numbers of telogen hairs suggest a diagnosis of a nutritional, endocrine, or metabolic disease.

Examination of the hair shaft follows bulbar evaluation. Hairs that are inappropriately curled, misshapen, and malformed suggest an underlying nutritional or metabolic disease or congenital or hereditary abnormality. Hairs with a normal shaft that are suddenly and cleanly broken or longitudinally split indicate external trauma from excessive licking or scratching or too vigorous grooming. Breakage of hairs with abnormal shafts can be seen in color dilution alopecia and other congenital hereditary disorders. Hair casts are seen in diseases associated with follicular keratinization abnormalities, such as sebaceous adenitis, demodicosis, primary seborrhoea, follicular dysplasia, and endocrinopathies.

Cytological examination

Cytological study is helpful in determining the relative depth of infection, determining whether the pustule contains bacteria or is sterile, identifying the type of inflammatory cells, various cutaneous neoplasms or other cellular infiltrate, or finding the acantholytic cells of the pemphigus diseases, the relative amount of protein or mucin and the presence of, yeasts, and bacteria.

Materials for cytologic examination can be gathered by a variety of techniques. Those most commonly used include direct smears for fluid-containing lesions, impression smears when lesions are moist or greasy, swab smears used to obtain specimens from draining tracts or sinuses, ear canals and interdigital webs, and dry crusty-surfaced lesions, acetate tape strips for collecting skin surface debris, scrapings to sample underneath crusts, vesicles, and peeling *Stratum corneum* and fine-needle aspiration to sample nodules, tumours, and cysts.

Collected materials are air-dried on the slide. Skin samples collected directly or with moistened cotton applicators can be heat-fixed before staining. Heat fixation is achieved by using a blow dryer for 2 to 5 seconds. Once heat-fixed and dried, the slide is stained and examined under a microscope. Common stains used in clinical practice include the modified Wright stain (Diff-Quik) and new methylene blue.

Examination for fungi

Direct examination of hairs-The hair, scales, and claw material may be cleared by placing specimens in several drops of 10% to 20% KOH on a microscope slide. A coverslip is added, and the slide is gently heated for 15 to 20 seconds. Arthroconidia occur as a sheath of tiny coccoid elements surrounding the hair, whereas hyphae grow within the hair shaft. Dermatophyte infected hairs appear swollen and frayed, irregular, or fuzzy in outline, and the clear definition between cuticle, cortex, and medulla is lost.

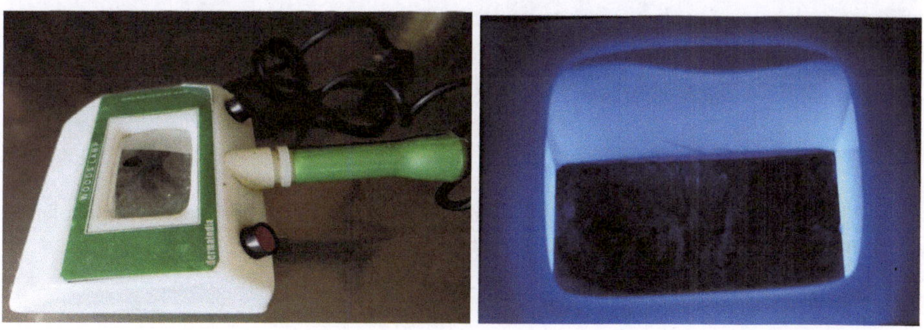

A B

Fig. 176: Wood's Lamp (A) and examination of lesions of a cat under Woods Lamp (B). Yellow-green fluorescence indicates infection with Microsporum spp.Figures are with the courtesy of Dr. J.P.Varshney (2025).

Wood's Lamp Examination-A Wood's lamp (Fig. 176A) is a UV light with a wavelength of 253.7 nm that is filtered through a cobalt or nickel filter. When exposed to the UV light, hairs invaded by *Microsporum canis* may result in a yellow-green fluorescence (Fig. 176B) in 30% to 80% of the isolates. The characteristic fluorescence is due to tryptophan metabolites produced by the fungus.

Fungal culture- Sabouraud dextrose agar and dermatophyte test medium (DTM) can be used for the isolation and identification of fungi.

Examination for bacteria

Cytological examination is the initial test of choice. When cytological study reveals rod-shaped organisms, or when cocci (Fig. 177) are seen but appropriate empirical therapy is ineffective, bacterial culture and susceptibility testing are indicated. Selection of appropriate lesions for culturing is critical. In cases with pustules, an intact pustule should be opened with a sterile needle. The pus collected on the needle should be transferred to the tip of a sterile swab. Papules may also be superficially punctured, and a relatively serous droplet of pus may be obtained. When papules or pustules are not present, cultures may be taken from under a crust or a skin biopsy may be taken for culturing. Disinfection of the skin overlying superficial papular and pustular lesions should not be attempted prior to culture, because negative cultures may result.

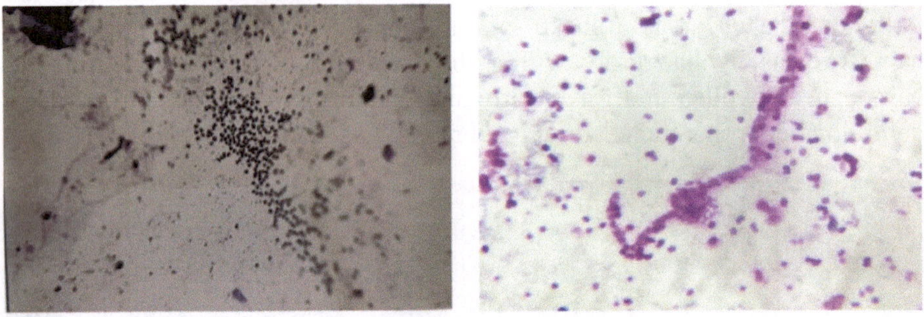

A B

Fig. 177: Impression smears showing organism (A) and long chain Dermatophilus congolensis (B) Figures are with the courtesy of D.J.P. Varshney (2025).

Allergy testing

History and physical examination findings and exclusion of other diseases provide the basis for diagnosis. Food allergies are most reliably diagnosed based on elimination diets followed by provocation trials. Contact allergies may be diagnosed through environmental restriction and provocation trials or via patch testing. Drug allergies are generally diagnosed by history and compatible clinical and dermato-histopathologic findings. Intradermal allergy testing and serologic tests are used to identify allergens that may be contributory factors in atopic dermatitis.

Biopsy

In general, skin biopsy should be performed for any dermatosis that is not responding to what appears to be appropriate therapy after 3 weeks. A 6-mm biopsy punch provides an adequate specimen; 4-mm punches are reserved for difficult sites such as near the eye, on the pinna, and on the nasal planum and foot pads of cats and small dogs. Biopsy should be performed on all obviously neoplastic or suspected neoplastic lesions, all persistent ulcerations, any case that is likely to have the major diseases most readily diagnosed by biopsy and dermatosis that is not responding to apparently rational therapy.

Further Readings

Budach, S.C. and Mueller, R.S. (2012). Reproducibility of a semiquantitative method to assess cutaneous cytology. Vet. Dermatol. 23(5):426–e80. doi: 10.1111/j.1365-3164.2012.01075.x. Epub 2012 Jul 19.

Jackson, P.G.G. and Cockcroft, P.D. (2002). Clinical Examination Of Farm Animals. Blackwell Science Ltd. Osney Mead, Oxford OX2 0EL, UK

Lundberg, A., Koch, S., Bierbrier, L. and Slater, M. (2022). Diagnosing Common Skin Conditions of Dogs and Cats in Community. Medicine Practice.Today's Veterinary Practice. Issue: May/June 2022

Miller, W.H., Griffin, C.E. and Campbell, K.L. (2021). Muller and Kirk's Small Animal Dermatology.8th Edition. Elsevier Science Publishing Co Inc,USA.

Moriello, K.A, Coyner, K., Paterson, S. and Mignon, B. (2017). Diagnosis and treatment of dermatophytosis in dogs and cats: Clinical Consensus Guidelines of the

Pereira, A.V., Pereira, S.A. and Gremião, I.D.F. (2012). Comparison of acetate tape impression with squeezing versus skin scraping for the diagnosis of canine demodicosis. Aust. Vet. J. 90(11):448-450.

Varshney, J.P. (2025). A Fully Illustrated Handbook. Clinical Diagnostic Processes in Canine Medicine"(NIPA, New Delhi

World Association for Veterinary Dermatology. Vet. Dermatol.28:266-e68. doi: 10.1111/vde.12440.

27

Blood Transfusion

Blood transfusion is one of the crucial life-saving procedures in veterinary medicine undertaken to increase red cell mass, volume replacement, replenishing coagulation factors, and protein replacement. The concept of blood transfusion was first introduced by Francis Potter in 1639. However, it was only in 1666 that Richard Lower reported the first successful direct transfusion from one dog to another. Blood is a biological drug and is not a definitive treatment, but it provides support until the disease is diagnosed or treatment is initiated.

Indications

Blood transfusion is indicated for the management of the following conditions:

- Anemia due to various diseases/disorders, like infection or acute/ chronic hemorrhage
- Bleeding disorders such as thrombocytopenia or coagulopathies
- Rodenticide toxicities
- Hypoproteinaemia usually of parasitic or infectious origin.
- Extensive burns
- Parasitism, toxicosis, or immune-mediated anemia
- Bone marrow disorders

Anaemia **:** Anaemia may be due to severe hemorrhage, hemolysis or decreased production of erythrocytes or hemoglobin.

Hemorrhagic Anaemia- It is caused by trauma leading to severe external / internal hemorrhage, bleeding disorders, chronic liver disease or severe parasitic infestations (strongyles, hook worm, Hemonchus sp.). In cattle, spontaneous pulmonary hemorrhage, abomasal ulcers, enzootic hematuria, pyelonephritis, rupture of middle uterine artery, cardiac tamponade, juvenile bovine angiomatosis and intraluminal hemorrahges are the major cause of hemorrhagic anemia. Nasal bleeding due to guttural pouch lesion, ethmoidal hematoma, excercised induced pulmonary hemorrhage, rupture of uterine artery following foaling, rupture of mesentric arteries owing to strongyle larva migration, splenic hematoma, hemangioma and hemangiosarcoma, and bleeding from vulva in association with ulcerated vericose veins are the main causes of severe hemorrhage in horses. In swine, major causes of severe bleeding are ulceration of esophagus and stomach, proliferative hemorrhagic enteropathy and congenital neonatal bleeding. In dogs, ancylostomiasis (in pups), chronic ehrlichiosis, colitis, disseminated intravascular coagulopathy, auto-immune thrombocytopenia, infectious canine hepatitis, parvo infection, pancytopenia and chronic kidney disease may lead to hemorrhagic anemia.

Hemolytic anemia- In farm animals, severe endoparasitic infestations (Haemanchosis, Oesophagostomiasis, *Fasciola hepatica* etc.), infectious diseases (babesiosis, anaplasmosis, theileriosis, nagana, trypanosomiasis, eperythrozoonosis, bacillary hemoglobinuria, leptospirosis), infestations (lice and ticks), production diseases (post-parturient haemoglobinuria), nutritional deficiencies, auto-immune hemolytic anaemia, and many poisonings may lead to hemolytic anemia. Chronic copper poisoning, eperythrozoonosis, babesiosis, poisoning by cruciferous plant, parasitic infestations (*Faciola hepatica, Haemonchus contortus*) are the most common cause of hemolytic anemia in sheep.In swine, isoerythrolysis, thrombocytopenia and coagulopathy leads to hemolytic anemia. The etiology of hemolytic anemia in horses varies from babesiosis, equine infectious anemia, phenothiazine poisoning, isoerythrolysis in foals, auto-immune hemolytic anemia, penicillin induced hemolysis to some snake envenomation.

Decreased production of erythrocytes or hemoglobin- Chronic diseases (chronic suppurative processes, radiation, bracken fern poisoning, intestinal parasitism, idiopathic hypoplastic anemia in foals,sudden movement to hgh altitude) and nutritional deficiencies (cobalt, copper, iron, potassium, pyridoxine or folic acid) lead to decreased production of erythrocytes or hemoglobin

Candidates for Blood Transfusion

- An animal exhibiting significant tachycardia, tachypnoea, lethargy, weakness, respiratory distress, pale or discolored mucous membranes, and prolonged capillary refill time.
- An animal having PCV of less than 10 percent.
- PCV has fallen rapidly to less than 20 percent in dogs or 15 percent in cats.

Blood Groups

The presence of glycolipids and glycoproteins on the surfaces of red blood cells allows classification into different blood groups. These antigens can elicit an immunological reaction in a noncompatible recipient by causing the formation of circulating anti-erythrocyte antibodies.There are many blood groups in different animals (Table 18).

Table 18: Blood groups in animals

Dog	13 (8 are labeled as DEA)
Cat	3
Horse and Donkey	7
Cattle	11
Sheep and goat	7

Canine blood groups

Karl Landsteiner, the pioneer of the human ABO system, first discovered canine blood types in 1910. Since then, over 13 canine blood groups have been described and eight DEA (Dog Erythrocyte Antigen) types have been recognized. The abbreviation DEA stands for "Dog Erythrocyte Antigen," which is followed by the numerical identification of the blood group containing polyclonal alloantibodies. The blood group system in dogs includes DEA 1.1, DEA 1.2, DEA 3, DEA 4, DEA 5, and DEA 7. DEA 1.1 and 1.2 are the most important blood groups and are found in 60% population of canines. DEA 1.1 is the most common blood type in dogs and, although naturally occurring antibodies to DEA 1.1 are rare, the determination of the DEA1.1 antigen is strongly recommended as this antigen is highly immunogenic and will result in antibody formation. Two newly identified blood groups are DAL (in Dalmatian dogs) and Kai.

Note: 15 ml of blood per kg BW can be collected from the dog every 6 weeks.

Feline blood groups

Cats have three main blood types (on the AB system) such as type A, B or AB.The antigens associated with these types are highly immunogenic and type B cats have high levels of naturally occurring antibodies. Type A blood group is the most common group and is found in 95% of American cats. The majority of Indian and 30% of cats in UK belong to blood group B. Blood group AB is extremely rare.

Note: Healthy adult cats can donate 45-60 ml every 6 weeks.

Equine blood groups

The seven blood groups in horses (A, C, D, K, P, Q and U) are internationally recognized with more than 30 erythrocyte antigens. Universal donor horse is not possible because of various possible antigenic combinations.

The cross-matching must be performed although impractical to minimize transfusion reaction.

Note: Adult horses can safely donate approximately 6-8 L of blood. Whole blood can be collected every 15-30 days and plasma collected every 7 days if the erythrocytes are returned to the donor.

Bovine blood groups

The internationally recognized blood groups in cattle are A, B, C, F, J, L, M, R, S, T, and Z. Out of these 11 groups, groups B and J are the most clinically relevant.

Note: Cattle can donate 8-14 ml of blood per kg of body weight.

Ovine and caprine blood groups

A, B, C, D, M, R, and X are the seven blood groups identified in sheep. The blood groups of the goat (A, B, C, M, J) are very similar to those of sheep with the B system equally complex. Many of the reagents used for blood typing of sheep are also used to type goat blood groups.

Blood Typing Technique

- The principle of blood typing methods is a visible hemagglutination reaction between patient RBC surface antigens and known reagent monoclonal or polyclonal antisera. Blood typing is significant in identifying the presence of dog erythrocyte antigens which are responsible for initiating approximately 70 to 80% of immune-mediated transfusion reaction
- A card-based agglutination assay, an immune-chromatographic cartridge, and a gel column diffusion assay are all commercially available blood typing kits. Antisera are lyophilized and placed in reaction wells on blood typing cards. Positive and negative control wells are included on the dog cards, while an auto control well is included on the cat cards. The availability of blood typing reagents for extended blood typing is limited. The procedure is simple, and the results can be obtained in less than 2 minutes without the use of any additional equipment. The auto control well on the cat typing card, as well as separate cards for dogs,

provides for auto-agglutination testing. Auto-agglutination resembles a positive reaction and may make correct typing impossible. Now blood typing kits are also commercially available for DEA 1.1 and 1.2. These kits are readily available for blood typing in practice and are easy to use and are accurate (Fig. 178).

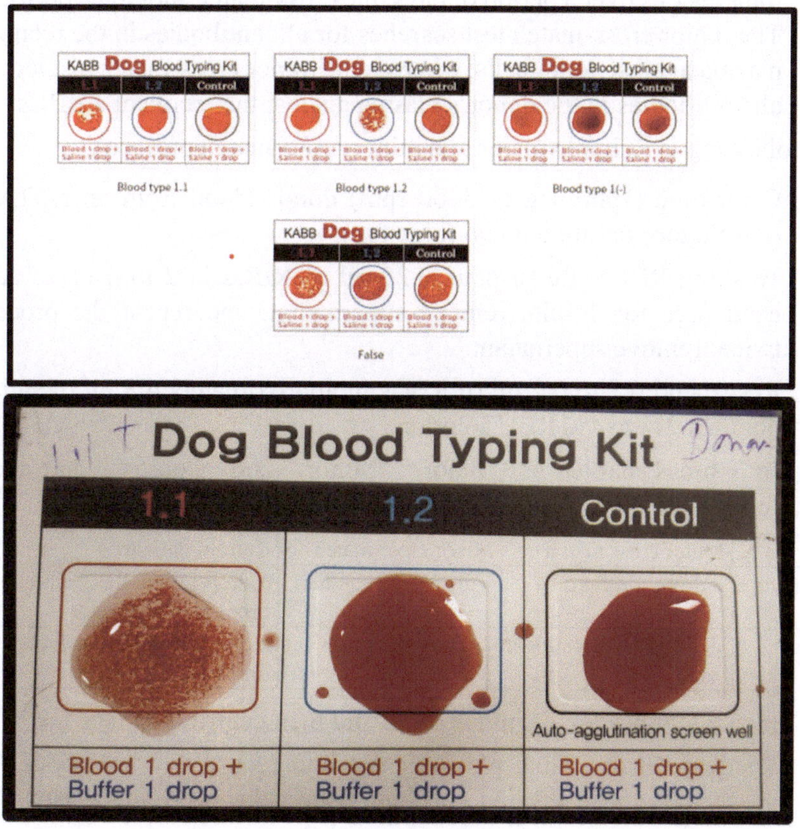

Fig. 178: Dog blood typing kit (DEA 1.1 positive).

Cross Matching

- Cross-matching reveals the serological compatibility or incompatibility between donor and recipient. Blood typing tests reveal the blood group antigens on the red blood cell surface. However, cross-matching reveals the serological compatibility or incompatibility between donor and recipient. Cross-matching, tests for the presence or absence of naturally and induced antibodies, do not replace blood typing. Cross-matching is similar to blood typing, except that specific antiserum are not used, and consists of a major and minor part. The major and minor cross-matching tests are performed for agglutination and/or hemolytic responses between donor and recipient. Animals with strong naturally occurring antibodies, such as cats, or those with induced antibodies, such as those from previous transfusions, should always have a substantial cross-match performed. Even if the same donor blood is used for multiple transfusions over a period of days, the cross-match should be performed. The major cross-match test searches for alloantibodies in the recipient's plasma against donor cells, whereas the minor cross-match test looks for alloantibodies in the donor's plasma against the recipient's RBCs.

The following is a simple major and minor cross-match protocol.

(a) Centrifuge (1000 x g or 3400 rpm) donor blood from an EDTA vial (purple top) or citrated vial (blue top).

(b) Washing RBCs: Re-suspend 0.25 ml of RBCs in 2 to 4 ml of saline; centrifuge for 1 min, remove supernatant, and repeat the procedure twice; remove supernatant.

(c) Resuspend 0.1 to 0.2 ml of RBCs in approximately 4.8 ml of saline to obtain a 2% to 4% RBC solution.

(d) In 3 tubes marked as "Major," "Minor," and "Control," and add the following:

	Major cross match	Minor cross match	Control Patient
Recipient	2 drops plasma	1 drop RBC solution	1 drop patient RBC solution + 2 drops patient plasma
Donor	1 drop RBC solution	2 drops plasma	1 drop donor RBC solution + 2 drops donor plasma

(e) Incubate tubes for 15 min at 37°C and then centrifuge tubes for 15 s.

(f) Reading results: Note plasma color and record any hemolysis. Then gently resuspend the red cell button into the overlying plasma layer, noting the presence of agglutinating clumps.

(g) Place a drop of re-suspended RBCs on a slide, cover it with a coverslip, and examine it under 10X and 40X. If the crossmatch is compatible, the RBCs are individually distributed (Fig. 179). While in incompatible match, hemolysis (compared to control) or agglutination is seen (Fig.180). Rouleaux, a physiological plasma-related phenomenon, may sometimes be observed. To distinguish this from agglutination, centrifuge the tubes again for 15 s, remove the serum, and add 2 drops of saline; then centrifuge the tubes once more and re-examine the RBC suspensions.

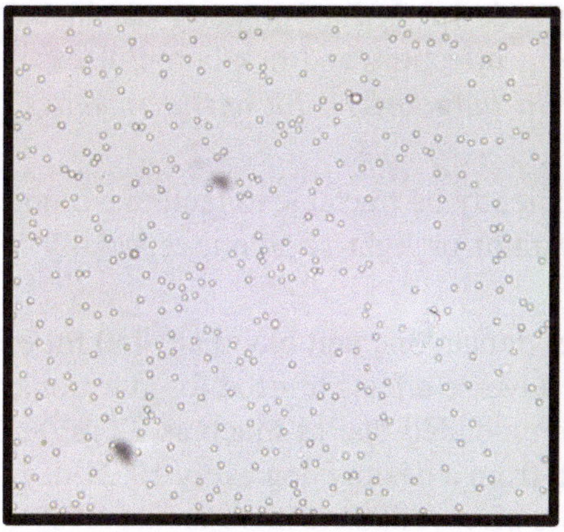

Fig. 179: Compatible cross-matching

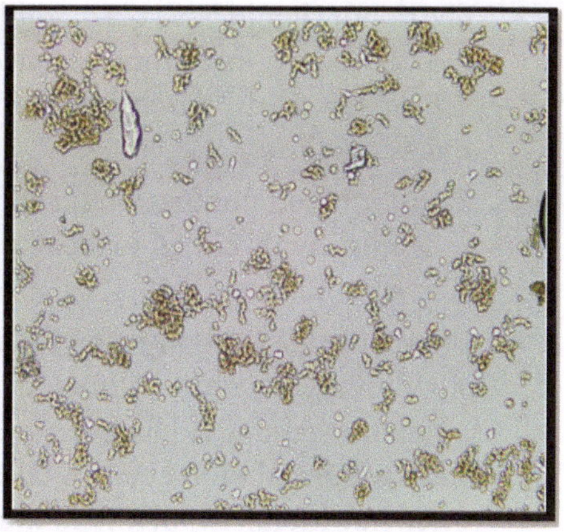

Fig.180: Incompatible cross-matching

- In general, the first transfusion is relatively safe because of the lack of naturally occurring alloantibodies so unlikely to experience an acute incompatibility reaction after first transfusion. In large animals, if crossmatching is not possible, compatibility of the blood can be judged by infusing 100 ml of donor blood slowly and observing for any adverse reaction for 15 minutes. If there is no clinical evidence, transfusion can be done safely. Subsequent transfusion without crossmatching may be repeated (same donor) within 2-3 days (never after the 5th day) of the first transfusion. However, during subsequent transfusions, when given at least 7 days after the first transfusion, dogs may experience acute, severe hemolytic transfusion reaction when given mismatched blood.

Selection of Donor Dog

- Factors such as age, general body condition, sex, and absence of diseases are to be taken into consideration while selecting a donor for blood collection so as to avoid transfer of unwanted diseases.
- The selected donor should be healthy male or non-pregnant female of 1 to 8 years age having no earlier blood transfusion.
- Donor dogs should weigh at least 23 Kg (in lean body condition) with a PCV > 40%. The best breed for blood donation is Grayhound (normal PCV>60%).
- To facilitate blood collection and maintain an aseptic phlebotomy site, donor should have readily accessible jugular veins.
- Donors should be calm in nature so that vascular trauma is minimized and blood collection is completed in less than 10 minutes with minimal restraint.

- The donor should have been regularly dewormed and properly vaccinated and should be free from heartworm infection, and tick-borne diseases *(Ehrlichia canis, Babesia, Borrelia burgdorferi, Rickettsia rickettssi,* and *Brucellosis*).
- Regular blood donors should sufficiently be tractable to facilitate blood collection with either manual restraint or light sedation, especially if they are client-owned.
- Dogs over 27 Kg body weight can donate one unit blood i.e. 450 ml of blood every 3 weeks without any adverse effects or need for nutritional supplementation. It has been recommended that as much as 22 ml/Kg body weight of blood can be taken from a healthy dog every 10-21 days when nutritional supplementation is given. All the donors should receive a physical examination and hematocrit check before each donation.
- Donors should be fed a balanced diet with adequate vitamins, minerals, and proteins. Oral iron can be supplemented at 0.5 mg iron/ml of blood taken for donors that are used every month.
- Adult horses and cattle can safely donate four liters of blood at a time. They should be tested negative for retroviruses
- If the donor is not cooperative during the collection of blood it may be sedated with either Butorphanol @ 0.1- 0.2 ± Diazepam @ 0.5mg/Kg IV or Acepromazine @ 0.04 mg/Kg IM (to avoid interference with platelet function, donors should not be sedated with acepromazine) or Diazepam @ 0.2mg/kg + Ketamine @ 5mg/Kg (or 1:1 in ml) IV.

Anticoagulants for Collection of Blood

- Two anticoagulants namely Citrate-phosphate-dextrose-adenine-1 (CPDA-1) and Acid- citrate-dextrose (ACD) are commonly used for the storage of blood. Citrate is present in the solutions used to collect blood and acts as an anticoagulant by binding calcium. These preservatives help RBCs to maintain their viability during storage by supporting their glycolytic energy metabolism (adenosine triphosphate synthesis).
- CPDA-1 is considered better anticoagulant because it maintains higher levels of 2,3-disphosphoglycerate (2,3-DPG) and adenosine triphosphate (ATP) in collected blood. One ml of CPDA is required for the collection of 7 ml of whole blood. In CPDA-1 blood can be stored for approximately 35 days at 4°C whereas, ACD allows the storage of blood for 21 days. One ml of ACD is required per 4 ml of whole blood collected.

- Heparin is not recommended for blood collection because it activates platelets that potentiate the activity of anti-thrombin, resulting in the inactivation of thrombin.

Blood Collection

- Blood collection (Fig.181) is done over a 5-10 minute period. Blood is drawn aseptically from the jugular vein by gravity and a bag is usually placed on the blood collection monitor. The maximal blood volume to be collected from a healthy dog is 20 ml blood/kg or one regular blood bag unit of 350 ± 45 ml per ≥ 25 kg dog. Fifteen to 20% of the estimated blood volume can be safely donated.

  Estimated blood volume (litres) = 0.08-0.09× Body weight (Kg)

- The donor is positioned in lateral recumbency on the table and the venipuncture site (jugular vein) is clipped and prepared with surgical scrub to minimize the chances of bacterial contamination. A local anesthetic cream can be applied.
- The donor is allowed to settle down and secured properly in a lateral recumbency.
- The jugular vein is punctured with a needle provided with the bag.
- The progress of blood collection is monitored via the blood collection monitor
- Once the desired amount is collected, the blood line is reclamped close to the needle and the needle is removed from the donor.
- Pressure is applied digitally or with a pressure wrap on the venipuncture site until hemostasis is achieved.
- The collected bag is labeled with the date of collection, product type, donor PCV (or Hb) and blood group, donor identification, and date of expiry.
- The donors are monitored further for any signs of hypotension or hypovolemia, such as weakness, pale

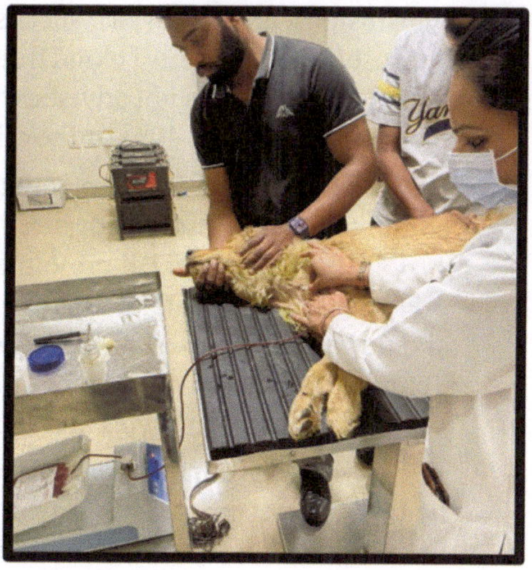

Fig. 181: Collection of blood from a donor in a CPDA bag.

mucous membranes, prolonged capillary refill time, weak pulses, altered mentation etc.

Calculating Blood Requirement

The amount of blood to be transfused can be determined by the following formula

Amount of blood required (ml)

$$= \frac{\text{Body weight. (Kg)} \times 90 \times (\text{Desired PCV} - \text{PCV of recipient})}{\text{PCV of donor}}$$

A basic guideline for small animals suggests that infusing 20 ml/kg of whole blood can raise the PCV by approximately 10%, assuming the donor's PCV around 40%.

For example, a recipient dog's body weight =10 kg, recipient's PCV = 10% and desired PCV=25%. Donor's PCV = 50%. Now, the amount of blood required for transfusion 10x90x (25-10)/50= 270 ml.

Blood Transfusion

- Blood should be infused through the cephalic or jugular vein of the recipient via an intravenous catheter (Fig.182). In severely hypotensive or paediatric patients, blood can also be administered through the intraosseous route (trochanteric fossa of the proximal femur). Marrow transfusions are nearly as successful as direct intravenous infusions because the absorption ofblood from the marrow cavity into the circulation is very efficient. It is not advised to infuse blood intraperitoneally as it is ineffective and only 40% of the administered blood is absorbed.
- Blood is administered intravenously through commercially available administration I/V sets with filters. Blood should be filtered using 150-170μm pore nonlatex filters either before or during administration to prevent clots and platelet aggregates from entering the recipient. Blood should be warmed to 37°C before administration to prevent hypothermia. Temperature should not be more than 37°C, because higher temperatures cause lysis of erythrocytes and inactivation of clotting factors.
- For the first 15 minutes, the rate of transfusion should be 0.5-1 ml/kg/hour, to detect immediate transfusion reaction. All vital signs should be monitored every 15 minutes in the first hour, followed by every 30 to 60 minutes afterward. Rapid transfusions can result in vomiting, or

muscle fasciculations. However, the rate of transfusion can be assessed individually depending on patient status (hydration status and severity of anaemia). For example, hypovolemic patients may require an infusion rate of 20ml/kg per hour, whereas patients with cardiac, renal, or hepatic disease or recumbent animals may require an infusion rate of only 1ml/kg per hour. The entire procedure should be completed within 4 hours to minimize bacterial contamination of the collected blood.

- In dogs, the half-life of erythrocytes transfused after matching is approximately 21 days. In cats, the half-life of erythrocytes transfused after matching is approximately 30- 38 days. In horses and cattle the survival time of compatible transfused erythrocytes is only 2-4 days.

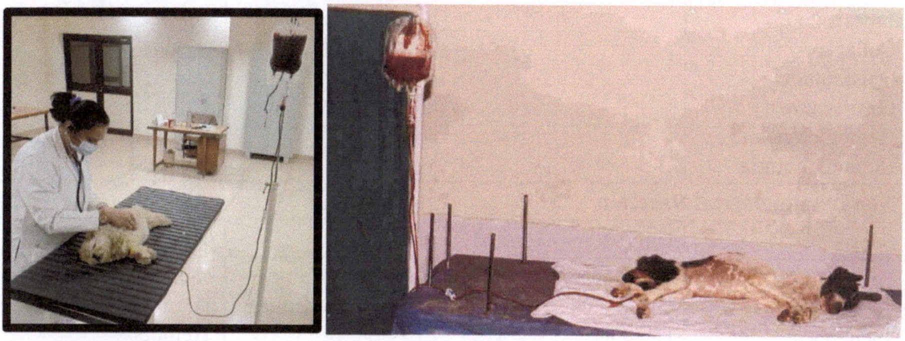

Fig. 182: Blood transfusion in dogs.

Hazards of Transfusion

- Even though transfusion is a lifesaving technique, potential risks are unavoidable. Transfusion reactions are classified as immunologic and non-immunologic reactions (Table 19).

Table 19: Common clinical signs of transfusion reaction

Immunological Reaction			
Haemolytic	**Hypersensitivity**	**Febrile**	**Delayed**
Fever Tachycardia or bradycardia Hypotension Dyspnea Cyanosis Salivation Lacrimation Urination Defecation Emesis Collapse Opisthotonus Cardiac arrest Hemoglobinemia Hemoglobinuria	Hives Edema Redness Vomiting Diarrhea Tachypnea Tachycardia Hypotension	Rise in temperature immediately or within 2 hours of the start of a transfusion	Anemia Fever Icterus
Non-immunological Reaction			
Sepsis	Circulatory overload	Hemolysis	Citrate intoxication
Fever Tachycardia Tachypnea Hypotension Hypoglycemia Hemoglobinemia Hemoglobinuria	Tachypnea Orthopnea Dyspnea Cyanosis Coughing Abnormal lung sounds Peripheral edema Jugular distention	Hemoglobinuria Hemoglobinemia	Face rubbing Vomiting Tremors Tetany Cardiac arrhythmia

Handling of Suspected Hemolytic Transfusion Reactions

- Stop the transfusion immediately.
- Maintain the intravenous line and administer crystalloid solutions.
- Monitor the animal for evidence of shock or disseminated intravascular coagulation,and check the patient's temperature. In addition, test serum and urine for evidence of haemoglobinemia and of haemoglobinuria.
- Administer supportive therapy. For anaphylactoid reactions immediately administer Adrenaline 5-10 ml of 1:1000 intravenously in large animals and 0.2-0.5 ml (1:1000) in small animals. Corticosteroids and antihistamines are given at standard dose rates.

- Check the blood bag for evidence of lysis by capillary tube centrifugation and collect a sample of blood for culture and sensitivity testing.
- Check blood typing or cross-matching.

Precautions

- Lactated Ringer's solution causes calcium chelation with citrate-containing anticoagulants and subsequent clot formation so its use should be avoided during blood transfusion.
- 5% dextrose in water causes swelling and lysis of erythrocytes and hypotonic saline fluids will lyse erythrocytes, so they are contraindicated.
- Excessive and rapid injection of blood or plasma can result in circulatory overload and heart failure.

Further Readings

Bell, G. (2006). Blood transfusions in cattle. UK Vet Livestock, 11(3), 39-43.

Hale, A. S. (1995). Canine blood groups and their importance in veterinary transfusion medicine. The Veterinary clinics of North America. Small Animal Practice, 25 : 1323–1332. https://doi.org/10.1016/s0195-5616(95)50157.

Helm, J.R., and Knottenbelt, C. (2010a). Blood transfusions in dogs and cats 1. Indications. In Practice, 32: 184 - 189.

Helm, J.R., and Knottenbelt, C. (2010b). Blood transfusions in dogs and cats 1. Practicalities of blood collection and administration. In Practice 32, 231- 237.

Lanevschi, A., and Wardrop, K. J. (2001). Principles of transfusion medicine in small animals. The Canadian Veterinary Journal 42: 447–454.

Yagi, K. and Holowaychuk, M. (2016). Manual of Veterinary Transfusion Medicine and Blood Banking, Wiley Blackwell.

28

Fluid Therapy

Fluid therapy is a critical intervention in Veterinary Medicine, requiring careful assessment and monitoring to ensure effectiveness and to minimize risks. Tailor made individualized approach is essential for optimal outcome.

Fluid Therapy in Small Animals

Indications

- Most common indication for fluid therapy is correction of dehydration and resuscitation from shock, hypokalemia, and metabolic acidosis.
- Less common indication for fluid therapy is in cases of acid-base disturbances (metabolic alkalosis or mixed acid-base disturbance), specific correction of other electrolyte disturbances such as sodium, potassium, calcium, magnesium.
- Other indication for fluid therapy is for the treatment of anemia and coagulopathies.

Body fluid compartmentalization

- Total body water (TBW) in adults is approximately 60% of body weight. While young animals (neonates 80% and pediatrics 75%) contain more water. Water is located in separate yet intertwined compartments within the body. The body water compartment is divided into intracellular and extracellular compartments.
- Intracellular fluid (ICF) is located within cells and contributes approximately two-thirds of TBW representing approximately 40% body weight on an average.
- Extracellular fluid (ECF) is located outside of cells and contributes one-third of TBW representing approximately 20% body weight on an average. The ECF comprises both the intravascular space and the interstitial space.
- The distribution of ECF between the interstitial and the intravascular spaces is about 75% to 25%, respectively. Thus, the interstitial fluid

comprises about 12% to 14% of the body weight, and the intravascular fluid comprises about 6% to 8% of the body weight.

- The intravascular space contains fluid within the blood vessel and contributes 8-10% of TBW.
- The interstitial extravascular compartment is the space located outside the blood vessel and contributes 24% of TBW.

Electrolyte distribution

- Sodium and chloride exist in high concentrations in ECF and low concentrations in ICF.
- Potassium, magnesium, and phosphorus exist in high concentrations in ICF and low concentrations in ECF.

Dehydration versus hypovolemia

- Dehydration refers to a decrease in TBW whereas hypovolemia largely refers to inadequate circulating intravascular fluid volume.
- Parameters used to determine hydration status should not be used to detect its intravascular volume status.
- Parameters for assessing dehydration are skin turgor, sunken eyes, dry mucus membranes and loss in body weight.
- Intravascular fluid volume and cardiac output largely determine organ perfusion. Parameters for assessing perfusion are capillary refill time (normally less than 2 second), mucus membrane color (moist and pink), arterial blood pressure, heart rate and urine output.

Estimation of dehydration

Dehydration and its Manifestations

The clinical signs in relation to degree of dehydration in dogs and cats are given in the table 20.

Table 20: Degree of dehydration and associated clinical signs in dogs and cats

Estimated degree of dehydration	Clinical signs
<5%	History of vomiting, diarrhoea or other fluid loss, normal mucus membrane.
5%	History of vomiting, diarrhoea or other fluid loss, tacky or dry mucus membrane, mild skin tenting.
7%	History of vomiting, diarrhoea or other fluid loss, dry mucus membrane, increased skin tenting, tachycardia, normal pulse quality and arterial blood pressure.
10%	History of vomiting, diarrhoea or other fluid loss, dry mucus membrane, increased skin tenting, tachycardia, decreased pulse pressure
12%	History of vomiting, diarrhoea or other fluid loss, dry mucus membrane, increased skin tenting, sunken eyes/dry cornea, tachy or bradycardia, hypotension, cold extremities, hypothermia and altered consciousness

Laboratory Assessment of Dehydration

Packed cell volume (PCV%) and total plasma protein (TPP, gm/dl) are helpful in evaluation of intravascular hydration and can be rapidly and inexpensively determined using microhematocrit tubes and a refractometer. Increased TPP and PCV provide documentation for intravascular dehydration.

Types of fluids

Fluids are broadly divided into two categories i.e., crystalloid and colloids

- Crystalloid fluids (Fig. 183) contain the solution of water with various forms of electrolyte, salt or sugar crystals, allowing them to redistribute quickly from the vascular space to the interstitial and intracellular fluid spaces. Thus, they exert their greatest effect in rehydration or replacement of prior fluid loss.

  Expansion of the vascular fluid space initially occurs, but with time, redistribution minimizes this effect.

  Crystalloid fluids can be classified according to osmolality by comparison to the animal's normal serum osmolality of approximately 300 mOsm/kg.

  Osmolality of commercial fluids is largely determined by sodium and glucose concentrations.

  Fluids with osmolality of less than 300 mOsm/ kg are hypotonic (e.g., 0.45% saline in water),

Fluids with osmolality of greater than 300 mOsm/kg are hypertonic (e.g., 5.0% dextrose added to 0.9% sodium chloride),

Fluids with osmolality near 300 mOsm/ kg are isotonic (e.g., lactated Ringer's solution, Plasma-Lyte 148, and Normosol-R).

Physiologic saline (0.9% sodium chloride) at 308 mOsm/kg is also within the isotonic range.

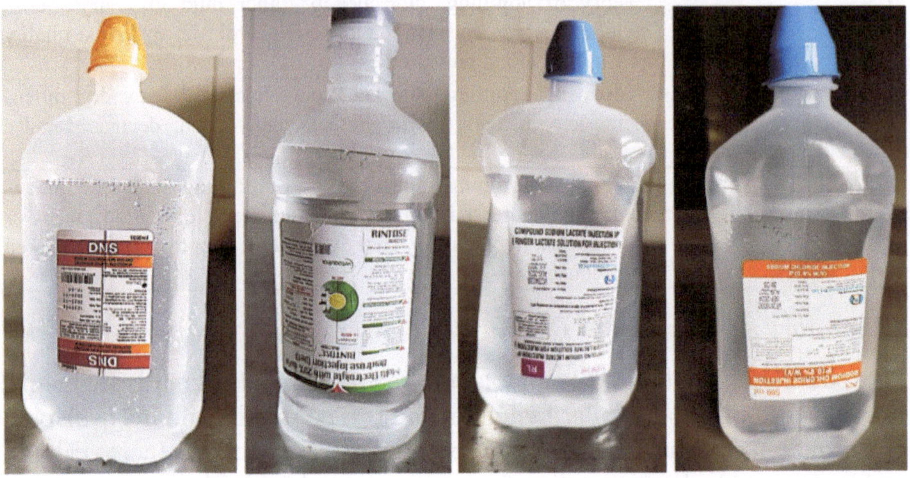

A B C D

Fig. 183: Different types of crystalloid fluid for parenteral infusion. DNS 5% (A), Multi electrolyte with 20% dextrose (B), Ringer Lactate solution (C) and normal saline solution (D).

A balanced crystalloid fluid is one that contains components very similar to extracellular space.

Crystalloid fluids can also be classified by their intended function for maintenance or replacement.

Replacement fluids contain component similar to extracellular fluid and in general, isotonic crystalloid fluids for replacement of fluid and electrolyte deficit have sodium concentration similar to plasma and the extracellular fluid compartment e.g. normal saline, lactated Ringers Solution, normosol-R, plasmalyte A.

Maintenance crystalloid solutions contain lower concentration of sodium and other compound than the extracellular space and used primarily to replace sensible and insensible fluid losses e.g. 0.45% sodium with 2.5% dextrose (half strength saline with dextrose)

- Colloidal solutions (Fig. 184) differ from crystalloids as they contain large macromolecules (natural or synthetic). These molecules, because of their size, do not readily pass through the normal endothelium, thus giving them oncotic pressure effects. This allows them to affect water movement from the interstitial fluid space.

In addition, the large net negative charge carried by these molecules attracts sodium and thus water from the interstitial spaces, which contributes to their volume-expanding effect. Depending on the clinical situation, colloid molecules exert most of their effect in the intravascular space, making them ideal for volume expansion of the intravascular space but less suitable for replacement of extravascular fluid losses.

Colloid solutions (Fig.184) are useful during the therapy of hypovolemia, septic shock, vasculitis, hypoproteinemia, and third spacing of fluid such as pleural and peritoneal effusions.

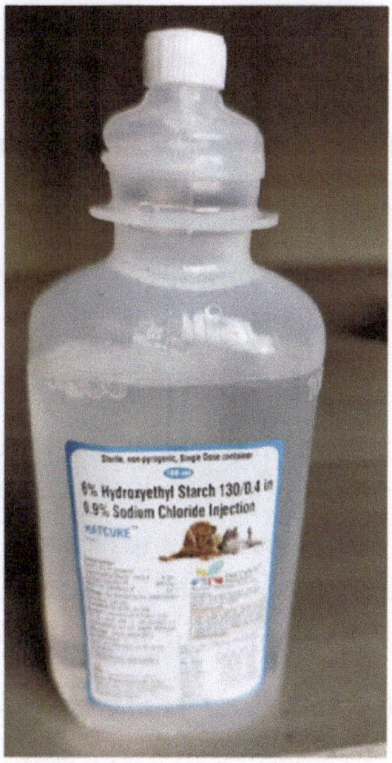

Fig. 184: Colloid fluid. 6% Hydroxyethyl Starch 130/0.4 in 0.9 % Sodium chloride solution.

Natural colloids are produced by the body, harvested from volunteer donors, and stored for later use. In small animal medicine, this group of fluids includes the blood transfusion components that possess albumin and other plasma proteins (whole blood and plasma).

Synthetic colloids include dextran 70, hydroxyethyl starch (hetastarch), pentastarch, and oxypolygelatin.

Calculation of fluid deficit

- After assessing the degree of dehydration, the volume of fluid to be administered to replace the fluid deficit can be calculated by the following formula:

  % Dehydration X Body weight (kg) X 1000 = ml fluid deficit

- The calculated fluid deficit is added to the animal's daily maintenance fluid requirement.

Calculation of maintenance fluid requirement

- Maintenance fluid requirement is defined as the volume of fluid (ml) and the amount of electrolyte (mEq or mg) that must be taken in daily to keep the volume of TBW and the electrolyte content normal.

- Obligatory losses of water and electrolytes occur daily as a consequence of normal metabolism. Water taken into the body in all of its forms is equal to the loss of water in a normal animal.
- Maintenance fluid requirement can be calculated by following formulas

  Cats – 80 x body weight in $(kg)^{0.75}$ per 24 hrs.

  Rule of thumb: 2-3 ml/kg/hr.

  Dogs

  a. 132 x body weight $(kg)^{0.75}$ per 24 hrs.

  b. 30 x body weight (kg) +70

  c. Rule of thumb: 2-6 ml/kg/hr.

Calculating fluid requirement for 24 hrs.

Fluid requirement for 24 hrs. = Deficit + Maintenance + ongoing losses

Sensible and insensible fluid losses

- Sensible fluid losses are those that can be measured and include fluid lost in the form of urine, feces, vomitus, and wound exudate.
- Urine output can be measured by collecting the urine in a closed collection system attached to the urinary catheter.
- Insensible losses are those that cannot be measured and include sweat, saliva, and excessive panting.

  Generally, insensible losses are estimated to be 20-30 ml/kg/day.

Rate of fluid administration

- The initial rate of fluid administration is dependent on the status of the patient. The guidelines for calculation of the required rate for various fluid therapies are as follows:

  Hypovolemic Shock- The initial rapid infusion in dogs @20-40ml/kg

  IV for 15 minutes. About half the volume for cats. Followed by 70-90ml/kg for dogs and 30-50ml/kg for cats over one hour. Then finally a maintenance rate of 10-12ml/kg/hr for dogs and 5-6ml/kg/hr for cats.

  Correction of Dehydration- Fluid volume needed (ml) = % Dehydration x Body weight (kg) x 1000.

  Maintenance rate of infusion is 2-6 ml/kg/hr. in dogs and 2-3 ml/kg/hr. for cats.

Fluids to be kept at hand

- Clinical correction of most fluid problems can be accomplished by maintaining only a few stock solutions, usually lactated Ringer's solution, Plasma-Lyte 148, or Normosol-R; 0.9% sodium chloride; and 5% dextrose in water.
- An alkalinizing basic electrolyte solution (lactated Ringer's solution, Plasma-Lyte 148, Normosol-R) is usually the fluid of choice in the absence of laboratory data until more information about electrolyte, osmolality, and acid- base status is available.
- A 0.9% sodium chloride solution is chosen when "extra" sodium or chloride is needed to maintain volume expansion or for the correction of metabolic alkalosis.
- Low-sodium fluids (0.45% saline in 2.5% dextrose, Plasma-Lyte 56, Normosol-M, or 5% dextrose in water) are chosen when treating sodium intolerant patients or patients with a water deficit (hypernatremia).
- It is useful to have one synthetic colloid solution available and should be administered in conjunction with crystalloids when there is lowered oncotic pressure or when rapid (and low volume) intravascular expansion is desired.

Routes of administration for parenteral fluids

- The route of fluid therapy depends on the nature of the illness, its clinical severity, its onset (acute versus chronic), the nature and magnitude of ongoing losses, and the composition of fluids to be given. The availability of personnel and equipment required for monitoring during IV therapy also influences decisions about the appropriate route of administration.
- *Subcutaneous Route*-This route (SC) is used for correction of mild to moderate dehydration using isotonic fluids. Subcutaneous route should not be opted for the emergency replacement of fluid in critically ill or severely dehydrated patients.

  Subcutaneous administration of 5% dextrose in water as an isotonic solution should be avoided in cases of severe dehydration.

  Subcutaneous fluid can be given by gravity flow through IV administration tubing or by direct injection from a large-volume hypodermic syringe. Subcutaneous infusion should be preferred on trunk so that fluid does not gravitate into limbs. Areas with surgical wounds should be avoided for subcutaneous infusion.The volume of fluid administered subcutaneously should not be more than 10-12 ml/kg per injection site.

- *Intravenous Route-* Intravenous fluid therapy is the preferred choice to treat severe hypovolemia, hypotension, and dehydration. Hypotonic, isotonic, and hypertonic fluids can be given by the intravenous (IV) route. Intravenous route for infusion should always be preferred when large volumes of fluid and rapid administration of fluid is desired.The jugular vein and cephalic vein are most commonly chosen for indwelling IV catheterization. The lateral saphenous and femoral veins also may be used. In cats and small dogs use 20 to 22-gauge catheters; in medium and large dogs use 14 to 20-gauge catheters.
- *Intraperitoneal Route-*This route of fluid infusion may be chosen in severely anemic puppies and kittens and for rewarming hypothermic animals. Isotonic to mildly hypotonic fluids may be used for rehydration. Hypertonic fluids worsen the dehydration. This route is not preferred if the patient has abdominal sepsis, ascites, peritonitis or recent abdominal surgery.
- *Intraosseous Route-* This route provides rapid access to circulation when venous catheterization is not feasible. Intraosseous (IO) fluid administration is comparable in effectiveness to the IV route. The bone marrow of the femur, tibia, or humerus is catheterized with a bone marrow or spinal needle and secure it in place. It is particularly useful in puppies and kittens and in emergency situations in which immediate intravenous access is difficult. Blood and crystalloid solutions can be infused safely by this route.

General guidelines for fluid administration

- Always prefer a new IV line and bag for each patient, regardless of route of administration.
- Ensure lines are primed to avoid air embolism.
- Fluid pumps and gravity flow systems require frequent monitoring. Check patients with gravity flow systems more frequently because catheter positioning can affect rate.
- If the infusion is being given by gravity flow, appropriate size bag (as per patient size particularly in small patients) should be selected to minimize risk of inadvertent overload.
- A buretrol (a fluid volume limiter that can be attached to an IV bag) can be used if frequent fluid composition changes are anticipated to reduce changing entire bag.
- Consider using T-ports to easily medicate a patient receiving IV fluids and Y-ports in animals receiving more than one compatible infusion.

- Consider using a syringe pump to either infuse small amounts of fluids or to provide a constant rate infusion.
- Consider a pressure bag for the delivery of boluses during resuscitation.
- Always perform aseptic catheter placement using wide clipping of hair over the vein and a standard surgical scrub. After securing the catheter in the vein, place a gauze sponge with an antimicrobial cream over the puncture site.
- Catheter complications are thrombophlebitis, thromboembolism, bacteremia, bacterial endocarditis, and catheter-fragment foreign body.

Catheter maintenance and monitoring

- The catheter should be placed aseptically and catheter site should be kept clean.
- The catheter site is checked once to twice daily. If there is no redness, swelling, pain, or patient discomfort (when injecting), the catheter may be left in place.
- Routine catheter changes at 72 hours are no longer recommended.
- Monitor the patient for fever, leucocytosis, and heart murmurs.
- Ensure that all injections made into the IV catheter or lines are done aseptically.
- Ensure that the IV line remains a closed and aseptic system. Do not disconnect the IV line unless absolutely necessary, and ensure aseptic technique is followed when changing IV fluid bags.
- When the catheter is not in use, flush regularly to prevent clotting.

Supplementation to parenteral fluid

- Supplementation of potassium (Table 4) to fluids is recommended if the serum potassium concentration is less than 3.5 mEq/L.The modified Sliding Scale of Scott recommends addition of potassium chloride to maintenance fluid volumes proportional to the degree of hypokalaemia as given in the table. When frequent monitoring of serum potassium is not practical, a fixed concentration of potassium from 20 to 30 mEq/L can be given in maintenance fluids. This concentration corresponds to the lower end of the Scott scale.
- Supplementation of potassium should not exceed 0.5 mEq/kg/hr to lessen the risk for development of hyperkalemia or cardiotoxicity. Electrocardiographic monitoring is needed during supplementation of fluids with potassium in animals.Animals with severe emaciation

should be less vigorously supplemented with potassium in their fluids. Less vigorous potassium supplementation and careful monitoring is indicated in reduced renal function and azotemia with oligouria cases to avoid hyperkalaemia.

Table 21: Supplementation of potassium to fluids

Serum K (mEq/L)	mEq KCl added to 250 ml	mEq KCl added to 1000ml	Maximum Infusion Rate (0.5 mEq/kg/hr.)
<2.0	20	80	6 ml/kg/hr
2.1–2.5	15	60	8 ml/kg/hr
2.6–3.0	10	40	12 ml/kg/hr
3.1–3.5	7	28	18 ml/kg/hr
>3.5 < 5.0	5	20	25 ml/kg/hr

- Supplementation of calcium to fluid is recommended in cases of hypocalcemia.Calcium gluconate and calcium chloride can be supplemented in replacement and maintenance fluids for the correction of symptomatic hypocalcemia. Calcium gluconate is the calcium salt of choice because it is non-irritating if the solution is injected outside of a vessel. In contrast, calcium chloride is extremely irritating to tissues but provides more elemental calcium in each milliliter of solution. Severe symptoms of ionized hypocalcemia (tetany and seizures) can be treated by administering 0.5 to 1.5 ml/kg of 10% calcium gluconate intravenously over 10 to 20 minutes while monitoring a continuous electrocardiogram. Calcium salts are not added to fluids containing lactate, acetate, bicarbonate, or phosphates owing to precipitation of calcium. Alkalinizing fluid therapy containing sodium bicarbonate decreases ionized calcium and may precipitate clinical signs of hypocalcaemia in patients with borderline hypocalcaemia. Therefore, alkalinizing fluids should be avoided in cases with borderline hypocalcaemia. Calcium salts should not be administered subcutaneously as their subcutaneous administration may cause severe skin necrosis and calcification
- Supplementation of phosphorus is recommended in cases of hypophosphatemia. Hypophosphatemia is present when the serum concentration is < 2.5 mg/dl. Mild hypophosphatemia (2.0– 2.5 mg/dl) is common and often transient. Hypophosphatemia often resolves quickly during therapy directed at the underlying cause. Severe hypophosphatemia (<1.5 mg/dl) is uncommon but can be life threatening (if phosphorus level is < 1.0 mg/dl) requiring phosphorus supplementation. Clinically significant hypophosphatemia that requires phosphate replacement therapy is most likely to be encountered in

patients with diabetic ketoacidosis 12 to 48 hours after initiation of insulin therapy or in patients in whom enteral nutrition is initiated after prolonged periods of anorexia. When necessary, sodium phosphate or potassium phosphate can be supplemented to replacement and maintenance fluids for the correction of symptomatic or severe hypophosphatemia. Phosphate salts should not be added to fluids containing calcium. Hypophosphatemia with its concurrent likely cause requires treatment. The phosphate supplementation can be done @ 0.01 to 0.03 mmol/kg/hour, intravenously, for 3 to 6 hours. Severe cases may require prolonged supplementation or higher dosages of phosphate (0.06–0.12 mmol/ kg/hour intravenously). Supplementation is continued only until the serum phosphorus concentration is greater than 0.65 mmol/L (2.0 mg/dl).

- Supplementation of magnesium is indicated in patients with documented hypokalemia especially resistant to supplementation efforts.As magnesium is primarily an intracellular ion, its serum measurements are often not reflective of total body stores. It is generally thought that low total serum magnesium is reflective of low total body magnesium status, but this is often a late finding. Patients with diabetic ketoacidosis, congestive heart failure, and ventricular arrhythmias are likely to get benefit from magnesium supplementation. Magnesium sulphate and magnesium chloride are the commonly available sources of magnesium. Magnesium sulphate is preferred to prevent excess of chloride when magnesium chloride is administered. The dosage of magnesium sulphate supplementation is 2.5 to 5 mg/kg/hour and can be added to most commercially available fluid types. A loading dose of 30 mg/kg of magnesium sulphate can also be administered over 20 minutes to 1 hour prior to beginning the lower rate of hourly supplementation. Magnesium supplementation can be extremely helpful in achieving and maintaining high normal levels of potassium when treating cardiac arrhythmias. Magnesium supplementation should not be done in oliguric patients as this is the primary route of excretion of excessive amounts.
- Supplementation of alkali to fluids is needed for the correction of metabolic acidosis. Supplementation of bicarbonate is most safely accomplished when blood gas analysis is present to determine the contribution of the respiratory system to the acid-base disturbance. Alkali replacement is calculated from either the bicarbonate value (available from blood gas analysis) or the total CO_2 value (serum biochemical profile). Total CO_2 (mEq/L) can be substituted for bicarbonate in calculations since the difference is usually of 1 to 2 mEq/L. Bicarbonate (HCO_3) deficit can be

assessed by substracting measured HCO_3 from normal HCO_3 level.The dose of bicarbonate is calculated using following methods.

(a) Method 1- HCO_3 deficit x 0.6 x body weight (kg) = "missing" mEq bicarbonate from TBW at the time of sampling. Administer total calculated replacement mEq over 8 to 12 hours or longer; it may take as long as 18 hours for administered bicarbonate to equilibrate inside cells.

(b) Method 2- HCO_3 deficit x 0.3 x body weight (kg) = "missing" mEq bicarbonate from ECF at the time of sampling. This formula is used when multiple injections of bicarbonate are anticipated, ongoing production of acid or loss of alkali is anticipated, and access to multiple HCO_3 measurements is possible. Without access to blood gas analysis or total CO_2, it is difficult to know with any certainty the magnitude of need.

(c) Empirical method-If the clinical diagnosis and suspicion for metabolic acidosis is high, bicarbonate may be administered empirically. Mild acidosis: administer 3 mEq/kg bicarbonate over 24 hours. Moderate acidosis: administer 6 mEq/kg bicarbonate over 24 hours. Severe acidosis: administer 9 mEq/kg bicarbonate over 24 hours

In emergency situations bicarbonate can be administered @ 0.5 to 1.0 mEq/kg of sodium bicarbonate over 5 to 10 minutes followed by re-evaluation of bicarnonate. Electrocardiographic monitoring is desirable during the injection.Precautions need to be taken during bicarbonate administration. Sodium bicarbonate should not be added to fluids containing calcium (lactated Ringer's solution), as it can precipitate calcium. Administration of sodium bicarbonate can result in hypernatremia, hyperosmolality, hypertension, overhydration, and volume overload if the administered sodium is not excreted.Overdosing with alkali can result in alkalosis, paradoxical cerebrospinal fluid acidosis, shifts of ionized calcium, a shift of the oxy-hemoglobin dissociation curve (decreasing oxygen unloading), and seizures.

- Supplementation of dextrose is indicated in hypoglycaemia due to sepsis, insulin overdosing, insulinoma, or liver disease. Dextrose supplementation at greater than 10% to 15% is necessary to provide sufficient calories for anorectic animals. To increase the dextrose concentration by 2.5%, 25 g of dextrose or 50 ml of 50% dextrose is added to 1000 ml of stock solution (e.g., lactated Ringers solution). Long-term supplementation of dextrose through a peripheral vein should not exceed concentrations over 5%, as the hypertonicity can induce significant phlebitis.

Fluid therapy in Large Animals (Ruminants and Horses)

- Under field conditions, laboratory evaluation of animals is not feasible. Polyionic crystalloid solutions are in general use. These solutions usually contain sodium, potassium, chloride, and calcium or magnesium at a concentration similar to the electrolyte composition of extracellular fluid. The solutions may also contain lactate or acetate as bicarbonate precursors. Dextrose may be added to the solution to make an initial mildly hypertonic solution.
- Polyionic crystalloid solutions are safe and can be used in large quantities without inducing electrolyte disturbances provided that circulating blood volume and renal function have been restored and are maintained. They can be used for most situations of dehydration and moderate academia or alkalemia and moderate electrolyte imbalances.
- Lactated Ringer's solution is the standard intravenous fluid for neonates and adult horses, because these animals tend to become acidemic when anorectic. Ringer's solution is the standard intra- venous fluid for adult ruminants because these animals tend to get alkalemic when anorectic.
- Isotonic saline (0.9% NaCl solution) is an isotonic crystalloid solution that has little merit in the routine treatment of sick ruminants, principally because ruminants usually develop hypocalcemia and hypokalemia when anorectic. The use of 0.9% saline solution should be confined to horses, the irrigation of surgical sites and wounds, or as a vehicle for adding other electrolytes and dextrose.
- Hypertonic saline solution and an oral water load represent a practical and inexpensive alternative to parenteral administration of large fluid volumes.
- The use of small volumes (4–5 ml/kg BW) of hypertonic saline solution, ranging in concentration from 7.0% to 7.5%, has been extensively evaluated for the treatment of various forms of haemorrhagic, septic, and endotoxic shock.
- Hypertonic saline solution is widely used for the treatment of dairy cattle with endo- toxic shock and endotoxemia associated with coliform mastitis. Affected cows are given 2 L of hypertonic saline (4–5 ml/kg BW) intravenously, followed by immediate access to drinking water and other supportive therapy. Cattle that do not drink water within 10

minutes of hypertonic saline should have 5 gallons of water pumped into their rumen. Hypertonic saline should never be given alone without providing the animal fresh water to drink or oro-ruminal administration of water.

- Hypertonic saline solutions are widely used for the initial resuscitation of critically ill horses that are undergoing abdominal surgery for colic.
- Hypertonic or isotonic sodium bicarbonates, such as 8.4, 5.0 (hypertonic), or 1.3% (isotonic) solutions of sodium bicarbonate, are used for severe strong ion (metabolic) acidosis and hyponatremia, particularly in dehydrated depressed calves with diarrhoea.
- Fifty percent dextrose solutions are commonly administered to ruminants with ketosis or hypoglycemia and produce a transient increase in cardiac contractility.
- Calcium borogluconate or calcium magnesium borogluconate (Fig.185) is the standard treatment for milk fever (hypocalcemia) in cattle.

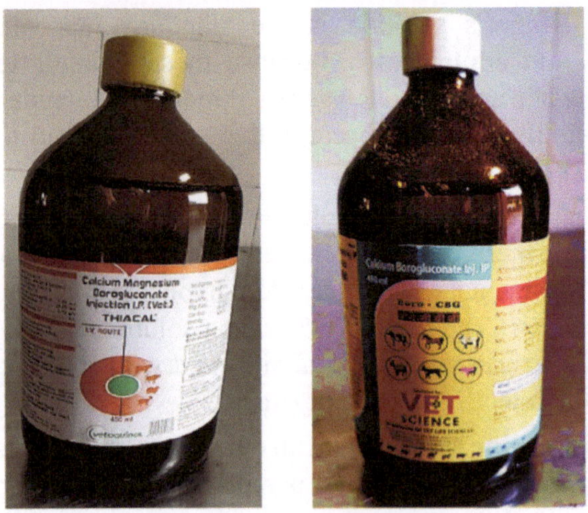

A B

Fig. 185: Calcium and magnesium solutions. Calcium magnesium borogluconate solution (A) and Calcium borogluconate solution (B) for the treatment of milk fever and magnesium deficit.

- Chloride-containing acidifying solutions, such as Ringer's solution, are used for treatment of strong ion (metabolic) alkalosis.
- Isotonic potassium chloride is used to treat severe hypokalemia in cattle (serum K^+ concentration <2.3 mEq/L) with severe muscle weakness or recumbency. This is usually seen in dairy cattle following the administration of isoflupredone acetate, a corticosteroid used to

treat ketosis in lactating cows. Recommended treatments involve the intravenous administration of isotonic potassium chloride (11.5 grams of potassium chloride per litre of sterile water) at a rate of 4 ml/kg/hour and combined with large doses of oral potassium salts (i.e. 200 grams of KCl per day), for correcting hypokalemia.

- The fluids are usually given for hydration or maintenance purpose. Hydration therapy is given in the first 4 to 6 hours @ 100 to 150 mL/kg BW intravenously. It is followed by maintenance therapy (a combination of continuous losses and maintenance requirements) that is given in the next 20 to 24 hours, depending on the severity and the course of the disease @ 60 to 80 mL/kg BW per 24 hours intravenously (approximately 3–4 mL/kg BW per hour). In some cases of profuse diarrhoea, the continuous losses and maintenance requirements will be about 150 ml/kg BW over a 24-hour period.
- The administration of large quantities of fluids intravenously to farm animals (Fig. 186) is best done with an indwelling jugular vein flexible catheter (10–14 gauge) or that is appropriately secured to the animal's neck to prevent withdrawal from the vein. Standard aseptic technique must be used. With a 14-gauge catheter, 20 L of fluids can be delivered as hydration therapy to a mature horse or cow over 4 hours.

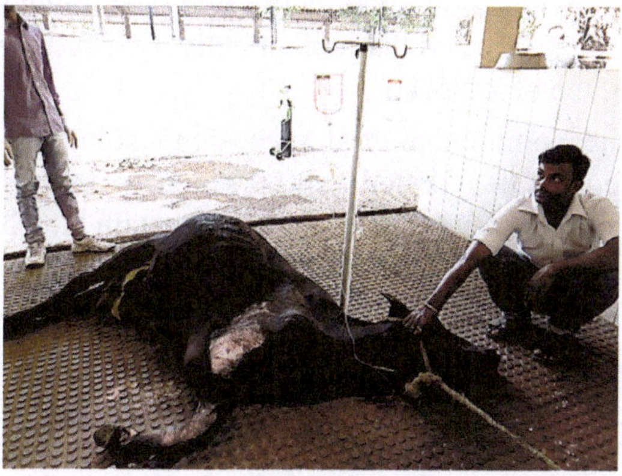

Fig. 186: A recumbent cow is being given hydration fluid therapy through jugular vein.

- Intravenous fluids are also administered to adult cattle and calves using the auricular vein. The auricular vein of neonatal calves can be successfully catheterized with an over the needle, 2.5 cm long, 22 gauge butterfly catheter, after clipping the external pinna and applying a tourniquet at the base of the ear to facilitate visualization of blood vessels and catheter advancement. The auricular vein of adult cattle can be successfully catheterized with an over the needle, 5 cm long, 14 gauge catheter, permitting 20 L of rehydration solution to be delivered over 4 hours.

- Mature cattle affected with metabolic alkalosis associated with diseases of the abomasum are usually hypokalemic, hypochloremic, and dehydrated. For such cases, a balanced electrolyte solution containing sodium (135–155 mEq/L), chloride (150–170 mEq/L), and potassium (10–20 mEq/L) is effective.
- Ringer's solution, 0.9% NaCl, or hypertonic saline along with oral potassium can be given for correction of strong ion (metabolic) alkalosis with hypochloremia and hypokalemia in adult anorectic cattle even without laboratory evaluation.
- Oral electrolyte solutions are used to replace fluid losses and correct electrolyte abnormalities in adult ruminants because they are cheap and easy to administer on farm. Generally, most dehydrated cattle have a metabolic alkalosis, it is important to use a non-alkalinizing oral electrolyte solution that does not contain bicarbonate, acetate, or propionate. In practice, this is accomplished with chloride-rich, high potassium solutions. By simply adding NaCl (7 grams/L), KCl (1.25 grams/L) and $CaCl_2$ (0.5 grams/L) to a liter of water (or 140 grams NaCl, 25 grams KCl, and 10 grams $CaCl_2$ in 20 liters or roughly 5 gallons of water), a non-alkalinizing oral electrolyte solution for adult ruminants can be created that will effectively rehydrate animals without alkalinizing the blood pH.
- Most of the commercial oral electrolyte solutions made for cattle are designed for dehydrated calves which typically have a metabolic acidosis. The majority of these products contain bicarbonate and are not indicated for use in adult ruminants.
- Oral fluid therapy provides an effective and inexpensive treatment in horses with impaction of the large colon and dorsal displacement of the colon. An absolute requirement for oral fluid therapy in the horse is that there is no gastric reflux. Generally, 6 to 8 L of water can be administered by nasogastric tube and funnel (gravity flow) every 15 to 20 minutes, and the administered fluid is rapidly transported to the large intestine. Fluid therapy recommendations for horses are given in the table (Crabtree and Epstein, 2021).

Table 22: Fluid Therapy Recommendations for adult horses (crabtree and epstein, 2021)

Route	Indication	Fluid type	Dose	Comments
IV	Resuscitation	Isotonic crystalloid	10–20 ml/kg bolus repeated as necessary to stabilize	Used in goal directed fashion with re-assessment of perfusion markers following each bolus. Caution with anuric/oliguric renal failure or heart disease.
		7.2% hypertonic saline	4 ml/kg bolus	Administration must be followed with isotonic crystalloids. Monitoring of electrolytes warranted with prolonged use.
		Hydroxyethyl starches	10 ml/kg/day	Higher rates associated with higher risk of coagulation derangements. Cannot monitor response with TP assessment due to falsely low reading on refractometer.
		Plasma	[(TPdesired – TPpatient)/ TPdonor] × 0.05 BW	Monitor for transfusion reactions.
	Maintenance	Isotonic crystalloid	40–60 ml/kg/day	This requirement likely reduced in an adult horse off-feed, although exact requirements in these cases unknown.
IG	Maintenance	Plain water or Electrolyte solution	40–60 ml/kg/day Administered as 4–6 L every 4–6 h or as a constant rate infusion	Typical electrolyte solution recipe: 5.27 g NaCl (table salt), 0.37 g KCl (lite salt), and 3.78 g $NaHCO_3$ in 1 L water. Volumes as high as 8–10 L per bolus reportedly well-tolerated. Volumes over twice maintenance for ingesta hydration not of any additional benefit.
	Fecal hydration	Plain water or Electrolyte solution	Up to 2× maintenance (i.e., 80–120 ml/kg/day) Administered as 4–8 L every 2–4 h or as a constant rate infusion.	
PR	Maintenance	Plain water	5 ml/kg/h *via* constant rate infusion	Plain water reportedly better tolerated than electrolyte solution.

Note. *IV- intravenous; IG- intragastric; PR- per rectum; GI- gastrointestinal; TP- total protein; PCV- packed cell volume; BW- body weight in kg.*

Table. 23. Summary of disturbances of body water, electrolytes, and acid-base balance in some common diseases of cattle and horses, and suggested fluid therapy.

Disease	Major abnormalities and deficits	Flue and electrolyte requirements
Neonatal calf diarrhea (including piglets and lambs)	Metabolic acidosis, low plasma bicarbonate, severe dehydration, loss of sodium, hyperkalemia when acidosis severe	Equal mixtures of isotonic saline and isotonic sodium bicarbonate with 5% dextrose, balanced electrolytes, intravenous and orally
D-lactic acidosis (carbohydrate engorgement of ruminants)	Metabolic acidosis, low plasma bicarbonate, severe dehydration	Sodium bicarbonate initially followed by balanced electrolytes, intravenously
Acute diffuse peritonitis	Dehydration, slight metabolic alkalosis caused by paralytic ileus	Balanced electrolyte solutions in large quantities intravenously for hydration and maintenance
Right-side dilatation/abomasal volvulus of cattle, abomasal impaction (dietary or vagal nerve injury).	Metabolic alkalosis, marked hypochloraemia, hypokalemia,severe dehydration	Balanced electrolyte solutions or high-potassium and chloride-acidifying solution, intravenously; may give acidifying solutions orally; can also use mixture of 2 L of isotonic saline (0.9%), 1 L isotonic potassium chloride (1.1%), and 1 L isotonic dextrose (5%)
Per acute coliform mastitis	Severe dehydration, mild electrolyte deficits including mild hypocalcaemia, metabolic acidosis if diarrhoea present	Balanced electrolyte solutions intravenously in large quantities for hydration and maintenance for 24—48 hours (100—150 mL/kg BW per 24 hours)
Acute diarrhoea in the horse (enteric salmonellosis)	Severe dehydration, marked hyponatremia, metabolic acidosis, hypokalemia occurs following bicarbonate therapy.	Hypertonic sodium bicarbonate (5%) 3–5 L/500 kg BW followed by high-sodium, high-potassium alkalinizing solution to correct hypokalemia following bicarbonate therapy, all by the intravenous route
Acute grain engorgement in the horse	Metabolic acidosis, dehydration, and shock	Hypertonic sodium bicarbonate (5%) 3–5 L/500 kg BW followed by balanced electrolytes intravenously
Water and electrolyte deprivation, oesophageal obstruction in horses	Moderate dehydration	Balanced electrolytes intravenously, when obstruction relieved, provide electrolyte solution orally
Acute intestinal obstruction	Metabolic acidosis or alkalosis dependent on level of obstruction, severe dehydration in horse, moderate in cow	Isotonic sodium bicarbonate initially, 3–5 L/500 kg BW followed by balanced electrolytes intravenously, horses may develop hypokalaemia following bicarbonate therapy and must be given potassium chloride

References

Article. :648774.doi:10.3389/fvets.2021. 648774.

Crabtree,N.E. and Epstein,K.L. (2021).Current concepts in fluid therapy in horses. Vet.Sci. 8. 1-16.Publisher: Frontiers Media S.A. Abstract; Full text

29

Ultrasonography for the Diagnosis of Various Diseases of Ruminants, Horses and Dogs

Ultrasonography, popularly known as sonography, is a non-invasive, radiation free and safe medical procedure being increasingly used for the imaging different body organs/ structures for the diagnosis of various diseases not only in humans but also in farm, companion and exotic pets. The technique involves ultrasound unit, patient and doctor/veterinarian (trained in sonography).

Ultrasound Unit

Ultrasound unit has processor (central processing unit commonly known as CPU), transducers, key board, data storage and printer (Fig. 187).

- The CPU (central processing unit) is basically a computer having microprocessor, memory, amplifiers and power supplies for the microprocessor and transducer. The CPU transmits electrical currents to the transducer to emit sound waves, and receives the returning echoes. Returning echoes are processed to form the image on the monitor. The processed data and/or image are also stored on the disk.
- Transducer or probe, other important component, has dual function of transmitting sound waves at a definite frequency as well as capturing the returning echoes. Linear and curved transducers are generally used. A linear transducer produces linear scan and a curved transducer produces an arc-shaped image. Probes are categorized according to ultrasound frequency produced as 2.5 MHz to 50 MHz and are selected as per species of animal to be examined, and area to be scanned.
- The returning echoes (received by the transducer) are converted into visible dots forming the anatomic image on the ultrasound screen.

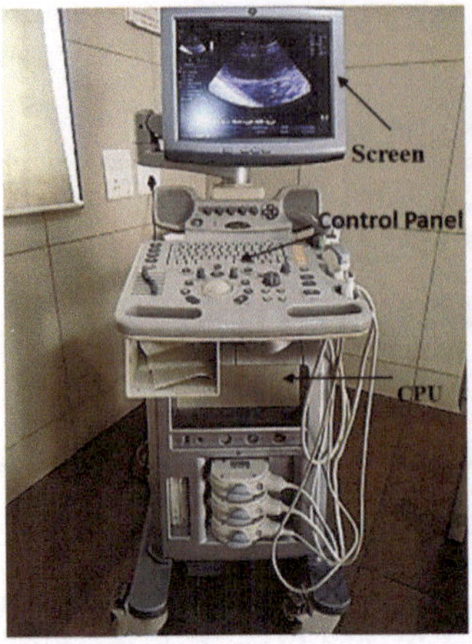

A

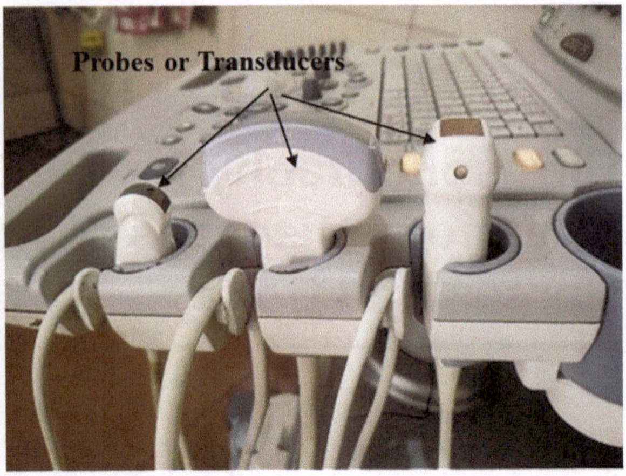

B

C

Fig. 187: Ultrasonography unit. Ultrasonography machine showing screen, control panel and CPU (A-B), probes (C) and printer.

Principles of Ultrasonography

- It uses the principle of "Doppler effect " or echoes to convert the reflected sound energy into image.
- The production of ultrasound waves is based on "pulse-echo-principle".

Piezo crystals, in transducer, are the source of ultrasound waves. A narrow beam of high frequency sound wave is sent into the soft tissues to be examined. The sound waves interact with the tissue interfaces. Some of the sound waves are reflected back to transducer as echoes. These echoes are processed by the computer of central processing unit and transformed into an image that is displayed on the screen of the ultrasound machine.

Patient Preparation

- For abdominal ultrasonography, a fast of 8-12 hrs is desirable to minimize abdominal gas.
- Drinking of the water should be encouraged prior to sonography of urinary bladder and prostrates so that urinary bladder is full.
- The site for sonography is prepared by shaving and cleaning the area in dogs and cats.
- In large animals for transrectal sonography, rectum needs to be cleaned off faecal material.
- Acoustic gel is applied on the area to ensure good contact.

Imaging Approaches

- Ultrasonographic imaging can be done through transabdominal, transthoracic, transurethral, transvaginal, transrectal and miscellaneous approaches. Of these, transrectal approach is commonly used in large animals and horses.
- Transabdominal approach (Fig.188) is taken for imaging abdominal wall, abdominal organs such as liver, gall bladder, spleen, stomach, intestines, pancreas, kidneys, urinary bladder, prostates, urethra, uterus in small animals.
- Transthoracic approach is taken for imaging heart, major vessels, lungs, diaphragm.
- Transurethral and transvaginal approaches are meant for imaging mucosal surface of urethra, bladder, vagina, cervix.
- Transrectal – This approach is used in large ruminants and equines for pregnancy diagnosis and evaluating reproductive and urinary tract.
- Miscellaneous – It is used for scanning tendons, ligaments, teat, udder, eyes.

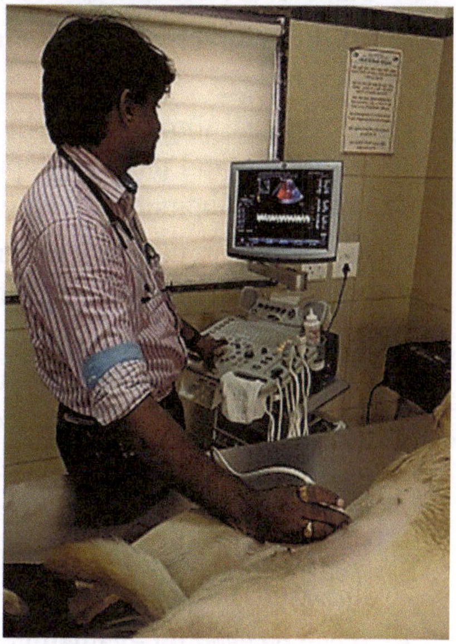

Fig. 188: Transabdominal ultrasonography in a dog.

Applications of Diagnostic Ultrasound

- It is also complementary to radiography and can provide more specific information.
- It can assist in the differential diagnosis of abscess, cysts, tumours and haematomas.
- Growth in stomach, intestines, liver, spleen, kidney, prostate, ovaries, and uterus can be detected.
- Identification of architecture (size, shape and location) of a mass or organ is of diagnostic value.
- Focal pathological area within parenchymal organs can be detected.
- Aspiration or needle biopsy is an important application of diagnostic ultrasound.
- Diagnosis of portosystemic shunts in liver of companion animals can be made that is otherwise not feasible.
- Ultrasonography has facilitated the diagnosis of gall bladder diseases.
- It can be used for the diagnosis of liths/crystals in kidneys, urinary bladder, urethra and gall bladder.
- Ultrasonography can be employed in evaluation of abdominal blood vessels.
- In females, ultrasonography can be used to evaluate uterine anatomy, involution and pathology, pregnancy, embryonic mortality, monitoring foetal development and identifying foetal sex or their abnormalities.
- M- mode or two dimensional echocardiography facilitates assessing cardiac size, thickness of ventricle wall, valves, septa, great vessels and cardiac contractility.
- Echocardiography can be used to diagnose pericardial effusions or masses (intra cardiac or pericardial).
- Ultrasonography can assist in differential diagnosis of ascites and haemoabdomen.
- It can also be used for evaluating soft tissues (muscles, tendons, ligaments).
- It can be used for evaluating adrenal glands and abdominal lymph glands
- Ultrasound guided nerve block is a new application of ultrasonography and intends to revolutionize the field of regional anaesthesia.
- It is the best technique to distinguish solids from fluid filled cavities.

- Transabdominal ultrasound has been used in evaluation of location, diameter of lesion, intestinal wall and motility, and intraluminal contents of the intestines in the goats with John's disease (Thrawat *et al.,* 2012).
- It has also been used in evaluation of animals with peritoneal or pleural fluids.

Limitations of Ultrasonography

- Examination of air-filled organs such as lungs or air sacs is difficult.
- Bone examination is not feasible.
- Examination of brain and spinal cord is also not feasible.
- In large animals, visualization of stomach and intestines is limited because of large amount of gas and ingesta.

Precautions

- It is operator dependent technology. Expertise is a prerequisite.
- Casual interpretation defeats the purpose.

Indications of Ultrasonography for the Diagnosis of Diseases in Dogs

- Liver Diseases: The liver is scanned for size, congestion, vessels, abscess, cyst, tumor, echotexture (hypoechoic, hyperechoic), and portosystemic shunts.
- Gall bladder Diseases: The gall bladder is scanned for detecting cholecystitis, sludge, mucocele, bile duct obstruction, calculi and mass.
- Splenic Diseases: The spleen is scanned for size, infarction, splenic vessels, mass (tumor), hypertrophy, and congestion.
- Esophageal Diseases- The esophagus is scanned for detecting mega esophagus, diverticulum, and foreign bodies.
- Gastric Diseases : The stomach is scanned for detecting foreign bodies, ulcers, dilatation, and volvulus.
- Intestinal Diseases: Intestines are scanned for foreign bodies, intussusceptions, obstruction, mucosal pattern, emptying time, intestinal wall thickness etc.
- Pancreatic Diseases: The pancreas is scanned for pancreatitis, and tumors.
- Renal failure and Renal Diseases: Kidneys are scanned for tumors, nephroliths, hydronephrosis, nephritis.
- Urinary Diseases: The urinary bladder is scanned for cystitis, cystoliths, bladder volume, tumor, cysts etc.

- Prostatic Diseases: The prostate glands are scanned for its size, inflammation, hyperplasia, tumor, abscess etc.
- Female Reproductive organs – Female reproductive tract is generally scanned for ovarian follicular dynamics, pregnancy detection, foetal wellbeing, foetal sexing, ovum pick up technology and detecting diseases of uterus and ovary.

Indications of Ultrasonography for the Diagnosis of Diseases in Ruminants

- Ultrasonography in cattle and buffaloes is mainly used in the field of reproduction and lameness. Per rectal sonography (Fig.189) is done for the diagnosis of pregnancy, diseases of reproductive tract, urinary tract, bony pelvis, portions of the small and large intestines.

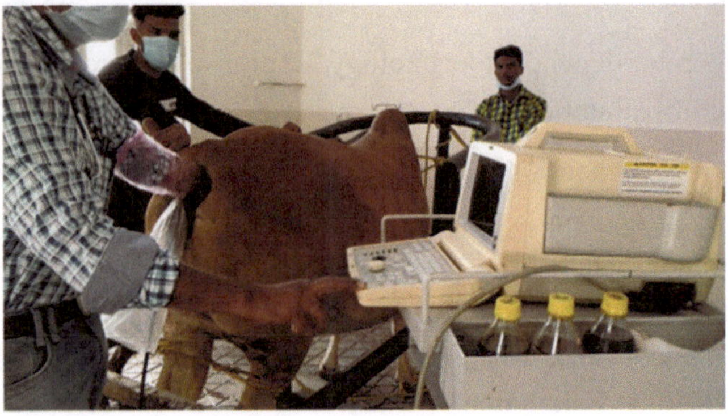

Fig. 189: Per rectal sonography in a cow.

- Its transabdominal use to visualize gastrointestinal tract is very much limited. Efforts have been made to use transabdominal ultrasound (Fig.190) in the diagnosis of peritonitis, abomasal displacement (left and right), abscesses, paralytic ileus and intussusception etc.

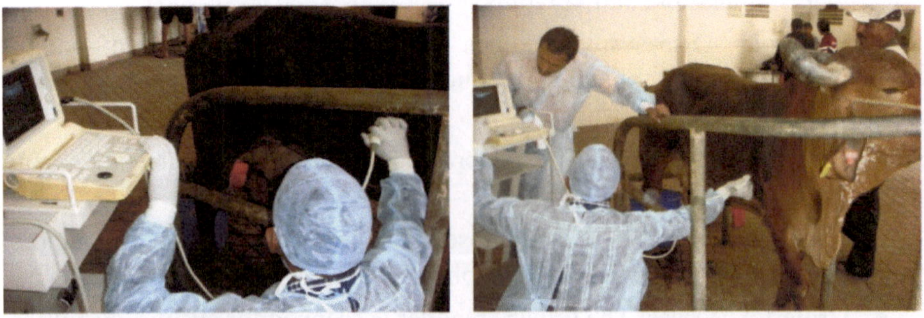

Fig. 190: Transabdominal sonography in cows

- Ultrasound can also be used in variety of non-reproductive imaging such as thoracic ultrasound can be used to investigate respiratory disease, teat ultrasound to investigate difficult milkers or reticulum ultrasound to investigate suspected hardware disease or motility disturbances, or top line ultrasound for a more objective measure of body condition score.
- Ultrasound examination can also be used to detect subclinical otitis in calves.
- The musculoskeletal system can be examined to image gross evidence of extension of deep digital sepsis or for differentiation between causes of swellings associated with lameness.

Small ruminants

- Use of ultrasound in reproductive management of small ruminants (goats and sheep) started in 1960.
- Its use in pregnancy diagnosis in goat and sheep has increased during last few decades.
- Transrectal approach for ultrasonographic examination is not feasible in caprine and ovine patients. In sheep and goat transabdominal approach (Fig.191) can be used.
- Images of transabdominal ultrasonographic examination is of poor quality due to excessive gas and ingesta in the gastrointestinal tract. Nevertheless, it can be used for identification of cystic ovaries and uterine pathology.
- Ultrasonography is also being used to detect uroliths in urinary tract of small ruminants. The uroliths at the level of the urethral process and / or at the level of penile sigmoid flexure are common and can be easily detected by transabdominal ultrasonography. Urinary bladder rupture or urethral rupture can also be detected.
- Both kidneys can be scanned by transabdominal approach through right paralumbar fossa.
- Liver and spleen can be visualized from lateral abdominal approach.
- Ultrasonography can also be used to evaluate the testis, epididymis and scrotum in male goats and sheep.

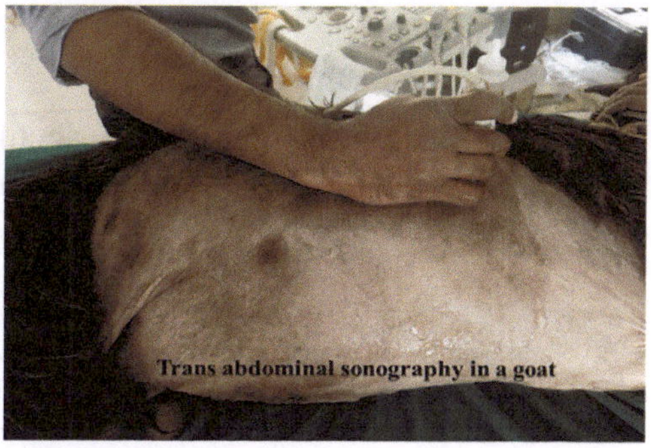

Fig. 191: Transabdominal sonography in a goat.

Indications of Ultrasonography for the diagnosis of Diseases in Horses

Ultrasonography (Fig. 192) is being increasingly used in horses for the diagnosis of following conditions.

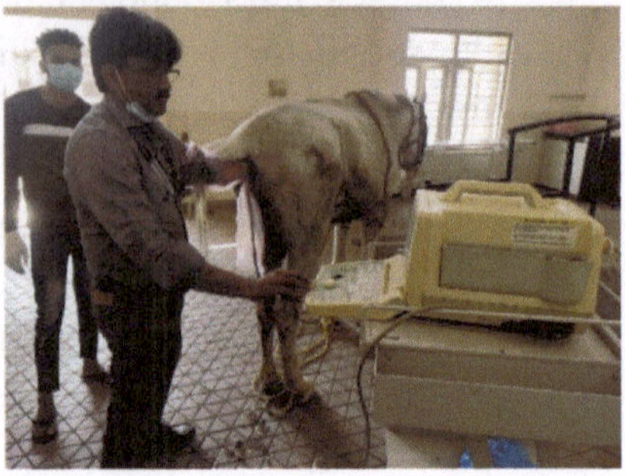

Fig. 192: Transrectal sonography in a horse.

- Lameness, assessing wounds, joint surface, fractures, soft tissue injuries, and detection of back and pelvic injuries.
- Affections of tendons and ligaments of distal limb e.g. SDFT.
- Fractures of ilial wing.
- Meniscal injury in stifle.
- Spine defects e.g. cervical facet joints.
- Pre breeding screening of uterus and ovaries.
- Pregnancy diagnosis, foetal sexing, viability of foetus and ascertaining the number of foetus (s).
- Checking foetal health.
- Diseases of liver, spleen and kidneys.
- Visualizing intestines in colic.
- Diagnosis of aortic-iliac thrombosis.
- Taking ultrasound guided biopsy.
- Cardiac evaluation to investigate cardiac murmurs, ataxia and sudden collapse.
- Evaluation of blood vessels.
- Evaluation of thoracic cavity including lungs.
- Diagnosis of lumps and bumps (haematomas or abscesses).
- Detection of foreign bodies.
- Examination of eye for periocular trauma, exophthalmos, corneal masses, corneal oedema, cataract, and anterior uveitis.

References and Further Readings

Thrawat, M., Alsobiyal, E., Hasahud, M and Buczinski, S. (2012). Trasnabdominal ultrasonographic findings in goats with paratuberculosis. Can. Vet. J., 53: 1063–1070.

Varshney, J.P. (2022) Ultrasound in Veterinary Medicine. Fundamentals and Applications. NIPA Genx Electronic Resources and Solutions.P.Ltd.New Delhi-110034.

Varshney, J.P.(2025). A Fully Illustrated Handbook on Clinical Diagnostic Processes in Canine Medicine. NIPA Genx Electronic Resources and Solutions.P.Ltd.New Delhi-110034.

30

Principles and Uses of Endoscopy in Animals

Endoscopy is a minimally invasive procedure that allows the clinician to look within a hollow organ or body cavity and obtain diagnostic information, which includes grossly evaluating an area of interest and obtaining tissue for histopathologic evaluation. Endoscopy is also used for therapeutic purpose and termed as interventional endoscopy.

Endoscope

An endoscope is basically consists of an insertion tube, hand piece and umbilical cord,

Insertion tube -Encases the mechanism for image transmission. It consists of fiberoptic glass bundles (fiber-scope) or charge coupled device (CCD) chip (video endoscope), biopsy/suction channel, irrigation/insufflation channel and deflection control cables.

Handpiece – It includes deflection control knobs,accessory channel entrance, irrigation/insufflation and suction valves.

Umbilical cord -It carries accessories for light transmission.

Types of Endoscopes

A number of different endoscopes available, depending on the area being examined. It can be classified as either flexible or rigid endoscopes, based on how they are designed.

Flexible endoscope

- Flexible endoscope (Fig. 193) uses either fibreoptics or video technology to generate an image and are referred to as fibrescopes or videoscopes, respectively. The fiberoptic endoscope is based on the principal of total internal reflection of light in tiny flexible glass fibers. Light entering one end of the fiber is transmitted by reflection along the walls of the fiber until it emerges at the other end. To minimize light loss and to

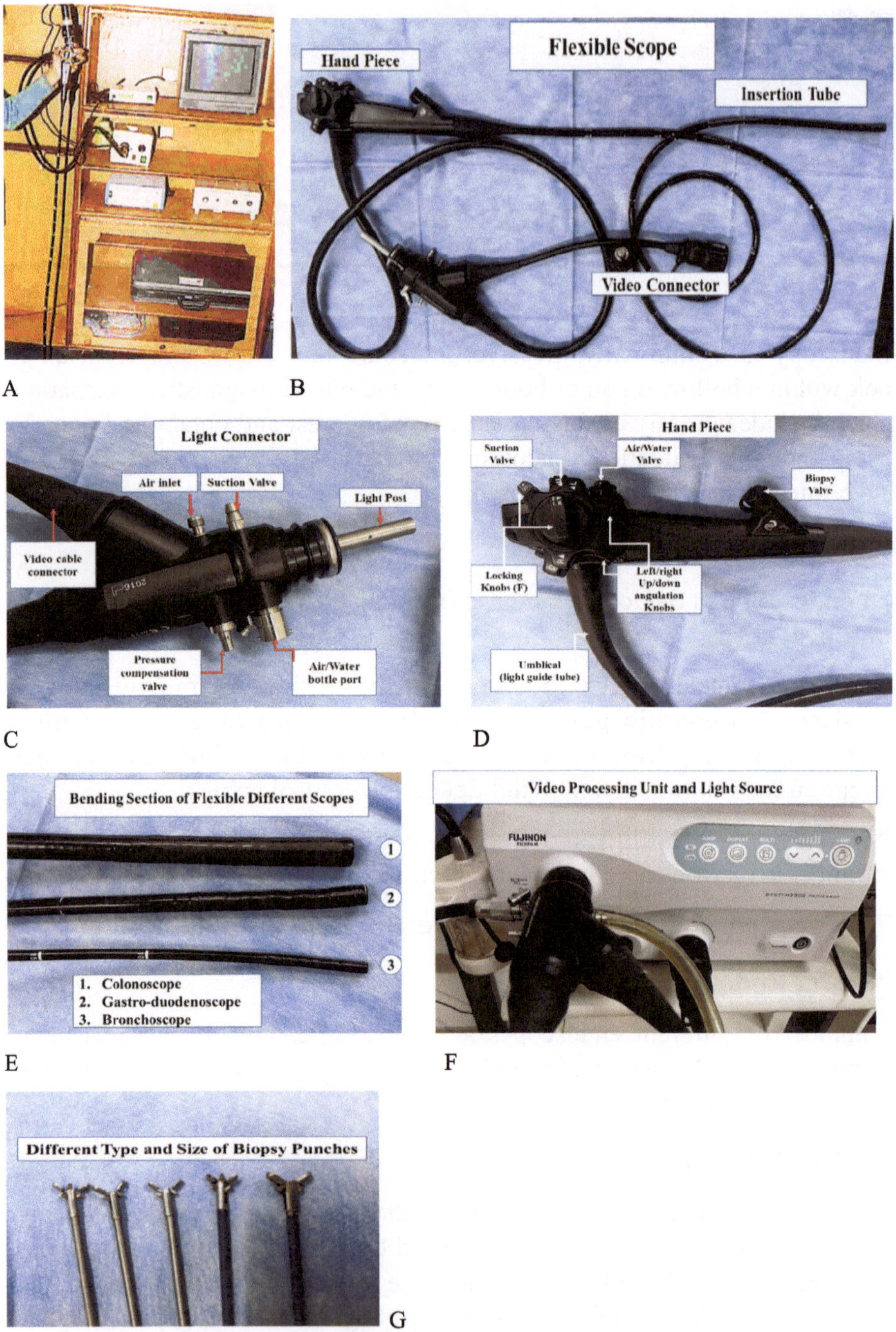

Fig. 193: Flexible endoscope (Gastroscope) unit. Images at B, C, D, E, F and G are with the courtesy of Dr. Gurpreet Singh Preet, Assistant Professor, TVCC, CoVS, GADVASU, Ludhiana.

prevent light from one fiber scattering into adjacent fibers, each fiber is wrapped with insulation, usually a glass coating of a different refractive index. Many of these insulated fibers are combined to form a bundle approximately 1/4" in diameter. There are two separate fiber bundles in flexible endoscopes, one for viewing and one for light transmission. The fibers in the visual bundle are spatially oriented so that the top of the object viewed at one end is oriented similarly at the other (a "coherent" fiber bundle). A lens system at the mucosal end focuses the image on that end of the bundle, while a lens system at the proximal, or viewing end, magnifies the image emerging from the bundle. Direct internal transmission of light through this flexible glass bundle enables the projection of images from one end to the other through the curves and coils that the endoscope adopts as it is passed into the gastrointestinal tract.

- Light for illumination at the mucosal end of the endoscope is transmitted from an external light source through a non-coherently oriented bundle. In most light sources light intensity can be varied for observation or photography.
- Endoscopes also contain a suction channel that can be used to evacuate mucus, fluid and blood from the lumen of the gut. Suction is controlled by a button or valve at the head of the instrument. A similar and adjacent mechanism allows control of a flow of air from an external pump into the lumen to dilate it and allow better visualization. Without such distention the walls would collapse around the tip of the endoscope impairing visibility. Another proximally controlled mechanism allows control of a fine spray of water over the lens to clean it of blood, feces or mucus.
- Most endoscopes also have a channel for the passage of instruments such as biopsy or grasping forceps, cytology brushes or irrigation catheters. The channel is usually the same as that used for suction. Biopsy catheters are passed through the channel so that the tip lies in the lumen and specimens are obtained under direct observation. Specimens are usually small (1-2 mm thick) and need both careful handling and an experienced pathologist for interpretation.
- Endoscopes also contain a pulley system whereby the tip end can be moved in 4 directions (right and left, and up and down) by control knobs on the handle. Complete 360 degree examination is achieved by slight rotation of the endoscope in combination with deflection of the tip.
- The main advantages of fiberscopes are reduced cost and increased portability. Limitations include inferior image quality and potential

breakage of individual fibers, leading to black spots that obscure the field of view.

- Videoscopes contain a video chip which converts the image to a digital signal. This is then transmitted through wires within the endoscope to a video processor, then to a viewing screen. An external camera does not need to be connected to a videoscope, but a special cable which transmits the video signal from the endoscope to the endoscope tower is required. The image quality of a video endoscope is far superior to that of a fiberscope and, therefore, carries a greater cost. The absence of fiberoptic bundles (and broken fibers) ensures that the image will not contain black dots.
- Several different types of flexible endoscope are used in practice, depending on the procedure being performed.
- A *gastroscope* is generally the largest, longest endoscope used in practice. It has an umbilical tube as well as an insertion tube, contains an air/water channel, larger instrument channel (usually 1.8-2.3mm across), and has 4-way angulation (the bending section can move left and right, as well as up and down). This function is unique to gastrointestinal endoscopes, as other flexible scopes have 2-way (up and down) angulation only. The gastroscope is used to examine the oesophagus, stomach, duodenum, ileum and colon.
- A *colonoscope*, has exactly the same functions as a gastroscope but it has a slightly longer insertion tube.
- A *bronchoscope* is a narrow endoscope, with a shorter insertion tube. This is used to image the nasopharynx, larynx and lower airways. They often do not have an umbilical tube and have a narrower instrument channel (1-1.8mm across). Most have suction facilities, but air and water channels are not necessary.
- The smallest flexible endoscope used in practice is the cystoscope. This is similar in structure to a bronchoscope but has a narrower insertion tube. This is used to examine the urethra and urinary bladder in male dogs.

Rigid endoscope

- The rigid endoscope is also known as a telescope, is used predominantly to image the rostral aspect of the nose, and the female lower urinary tract. Rigid scopes are also used surgically for arthroscopic and laparoscopic (Fig. 194) procedures (the minimally invasive examination of joints, and the abdominal/thoracic cavity).

The telescope allows light to be directed to the area of interest. It contains glass lenses and fiberoptics but no flexible materials.

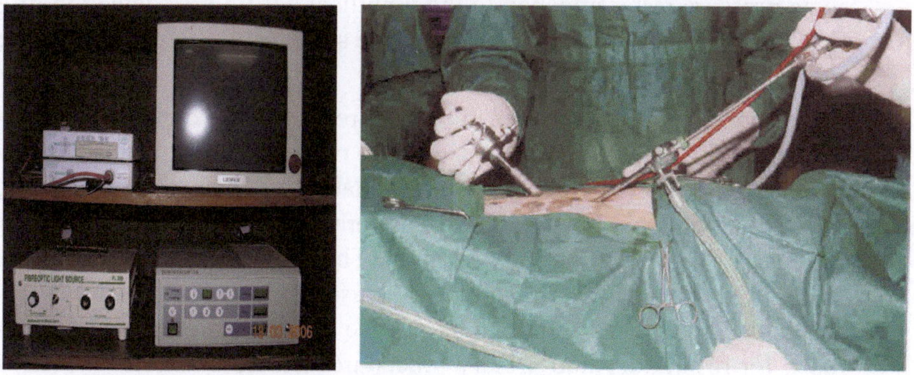

Fig. 194: Rigid endoscope (laparoscope) unit.

- Rigid endoscope is delicate and need to be housed within an outer working sheath during use, and stored within a plastic protective cover whenever they are not in use (including during cleaning). The working sheath contains ports for the introduction of instruments and the administration of irrigation fluid. Fluid administration is often necessary during the use of rigid endoscopes, as the image quality is greatly improved when visualizing through a column of fluid. The working end of the telescope may be angled, depending on the intended use; the angle will change the clinician's field of view. The operator end of the telescope contains an eyepiece (to which a camera is usually attached), and light guidepost (to which the light source is attached).

Endoscope tower

- For the proper functioning of endoscope, a number of external components are required. These may be housed together in one unit (e.g. a unit containing a viewing screen, air pump, light source and video processor) or be separate units combined to make up an endoscopy tower. These contain a light source, air pump, video processing unit, image capture device and viewing screen. If tower is used for surgical endoscopy (laparoscopy/arthroscopy), a carbon dioxide insufflator and an arthroscopic fluid pump may also be present. In addition to the equipment listed above, an endoscope water bottle and a suction unit will be required.

Care of endoscope

- Following manufacturer guidelines about safe and appropriate cleaning is the best way to care for an endoscope.

- Leak Testing- Prior to and after each procedure, perform a leak test to check for internal leaks. This test is performed using a manometer type pressure tester attached to the umbilical cord. Early detection of a leak, prior to submerging a portion of the scope into water or enzymatic cleaner, may prevent costly damage.
- Between Procedures-After each procedure, wipe the scope down using a soft cloth to remove residual organic material. Flush the accessory channel with an enzymatic cleaner; nonabrasive, neutral pH cleaners. Rinse the accessory channel with demineralized water; then dry with air (injected via a large luerlock syringe). This "quick clean" approach is appropriate between GI procedures because sterility is not crucial unless heightened concern for infectious disease exists. However, endoscopes used for respiratory and urinary procedures should be sterile, which is accomplished by cold disinfection using a glutaraldehyde solution or gas sterilization. Following cold disinfection, rinse the scope thoroughly in sterile water and dry with sterile towels.
- Thorough Cleaning-Perform comprehensive cleaning after a procedure, if needed, or at the end of the day.

  Immerse the scope in a dilute bath of enzymatic cleaner for 10 to 15 minutes, during which time the channel can be flushed and cleaned with channel cleaning brushes.

  Rinse the scope using demineralized water.

  Hang the scope to allow it to dry.

  Store flexible scopes by hanging them; rigid scopes are best stored in protective cases once they are thoroughly dried.

Diagnostic Utility of Endoscope in Small Animals

Endoscopy plays a great role in the diagnosis of diseases of upper and lower gastro-intestinal tract, nasal cavity, bronchi, urogenital system, ear canal, joints and abdomen.

Gastro-intestinal endoscopy

- Before performing an endoscopy, routine investigations are essential to confirm the procedure's indication. This includes a complete history, a thorough physical examination, and a fecal examination if diarrhea is present. Routine laboratory tests are crucial to determine whether systemic disease is causing gastro-intestinal symptoms and to ensure patient's safety to anesthesia. Plain thoracic and abdominal radiographs and an abdominal ultrasound should always precede endoscopy to check for unexpected foreign bodies or masses/lesions beyond the endoscope's reach. Radiographs and ultrasonography can also detect extramural diseases, whereas endoscopy is more sensitive for detecting mucosal disease and offers the advantage of obtaining definitive diagnoses through biopsies.

- *Upper GI tract endoscopy*- It is encompassing the examination of the esophagus, stomach, and proximal small intestine (Table 24), is one of the most common endoscopic procedures in companion animal practice. While a rigid endoscope can examine the esophagus and aid in foreign body removal, a flexible endoscope is preferred for comprehensive diagnostic investigation of the esophagus and is essential for a complete examination of the stomach and intubation of the small intestine. Modern gastroscopes allow for tissue and liquid sampling and even some therapeutic procedures, unlike early flexible endoscopes that only permitted observation.

  Endoscopy is often indicated for investigating suspected esophageal, gastric, and small intestinal diseases (Fig 195) if the lesion is accessible from the GI tract lumen. Endoscopy provides the safest method for obtaining a definitive diagnosis, particularly in patients with protein-losing enteropathies who have impaired healing.

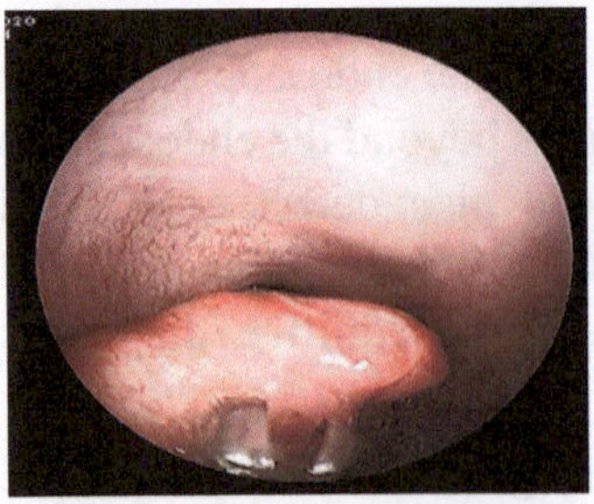

Endoscopic image showing oesophageal nodule.

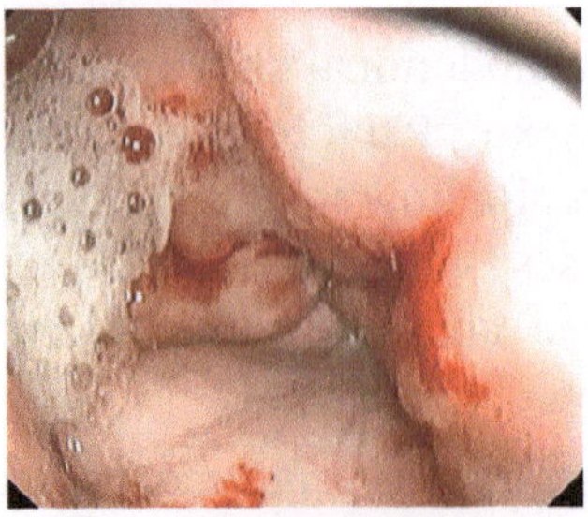

Endoscopic image showing ulceration at cardiac end of esophagus.

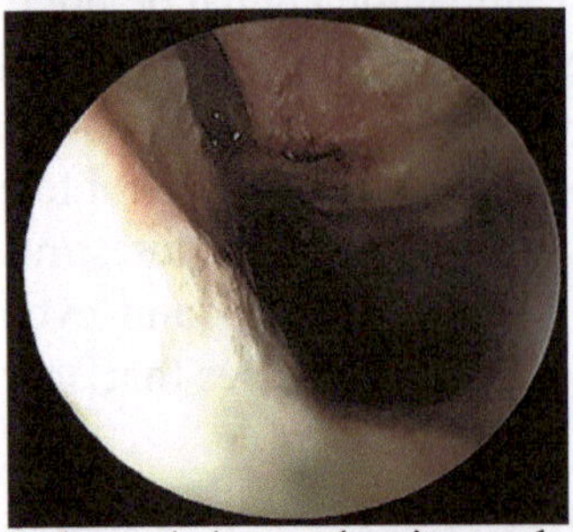

Endoscopic image showing pocketing Esophagus suggesting esophageal

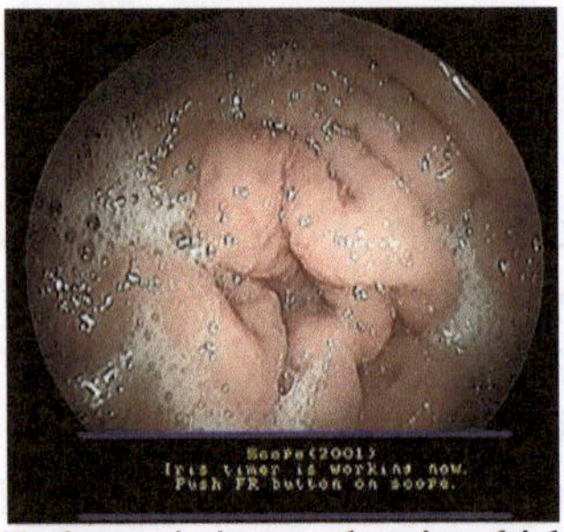

Endoscopic image showing thickened gastric rugal of folds in chronic gastritis diverticulum

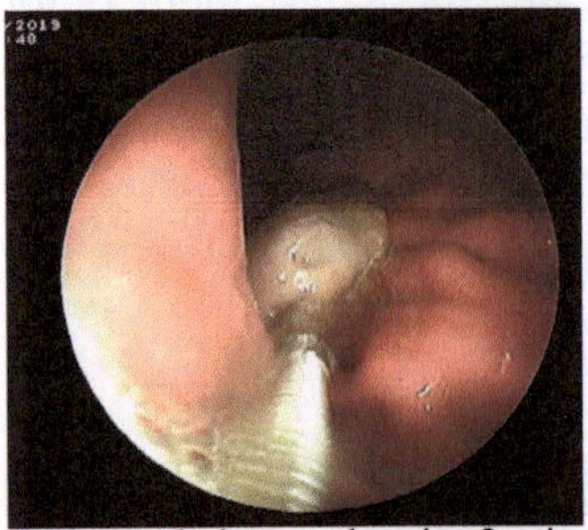

Endoscopic image showingforeign body in stomach and its retrieval via forceps

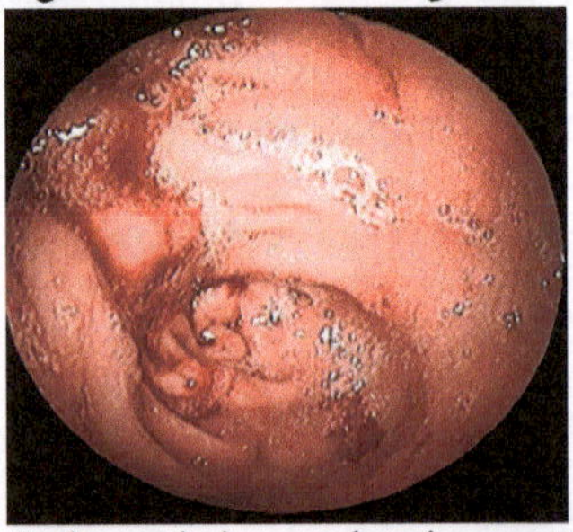

Endoscopic image showing gastric Endoscopic image showing gastric Hemorrhage

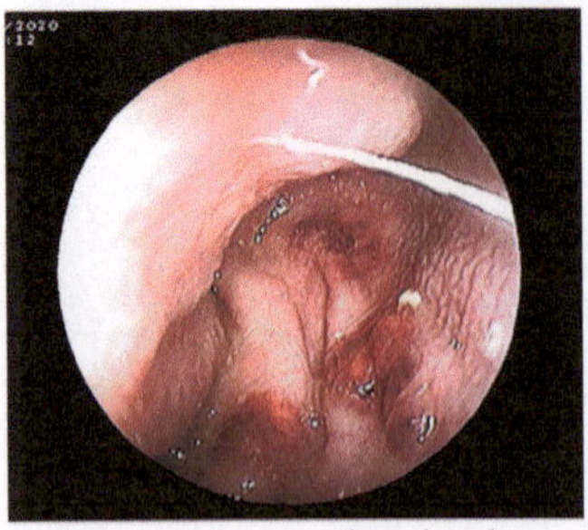

Endoscopic image showing hyperimia and ulceration of colonic wall.

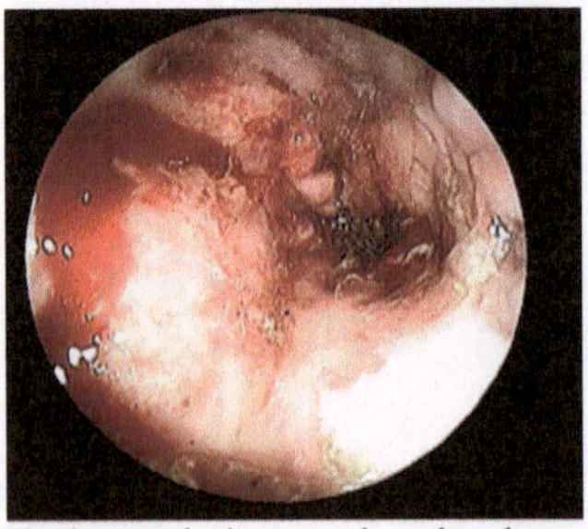

Endoscopic image showing hemorrhage and ulceration of colon wall.

Fig. 195: Showing endoscopic view of different gastrointestinal lesions. Images at with the courtesy of Dr.Gurpreet Singh Preet, Assistant professor, TVCC, CoVS, GADVASU, Ludhiana.

Diagnostic endoscopy is indicated for clinical signs related to esophageal diseases (regurgitation, dysphagia, painful swallowing, excessive salivation/inability to swallow), gastric diseases (nausea and salivation, chronic vomiting, haematemesis and/or melaena, unexplained anorexia), and intestinal diseases (chronic vomiting, chronic diarrhoea, melaena, haematemesis when no gastic lesion, change in appetite, unexplained weight loss and abdominal pain).

Table 24. Indications for upper gastro-intestinal tract endoscopy in dogs and cats

Oesophagus	Stomach	Intestine
Foreign body	Foreign body	Focal causes of haemorrhage
Megaoesophagus	Ulcer	Focal neoplasia
Hiatal hernia	Neoplasia	Patchy disease
Gastro-oesophageal intussusception	Chronic gastritis	Inflammatory bowel disease
Neoplasia	Pyloric stenosis	Lymphangiectasia
Stricture		Alimentary Lymphoma
Vascular ring anomaly (endoscopy rarely necessary) unless foreign		

- *Lower GI tract endoscopy*- The lower gastrointestinal (GI) tract of dogs and cats is relatively simple in anatomical structure compared to many other species. It consists of a straightforward tube divided into the cecum, colon, rectum, and anus. This simplicity and the accessibility of the lower bowel make it particularly well-suited for examination via flexible endoscopy. Colonic disease is common in small animal practice. Endoscopy now offers a straightforward and readily available method for examining the entire mucosal surface of the lower bowel (Table 25) and collecting biopsy samples for definitive diagnoses, revolutionizing the diagnosis of lower bowel disorders.

  Clinical signs of lower GI disease typically include diarrhea with or without fresh blood (hematochezia), mucus, tenesmus, dyschezia, and constipation.

  Although many of these conditions can be diagnosed endoscopically, it's crucial to begin with a detailed history and a thorough physical examination.

Table 25: Indications for lower gastro-intestinal tract endoscopy in dogs and cats.

Disorder	Comments
Colitis	Lymphocytic–plasmacytic; eosinophilic; histiocytic; granulomatous
Infection	Salmonella spp.; Campylobacter spp.; Yersinia spp.; Clostridium spp.; Trichuris vulpis; Giardia; Uncinaria spp.
Caecal disorders	Typhlitis; caecal inversion; caecal abscessation; perforation; neoplasia
Intussusception	Ileocaecal; caecocolic; colocolonic
Neoplasia	Adenocarcinoma; lymphoma; leiomyosarcoma
Rectal disorders	Stricture; adenomatous polyps; adenocarcinoma; foreign body

Rhinoscopy

- It is used for examination of nasal passages for diagnostic and therapeutic purposes. It provides visualization of nasal cavity, choanae, nasopharynx and frontal sinuses for diagnosis of tumors, nasal foreign bodies, fungal rhinitis, and anatomic nasal abnormalities. It is also used to treat nasal, sinus or nasopharyngeal problems, such as removal of foreign body, ectoparasites, nasal polyps, and introduction of medication into the nasal cavity or sinus to eradicate fungal infections.

  Indications for rhinoscopy include sneezing, reverse sneezing, nasal discharge, epistaxis, unexplained nasal discharge and stertor or nasal airflow obstruction, rubbing or scratching at the nose or face, ulceration of the rhinarium, facial or nasal pain or sensitivity, facial swelling or distortion, nasal or oral odor, and knowledge of foreign body inhalation or ingestion.

Bronchoscopy

- Bronchoscopy may be used for diagnostic, therapeutic, and prognostic purposes.

  Diagnostic bronchoscopy is used to obtain visual information concerning the airways (e.g., compression, dynamic collapse, and dilation) as well as to obtain samples (cytology, culture, and occasionally biopsy) to help establish a specific etiologic diagnosis.

  Bronchoscopy is useful therapeutically, especially in the removal of airway foreign bodies. It is also helpful in determining prognosis when nonreversible anatomic or mucosal changes are recognized in the airways. The indications of brochioscopy are abnormal breathing pattern, acute and chronic cough, airway collapse, inflammatory conditions, neoplasia, unexplained radiographic infiltrates.

Cystoscopy

- It is a minimally invasive diagnostic and treatment tool for the urogenital system. Cystoscopy provides minimally invasive direct visualization of the vagina, urethral opening, urethra, bladder, and ureteral openings. Indications for cystoscopy include chronic or recurrent urinary tract infection (UTI), hematuria, dysuria, stranguria, pollakiuria, trauma, calculi, incontinence, neoplastic cells in urine sediments, abnormal findings in urine analysis, radiographs and ultrasound etc. A wide variety of urinary tract pathology can be assessed by cystoscopy. Tumors that originate in or penetrate the mucosa can be visualized, their extent evaluated, and biopsies done. Chronic inflammation can be defined and biopsy samples can be obtained for histopathology and mucosal culture studies. The extent of inflammatory involvement can be evaluated and the ability of the bladder to distend and contract can be assessed. Small cystic and urethral calculi can be removed therapeutically and for stone analysis. Transurethral cystoscopy (TUC) is the procedure of choice for assessment of urinary trauma in that the entire urinary tract can be easily and quickly assessed. TUC is also the procedure of choice for diagnosis and definition of ectopic ureters. Bladder diverticula are assessed effectively by cystoscopy. Use of cystoscopy is limited to lesions that are visible within the lumen of the urethra and bladder, lesions that involve the mucosa, and the ureteral openings.

Otoscopy

- It allows visualization of the ear for diagnosis and management of ear diseases with the help of the hand-held otoscope or video otoscope. The various indications are head shaking, ear scratching, ear discharge, foul odor from ear and pain on manipulation, neurologic signs, deafness etc. In addition to diagnosis of ear affections, it also serves to perform deep ear flushing, myringotomy, biopsy and foreign body retrieval.

Arthroscopy

- Arthroscopy is indicated whenever history, physical findings, radiographic changes, or laboratory results suggest joint disease. A history of lameness and stiffness; difficulty or reluctance to get up, to go up or down stairs, to get up and down off the couch or favorite chair, or to get into or out of the car or truck; joint pain, swelling, thickening or crepitus; reduced range of joint motion; and joint instability on physical examination are potential reasons to perform arthroscopy. Radiographic abnormalities of increased joint fluid or joint capsule thickening, periarticular osteophytes, periarticular sclerosis, osteochondritis dissecans (OCD) lesions, ununited anconeal processes (UAPs), ununited caudal glenoid ossification center (UCGOC), intraarticular fractures or chips, periarticular bone lysis, or any other radiographic abnormalities involving a joint are also indications for arthroscopy.

  Arthroscopy is most commonly performed in the shoulder, elbow, and stifle in dogs. It is less commonly performed on the radiocarpal, hip, and tibio-tarsal joints. Conditions diagnosed with arthroscopy include osteochondritis dissecans (OCD) of the shoulder, stifle, elbow and tibiotarsal joints; partial and complete cranial and caudal cruciate ligament ruptures; meniscal injuries; MCPP (FCP); UCGOC, UAP, ununited supraglenoid tubercle, degenerative joint disease (DJD); intraarticular fractures; immune mediated arthritis; synovitis; bicipital tendinitis and partial or complete bicipital tendon rupture; intraarticular soft tissue injuries of the shoulder, radiocarpal joint, stifle, and hip; septic arthritis; and neoplasia.

Operative procedures currently being performed with arthroscopy include removal of OCD cartilage flaps and debridement of the cartilage defects, coronoid process fragment removal and coronoid process revision, free joint body removal, bicipital tendon transection, carpal chip removal, partial and total meniscectomy, cruciate ligament debridement, meniscal release, UCGOC removal, ununited supraglenoid tubercle removal, UAP removal, screw fixation of UAP fragments, osteophyte removal in chronic DJD of the elbow and tarsus, thermal modification for stabilization of medial soft tissue injuries of the shoulder, intraarticular repair of ruptured cranial cruciate ligaments, fixation of avulsed ligament attachments, and repair of lateral humeral condyle and acetabular fractures.

Laparoscopy

- It is also called as celioscopy/pelvioscopy. It involves the visual examination (Fig.196), surgical procedures and biopsy of the peritoneal cavity and its organs. Diagnostic indications include exploration and biopsy collection of liver, kidney, prostate, lymph nodes, spleen, intestine, urinary-reproductive tract, cholecystocentesis, cancer staging, laparoscopic surgeries for gastropexy, ovariectomy etc.

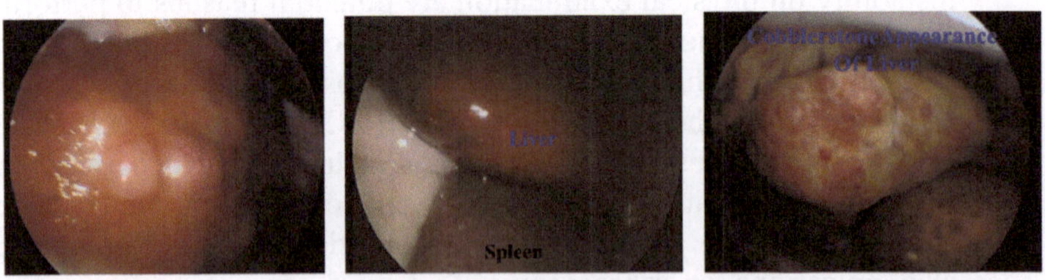

Hepatic tumor — Enlarged liver and spleen — Cobblerstone appearance of liver

Fig. 196: Laparoscopic view of the abdomen of a dog showing hepatic tumor, enlarged liver and spleen, and cobblerstone appearance of the liver. Courtesy S.K.Maiti, J.P.Varshney,S. Chaudhary, A. Pawde,Naveen Kumar and Ninu, A.R.(2014).

Diagnostic Utility of Endoscope in Large Animals

Use of endoscopy in bovines

- Theloscopy in bovines offers a precision approach for diagnosing and treating teat problems. Theloscopy is a minimal invasive procedure proved to be a valuable imaging modality to investigate, document and categorize the diverse alterations of the inner teat cistern lining. It can diagnose from teat sphincter constrictions to foreign bodies in the teat canal. The main indication for theloscopy turned out to be treatment of distal teat stenosis under visual control. Mucosal detachment can be removed using cutting instruments or by coagulating the mucosal lesion using the theloresectoscope.
- Specific clinical signs, such as regurgitation, salivation, recurrent tympany, or swelling in the neck region are the indications of endoscopy of upper digestive tract in cattle.

  Esophagoscopy is used for the diagnosis of esophageal foreign bodies, megaesophagus, esophageal diverticulum, rupture, esophageal obstruction and stricture, inflammation, or neoplasia.

  The potential as a diagnostic tool for detection of infectious diseases such as Malignant Catarrhal Fever or Bovine Virus Diarrhoea in cattle is demonstrated by evaluation of the oesophageal mucous membrane for typical signs of inflammation and biopsy for detection of BVDV Ag by PCR technique in persistently and transiently infected animals.
- Rumenoscopic diagnosis of polyethylene bags in the rumen have been reported in adult cow showing clinical signs such as anorexia and recurrent tympany.
- By visual inspection of the surface of the mucous membrane, different types of cystitis (catarrhalic cystitis, haemorrhagic cystitis, fibrinous-purulent cystitis) can be diagnosed. The endoscopic assessment of mucosal lesions is also demonstrated in cows with bovine enzootic haematuria ranging from reddening over ecchymotic haemorrhages to visualisation of a tumour. some case reports demonstrate the potential of cystoscopy in cows for diagnosing rupture of the urinary bladder or urachal rupture.
- Correction of left displacement of the abomasum (LDA) and right displacement of the abomasum (RDA) is the most popular application of laparoscopic endoscopy in cattle.

- Gastroscopic examination is limited to monogastric animals and has particular utility in horses and foals. It is useful in confirming the presence or absence of gastric ulcers, impaction, neoplasia, and inflammatory disease.
- In general, clinical signs such as strides, pathological nasal discharge, swelling in the retropharyngeal area, coughing, dyspnoea are described as indications for endoscopy of the upper respiratory tract in cattle. Pharyngeal diseases are diagnosed by using endoscopy alone or in combination with other imaging techniques such as ultrasonography or radiography. The endoscopic diagnosis of retro- pharyngeal abscess, nasal neoplasia, laryngeal cyst, nasal abscess, nasal schistosomiasis and tracheal collapse have been reported. Aside from visual inspection, biopsy under endoscopic control lead to exact diagnosis (nasal schistosomiasis, squamous cell carcinoma, chondrosarcoma).
- Another indication of bronchoscopy is the bronchoalveolar lavage (BAL), which is performed in animals either with lung disease or as sampling technique for the lower respiratory tract.
- Thoracoscopy is used in the diseased calves for the diagnosis of small lung abscesses, emphysema, edema, and lung adhesions.

Use of endoscopy in horses

- The broad utility of endoscopy is extremely helpful to equine clinicians to diagnose dorsal displacement of soft palate, epiglottic entrapment, rostral displacement of the palatopharyngeal arches, arytenoid chondritis, tracheal collapse or stenosis, and pharyngeal disease that may include narrowing. Developmental or congenital abnormalities may be identified with upper airway endoscopy such as cleft palate, subepiglottic cyst, or choanal atresia. When exudate or hemorrhage is present, endoscopy can help determine its origin. Examples include guttural pouch mycosis, guttural pouch empyema, progressive ethmoidal hematoma, retropharyngeal lymphadenopathy, pulmonary hemorrhage, or exudate or mucus.
- Dynamic endoscopy is the best modality in horses showing exercise intolerance or excessive respiratory noise that has remained undiagnosed at rest. Conditions commonly identified with dynamic endoscopy include dorsal displacement of the soft palate and recurrent laryngeal neuropathy (RLN).
- Endoscopic examination of paranasal sinuses is indicated in horse in which diagnosis of the disease requires collection of tissue from

the sinus. Therapeutic interventions that can be performed during endoscopic examination of the paranasal sinuses include lavage, removal of accretions of inflammatory material, drainage of cysts, and creation or enlargement of drainage holes.

- Common non-infectious causes of lower-airway disease, such as exercise-induced pulmonary hemorrhage (EIPH) and equine asthma syndrome (formerly known as inflammatory airway disease, recurrent airway obstruction, heaves, and chronic obstructive pulmonary disease), are marked by the accumulation of blood (EIPH) or mucus (equine asthma syndrome) in the trachea, both of which can be visualized endoscopically. Diagnostic sampling may include a collection of tracheal or bronchial secretions, depending on the purpose of sampling. Tracheal secretion collection is most appropriate for culture purposes, whereas bronchoalveolar lavage (BAL) washing of distal airways can be performed for examination of lower airway cellularity.
- Pleuroscopy, using a rigid or flexible endoscope, enables direct visual inspection of the pleural cavity for the diagnosis of pleural disease. The technique is particularly valuable in diagnosis of diseases of the thorax that extend to the pleural surface and do not exfoliate large quantities of cells, thereby making diagnosis by examination of fluid obtained by pleurocentesis unlikely. The procedure is useful in collection of tissue samples, such as from suspected thoracic neoplasia 10-12 or in therapeutic procedures including relief of pleural adhesions and resection of lung sections, and in collection of large biopsies samples from the lung.
- Gastroscopic examination has particular utility in horses and foals for confirming the presence or absence of gastric ulcers, impaction, neoplasia, and inflammatory disease.
- Laparoscopy in standing horses is a valuable diagnostic aid for the examination of the structures in the dorsal regions of the abdomen. In this procedure a laparoscope is passed through an incision in the abdominal wall of either the left or right paralumbar fossa. In the standing horse, the anatomic structures of importance that can be viewed in the left half of the abdomen are the hepatic duct, left lateral and quadrate lobes of the liver, stomach, left kidney with associated nephrosplenic ligament, segments of the jejunum, descending colon and ascending colon, left side of the male and female reproductive tracts, urinary bladder, vaginal ring, and mesorchium. The important structures observable in the right side

of the abdomen are the common hepatic duct, left lateral, quadrate and right lobes of the liver, caudate process of the liver, stomach, duodenum, right dorsal colon, epiploic foramen, omental bursa, right kidney, base of the cecum, segments of jejunum, descending colon and ascending colon, urinary bladder, right half of the male and female reproductive tracts, and rectum.

- Endoscopy of the urinary tract is a useful diagnostic aid when the complaint is abnormal urination and hematuria. Cystoscopy is useful in diagnosis of calculi, inflammation, and masses. Viewing the ureteral openings in the dorsal aspect of the trigone can help determine the source of hematuria or pyuria.
- When conservative hormonal treatment of the ovarian cyst fails, aspiration of the cystic content can be carried out under the guidance of endoscopy.

References and Further Readings

Franz, S., Hofer, L. and Dadak, A.M. (2024). The role of endoscopy in bovine internal Medicine-A review of current indication fields. Ve.J.. https://doi.org/10.1016/j.tvjl.2024.106093.

Maiti, S.K., Varshney, J.P., Chaudhary,S., Pawde,A., Kumar,N. and Ninu, A.R.(2014). Hepatic Liposarcoma in dog. A rare clinical case report. Indian.J.Vet.Pathol. 38:197.

Reed, S.M., Bayly, W.M. and Sellon, D.C. (2018). Equine Internal Medicine. 4th Edition, St. Louis, Missouri

McCarthy, T.C. (2005). Veterinary Endoscopy for The Small Animal Practitioner. Elsevier Saunders, St. Louis, Missouri

Varshney, J.P. (2025). A Fully Illustrated Handbook. Clinical Diagnostic Processes in Canine Medicine"(NIPA, New Delhi,

31

Herd Medicine

Herd

A large number of animals that live and feed together in a group (Fig.197) is called as herd. In herd, all animals are considered as a unit rather than individual. Herd medicine, also known as herd health or production medicine, is the discipline of veterinary medicine that deals with the diagnosis, treatment and prevention of the diseases in the herd as a whole.

Fig. 197: A herd of animals.Where a large number of animals live and feed together

Concept and Significance of Herd Health

Herd health is a combination of regularly scheduled veterinary activity and management practices undertaken to maintain optimum health to achieve optimum production in the farm. Traditionally, the aim of the herd health is to control or eliminate diseases and rectify management inefficiencies impacting the welfare and productivity in the frame work of risk monitoring and assessment in conjunction with best farming practices. The main emphasis is on identifying the disease processes affecting the herd as a whole; treating them to stop the diseases and restore herd health; and preventing the occurrence of disease in the herd employing control measures so that production is not affected. The practice of herd medicine or production medicine is essential to contain or halting losses in production owing to sick animals as well as animals with subclinical sickness or disease. In simple words, the objective of herd medicine is to identify risk factors for disease(s) to control them so that production losses associated with subclinical disease are avoided and overall herd production improves. Herd health program must be implemented in a practical as well as economical way. Reducing pathogen load on the farm increases food safety along the food chain. Improved herd health not only reduces antimicrobial requirement but also slows down the development of antimicrobial resistance. Hygiene and nutrition are the very important factors in reducing the occurrence of diseases and increasing the herd productivity.

In dairy animals, milk production depends on nutritional balance and udder health therefore production diseases and mastitis are of great concern in these animals. As cows cannot produce milk without reproducing, reproductive problems are another area of great concern. Dairy animals kept confined with cement floors are predisposed to lameness. Another important area of concern in dairy animals is their gut health and availability of nutrients through feed and fodder. Any imbalance, deficiency or excess in nutrient supply may adversely affect production.

Herd performance also depends on genetic and management practices. In the herd a few animals may be affected by genetics at a time. While entire herd or group of animals is affected by management practices. Management practices include nutrition, hygiene, floor space, and shed conditions. Any specific practice altering normal behavior of some animals may make these animals more susceptible to disease than others. For example, cows in the herd not liking the bedding provided in the free -stalls remain standing for longer periods putting more stress on their feet and legs (altered behavior), have higher incidence of lameness.

Scientific management practices in the herd are not only cost effective in reducing the incidence of preventable diseases but also reduces the requirement of pharmaceuticals. Diseases such as caprine arthritis-encephalitis (CAE), *Mycoplasma*, caseous lymphadenitis, and Johne's disease are unresponsive to treatment regimen but can be prevented through adoption of specific pathogen prevention programs and culling of affected animals.

Objectives of Herd Health Programme

- To maintain health of the herd as well as production at the most efficient level to get maximum economic returns.
- To control and manage animal health and production at a high level of efficiency adopting new techniques for continued improvement.
- To make provision for comfortable housing commensurating with reasonable animal welfare, minimum pollution of environment by animal waste and to prevent zoonotic diseases.

Practices for Herd Health Improvement

The following practices can improve herd health.

- A clean environment
- Adequate nutrition program
- Good reproduction (A.I. and Calving) practices
- Quality vaccination management
- Keeping records

Biosecurity

- Biosecurity is a set of management practices that prevent infectious diseases from being carried into a herd.
- It plays very important role in decreasing disease transmission, reducing death loss, preventing production loss, improving cost benefit ratio, and preventing premature culling of animals.
- It is critical to global trade, food safety and antibiotic resistance.

Biosecurity Measures on Farm

- Good sanitation.
- Stock and disease free boundaries.
- Isolation and acclimation of new animals at least for two weeks.
- Regular disease testing and monitoring.
- Regular vaccination against important prevalent diseases in the area.
- Foot bath.
- Close herd policy (no cattle movement) prevents direct transmission of the disease onto a farm.
- Maintaining proper good records of all activities on the farm.

Herd Health Programmes

A. *Preventing disease exposure*

Biosecurity practices and sanitation plays an important role in preventing diseases exposure in the herd. Introduction of new animals can be a source of infection in the herd. This is more important in breeding herd. Purchase of new animals should be from known source with a proper vaccination history. As an additional measures purchased animals should be tested for high risk diseases prior to mixing in the herd and should be strictly quarantined and a period of 21 to 30 days of diseases surveillance is recommended. Fence-line contact should also be avoided. All herd health programs should consider testing calves for bovine viral diarrhea virus persistent infection (BVD-PI). Vaccination against brucellosis (Bangs) in replacement heifers should be ensured before taking the animals to other places. Additional testing against diseases like bovine leukemia virus (BLV), Johne's disease, or any other may be considered as per diseases prevalent in a location.

Sanitation in the herd is vital in reducing/ preventing disease exposure. Maintaining sanitation in dairy herds is easy as compared to beef animals. In beef or pasturing animals efforts should be made to reduce excessive exposure to waste, mud, and other environmental sources. Equipment used on animals should be sanitized and all handling should also be done under sanitized conditions. Always fresh syringe and needle should be used when drugs are given parenterally. Cleaning of vulva prior to insemination reduces the risk of infection. During calving, good sanitation is to be maintained in calving pens.

B. *Parasite control*

Internal and external parasites adversely affect health and reproductive efficiency. Subclinical parasitism can lead to diminished immunity and nutritional deficits. Parasitic load in animals is affected by location, season necessitating effective surveillance, and modulating strategies along with anthelminthics used for parasitic control. Rotational grazing and keeping used pastures unstocked for reasonably sufficient period, proper and timely disposal of dung, eliminating sources of standing water, selective deworming of young and vulnerable animals, and timing deworming as per seasonal fluctuations in the presence of parasite are important effective parasitic control strategies. Frequent change in anthelmintics carries advantage of avoiding the problem of parasite resistance.

C. Vaccination

Vaccination is a must in herd health programs. Some vaccinations are given annually to entire herd. Vaccination against brucellosis, require only a single dose early in life. For an effective development of herd immunity in all age groups in the farm, a protocol is to be carefully followed keeping in view vaccine type, cost, stage of production, and timing of administration.

Modified live vaccines (MLV)-These vaccines provide superior immunity but may not be appropriate for all animals due to the possibility for developing active disease, risk of abortion.

Killed vaccines -Immune response and protection by killed vaccines is inferior to MLV.

Handling of vaccine and Vaccination: The efficacy of the vaccines depends on proper handling and administration of the vaccine maintaining cold chain and sanitary conditions. Expiry date vaccines should not be used. Poor compliance with recommended handling instructions such as maintain cold chain, using new sterilized separate syringe and needles, avoiding exposure to sunlight, or inaccurate dosing may decrease efficacy of the vaccine resulting into diminished protection of the herd.

Calves

Development of immunity in calves depends on both vaccination and adequate colostrum. Neonatal calves can be prevented from scours by vaccinating their dams prior to calving and relying on the transfer of maternal antibodies in colostrum. Prior to weaning, calves may be vaccinated against infectious bovine rhinotracheitis (IBR), bovine viral diarrhea (BVD), parainfluenza-3 (PI3), bovine respiratory syncytial virus (BRSV), and clostridial disease followed by a booster vaccination at weaning, Replacement heifer calves should be vaccinated against brucellosis.

Cows and heifers

Pre-breeding vaccinations should be administered to cows and heifers entering the breeding herd 30 to 60 days prior to breeding.

Bulls

Breeding bulls needs to be examined for their breeding soundness and semen quality during pre- breeding period. Testing for trichomoniasis and annual vaccinations against IBR, BVD, PI3, BRSV, leptospirosis and diseases prevalent in the area should be done.

Factors Affecting Herd Health Programmes

- Failure to appreciate the existence of subclinical diseases and their potential to cause economic losses that can be reversed by taking positive steps for improving performances of the animals.
- Farmers traditional willingness to pay for emergency veterinary advise and care but reluctancy to pay for preventive veterinary advise when results are not immediately visible.
- Farmers perception of a veterinarian as primary provider of clinical service only.
- Failure of large dairy owners to hire a veterinarian for integrated service.

32

Peri-Parturient, Calf and Heifer Health Care

Peri-parturient Health Care

Periparturient period is defined as a period immediately around calving. It is the most critical time when high yielding cows are prone to nutritional deficits due to great metabolic alterations from dry to milking. Reduced feed intake and depressed immune system at this time make cows vulnerable to periparturient diseases affecting milk yield adversely and increasing mortality risk. Health care during this critical period aims at minimising deficits particularly in relation to dry matter intakes (DMI's) and energy and mineral balance.

- Milk fever is the most common and important periparturient disease affecting high yielders. Due to the sudden onset of lactation, there is a huge demand for calcium resulting in dramatic decrease in blood calcium level during calving causing milk fever.
- Retained placenta and subsequent metritis is seen in many cows with clinical or subclinical milk fever affecting the milk production at the beginning of the lactation and delaying estrus.
- Subclinical hypocalcaemia (milk fever) also increases the risk of developing left abomasal displacement.
- Cows with milk fever have increased chance of developing coliform mastitis in following lactation.
- The use of dietary cation-anion balance (DCAB) or more often partial DCAB diets in the last ten days of the dry period can assist in controlling the occurrence of milk fever. These diets are now a vital part of management of the periparturient cow.
- Energy deficit (from insufficient voluntary dry matter intake) and urge of high production in cows during transitional period leads to negative energy balance (NEB), for a period of days to weeks after calving, making them susceptible to ketosis. The cow undergoes to various metabolic adaptations (hepatic gluconeogenesis, mobilizing body fat)

during lactation to sustain milk production. Subclinical hypocalcaemia contributes to development of ketosis by lowering smooth muscle function.

- Nutritional management of dry cows, transition cows, and fresh cows during periparturient period requires meeting of their nutritional demands as well as promoting appetite and a balanced dry matter intake (DMI) following calving. Dry matter intake after calving can be promoted by (a) minimizing environmental stressors and keeping the cows comfortable, (b) optimizing body condition score (BCS around 3 at dry off), and (c) feeding management during dry period (to limit nutrient intakes to the requirements of the cows rather than volume and controlling energy intake).

  For fresh cows (postpartum) a diet with 23 percent to 25 percent dry matter with moderate fermentability (ground dry corn rather than high-moisture corn or ground barley) and adequate, effective forage fibre is a best option. Dietary fat may contribute in lowering NEFA levels and preventing ketosis.
- Dairy cows are vulnerable to production and other diseases during the peripartum and early lactation. Disorders like fatty liver, ketosis and impaired reproductive performance are associated with postpartum negative energy balance (NEB). Increase in the incidence of metritis, retained placenta, milk fever, ketosis, abomasal displacement and longer interval to first breeding has been reported in cows who lost > 1 BCS unit (on 5 point scale) as compared to those that lost <1 BCS unit during the transition.

Calf Health Care

Calf health management is crucial for ensuring an overall productivity and profitability of a dairy operation. Effective management practices can significantly reduce morbidity and mortality rates among calves. Key aspects of calf health management are as follows:

Care of the new-born calves

- When the calf is born, the mucous and phlegm should be cleaned from the nose and mouth.
- If the calf does not breathe, it should be held from the rear legs and lifted from the floor with the head down or alternate compression and relaxation of the chest will often start the calf breathing.
- The navel should be dipped in tincture iodine.

- The calf should be wiped with a clean coarse cloth to hasten drying.
- All wet bedding from the cow pen should be removed and cow's udder should be cleaned with clean water and dilute solution of the potassium permanganate.
- Healthy and normal calf will stand on its leg and suck within 30 minutes. If calf's fails to suck, assistance should be provided in getting the first feeding.
- Normally meconium (dark material that has been collected in the intestine of unborn calf) passes within 2 hours after the first feed. If it does not pass in due time, an enema consisting of one and a half teaspoonful of sodium bicarbonate in a litre of warm water should be given.

Colostrum feeding to calves

- Colostrum is the first milk produced by the dam following birth. It is rich in antibodies, growth factors, and nutrients essential for the calf's immunity and development. Since calves are born with an immature immune system and lack antibodies, colostrum feeding (Fig. 198) is crucial and essential for passive immunity transfer.
- There are three Q (quantity, quality, quickly) principle of colostrum feeding. The quantity of colostrum to be given as a first feed should be 3 litres or 10% of body weight within the first 2 hours after birth. Early intake maximizes the absorption of antibodies. This should be followed up by another similar-sized feed within 12 hours of birth. The colostrum should be fed at body temperature (38°C).

  The quality of the colostrum should be good. Good quality colostrum contains at least 50g/L of IgG. Any colostrum containing <20g/L of IgG should not be used. Colostrum quality declines when it is held in the udder for a longer period as it becomes more diluted with time. If the dam's colostrum is not of sufficiently good quality, the supply of good quality frozen colostrum from cows of known health status should be ensured.

  The first colostrum feeding should be done as quickly as possible within 2 hrs of birth. The efficiency of antibody absorption from colostrum declines rapidly from over 40% at birth to less than 5% by 20 hours.
- Colostrum can be stored in refrigerator if intended to be used within 24 hrs. For longer period storage, colostrum can be frozen in sanitized containers.

Frozen colostrum is gradually thawed in warm water (not hot, to prevent denaturing antibodies). Colostrum should not be microwaved, as it can damage antibodies.

Colostrum Feeding Techniques

i. Bottle feeding can be done to feed colostrum to calves. It has an advantage of monitoring of the exact amount consumed. Use a clean bottle and nipple, and ensure the calf nurses vigorously. Hold the bottle to mimic natural nursing height.

ii. Another way of feeding colostrum is through esophageal tube. Weak calves and those unable to suckle effectively, can be fed through esophageal tube. Use a clean, sanitized esophageal tube feeder. Ensure proper placement in the esophagus to avoid aspiration into the lungs. Administer the colostrum slowly and gently.

Monitoring Colostrum Intake

i. Clinical signs such as weakness, lethargy or failure to stand in new born calves may indicate that not enough colostrum is being consumed by the calf.

ii. For monitoring colostrum intake, blood samples can be tested either for IgG or serum total protein tests within a week after birth to assess the success of passive transfer. Levels above 12 g/l (IgG) and 5.5 g/dL (serum total protein) generally indicate adequate colostrum intake.

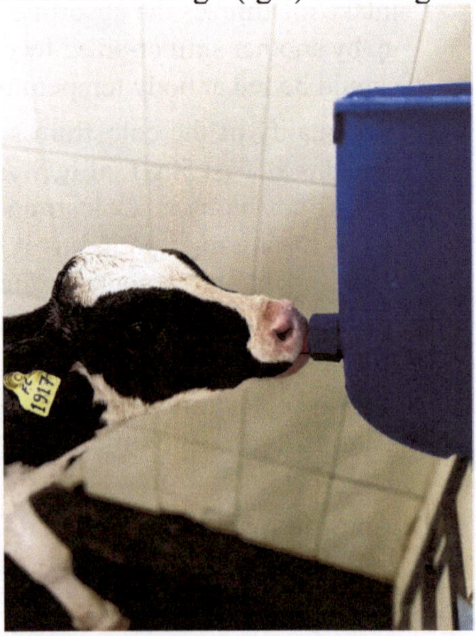

Fig. 198: Colostrum feeding to a calf.

Naval care

- Proper navel care is crucial in newborn calves to prevent infections and promote overall health. The navel, or umbilical cord, is a potential entry point for pathogens immediately after birth.
- The navel provides a direct pathway for bacteria to enter the bloodstream, potentially leading to serious infections such as septicemia, arthritis, and umbilical abscesses. Ensuring

proper navel care helps to prevent these infections and promotes healthy growth and development in calves.

- Ensuring a clean and dry birth area can minimize the exposure of the navel to pathogens.
- Handling a new born calf with clean hands or with gloves is very important in reducing the risk of introducing bacteria.
- Disinfect the navel as soon as possible after birth, ideally within the first 15 minutes. Use a 7% tincture of iodine solution for effective disinfection. Lower concentrations or alternative products like chlorhexidine can also be used. Fully immerse the navel in the disinfectant using a navel dip cup or spray bottle. Ensure complete coverage, allowing the solution to penetrate the navel stump and surrounding area.
- The navel should be monitored over the first week of life to ensure it is drying properly and healing without signs of infection.
- The calf should be kept in a clean, dry environment with fresh bedding to reduce the risk of contamination during the healing process.

Signs of Navel Infection

- Be vigilant for signs of navel infection such as swelling, redness, discharge, pain, fever and lethargy.
- Swelling around the navel may be persistent or increasing.
- Redness around the navel area is suggestive of inflammation.
- Pus or other discharge from the navel is also indicative of navel infection.
- Discomfort on touching the navel area is due to pain and suggestive of infection/inflammation.
- Elevated body temperature (fever) is indicative of a systemic infection.
- Navel infection may lead to weakness and calf appears less active.

Treatment of Navel Infection

- If a navel infection is suspected, seek prompt advice from a qualified veterinarian.
- Antibiotics need to be administered as per the advice of the veterinarian.
- If there is abscessization, it needs surgical drainage.
- Additional supportive care with adequate nutrition and hydration helps in rapid recovery of the calf.

Calf starter

- Calf starter feed is essential for promoting rumen development, providing necessary nutrients, and ensuring healthy growth in young calves. Introducing calf starters early in a calf's life encourages solid feed intake, which is critical for the transition from a milk-based diet to solid foods.
- Calf starter should be energy-dense to support rapid growth. Carbohydrates from grains like corn or oats are commonly used. Calf starter should have protein content of 18-22% to support muscle development and overall growth. Calf starter should be highly palatable to encourage intake. Molasses is often added to enhance flavour. Fiber content in calf starter should be moderate to promote rumen development without causing digestive issues.Calf starter should have balanced levels of essential vitamins and minerals, including calcium, phosphorus, and vitamins A, D, and E, to support overall health.
- Example of calf starter (approximate percentage) -Maize- 52 %; Oats - 20 %; Soya bean meal - 20 %; Molasses - 5 %; Salt - 0.5 %; Minerals (Macro & Micro) - 1.5%; Vitamins - 1%.
- Introduce calf starter at 3-5 days of age. Ensure starter feed is always available so calves can begin nibbling as soon as they show interest.
- Keep the feed (calf starter) fresh and clean by providing it in small amounts and replenishing it regularly.
- Use shallow pans or troughs to make it easy for calves to access the feed.
- Encourage calves for taking calf starter by placing some starter in mouth to stimulate interest if they are slow to start eating.

Monitoring and Management

- Amount of starter feed consumed by each calf should be monitored. Gradually increase the amount of starter feed offered as intake rises. A healthy target is for calves is to consume 1-2 pounds of starter feed per day by three weeks of age.
- Ensure availability of fresh, clean water at all times. Water intake is crucial as it aids in digestion and encourages starter feed consumption.
- Regularly monitor health and growth by regular weighting to ensure that they are gaining weight.
- Monitor regularly for any signs of digestive issues, such as diarrhea or bloat and adjust feeding practices accordingly.

Weaning Process

- Begin weaning when calves are consistently consuming at least 2 pounds of starter feed per day for three consecutive days, typically around 6-8 weeks of age.
- Gradually reduce milk feedings to encourage greater starter intake.
- After weaning, continue providing calf starter until calves are fully transitioned to a grower ration.
- Ensure a smooth transition by mixing starter feed with the grower ration initially.

Parasitic infestation and deworming in calves

- Parasitic infections in calves can lead to poor growth, reduced weight gain, anemia, diarrhea, and overall weakened immune function. Parasitic infestation in calves can be due to round worms, coccidia, tape worms and lung worms. Effective deworming is essential to maintain calf health, improve feed efficiency, and enhance overall productivity.
- Roundworms (Ascarids) are common endo parasites in young calves, They can cause weight loss and diarrhea.
- Coccidia (protozoan parasites) are causing coccidiosis in young ones leading to bloody diarrhea and dehydration.
- Tapeworms are less common in young ones but can cause intestinal blockages and nutrient malabsorption.
- Lungworms can affect the respiratory system, leading to coughing and respiratory distress.

Deworming strategies

- First deworming should be done at 3-4 weeks of age to target roundworms.
- Follow-up deworming should be done at regular intervals based on the farm's parasite load, typically every 4-6 weeks until weaning.
- In some regions, deworming is timed with specific seasons to target peak parasite activity.
- Benzimidazoles anthelmintics (e.g., Fenbendazole) are effective against a broad spectrum of nematodes.
- Macrocyclic Lactones (e.g., Ivermectin) is effective against both gastrointestinal and lungworms.
- Levamisole is effective against nematodes, used in rotation to prevent resistance.

- Coccidiostats (e.g., Amprolium) should be used to specifically target coccidia.

Administration Methods of deworming

- Oral Drench- They are administered directly into the mouth using a dosing gun. Precise dosage is to be ensured.
- Injectable-They are administered subcutaneously or intramuscularly. They provide systemic protection.
- Topical Pour-on- They are applied along the backline. Easy to administer but ensure even application to avoid under-dosing.
- Accurate weight measurement avoids under or over-dosing.
- Manufacturer's guidelines need to be followed. Always follow the dosage and administration instructions provided by the dewormer manufacturer.
- Rotation of dewormers is advisable at farms. Rotate classes of dewormers to prevent development of resistance.

Vaccination in Calves

- Vaccination is a critical component of calf health management, aimed at preventing infectious diseases that can cause significant morbidity and morbidity. Effective vaccination protocols help ensure the long-term health, productivity, and welfare of calves. The vaccination schedule of calves is given in the table 26.

Table 26: Vaccination schedule in calves

Sl. No.	Name of Disease	Age at first dose	Booster dose	Subsequent dose(s)
1	Foot and Mouth Disease (FMD)	4 months and above	1 month after the first dose	Six monthly
2	Haemorrhagic Septicaemia (HS)	6 months and above	-	Twice/thrice in a year in endemic areas.
3	Black Quarter (BQ)	6 months and above	-	Annually in endemic areas.
4	Brucellosis	4-8 months of age (Only female calves)	-	Once in a lifetime
5	Theileriosis	3 months of age and above	-	Annually. Only required for crossbred and exotic cattle.
6	Anthrax	4 months and above	-	Annually in endemic areas.

Heifer Health Care

For high quality replacement stock in the farm, heifers should be provided with an excellent health care and management services enumerated below.

- Heifers should be fed with a good quality feed in sufficient amount to sustain normal growth. They need relatively a more protein feed rather than energy during the early stage. Growth depends upon the quality of forage fed. They may grow well on ad libitum feeding of an excellent hay.
- The heifers may be kept in a loose housing system with a dry shelter free from drafts and open to one side.
- The size of the heifer is more important rather than age at breeding as small heifers may have difficult parturition.
- The heifer should be growing and in good flesh at calving time to produce milk at the most profitable level.
- Heifer should be placed in a separate shed 6-8 week before calving.
- A heifer should be fed 2 - 3 kg concentrate daily along with ad libitum good quality forage.
- Heifers should be made accustomed to milking procedure before calving and should be handled gently with kindness.
- Maintaining good health is necessary for the proper growth of the heifer. Hygienic housing, ad libitum drinking water, nutritional balanced feeding and preventive health care practices are necessary for the maintenance of good health of the heifers.
- The heifers in the herd should be checked periodically for growth and wellbeing. Heifers lagging behind below the required standards need to be removed from the herd.
- Heifers getting close to reaching reproductive age should be checked for estrus cycle on regular basis. On detection of heat, the heifer should be inseminated at the proper time. A proper breeding record is to be maintained.
- First calving may be difficult for some heifers. It requires an extra care during this period.
- Heifers health care programme has five important components viz. biosecurity process, vaccination programme, parasitic management, mineral (macro and micro), and farm specific or regional issues.

- As a biosecurity process, all animals entering the farm should be disease tested (tuberculosis, paratuberculosis, IBR, brucellosis etc.) and vaccinated as per the practice in the area. Reproductive diseases pose a threat to the success of pregnancy, heifer health and calf health. Heifers may be affected bovine viral diarrhoea virus (BVDV), infectious bovine rhinotracheitis virus (IBR), leptospirosis and vibro infections. Heifers should also be vaccinated against illnesses such as H.S., B.Q F.M.D. as per the prevailing vaccination policy in the area.
- Parasites may be farm specific. Quarantine drenches should be administered to the heifers on arrival and then retained in the quarantine paddocks.
- In heifers the efficiency of uptake of trace mineral from pasture changes as the rumen develops. Therefore supplementation of copper, cobalt, iodine, selenium etc. should be given.

Further Readings

Atkinson, O. (2016). Management of transition cows in dairy practice. In Pract. 38, 229–240.

Constable, P.D., Hinchcliff, K.W., Done, S.H.and Grunberg, W. (2017). A textbook of the diseases of cattle, horses sheep, pigs and goats and 11th edn., Elesvier, St. Louis, Missouri, pp137-152.

O'Boyle, N. (2008). Nutrition of the periparturient dairy cow. In: Pract. 30, 495–500.

Redfern, E. A., Sinclair, L.A. and Robinson, P. A. (2021). Dairy cow health and management in the transition period: The need to understand the human dimension. Resch. Vet.Sci. 137: 94–101.

33

Udder Health, Foot Health and Heat Stress Management

Udder Health

Mastitis is a significant and costly disease of udder affecting dairy cows and buffaloes, leading to reduced milk production, increased veterinary costs, and compromised animal welfare. Effective udder hygiene practices are crucial in preventing and controlling mastitis, maximizing milk production, ensuring animal welfare, minimizing costs, and maintaining consumer trust. By focusing on udder health management through preventative measures and the best practices, dairy farmers can enhance the overall sustainability and profitability of dairy operations.

There are six major steps that can lead to a better udder health.

- Farm hygiene and udder hygiene
- Hygienic milking procedure
- Dry cow management and therapy
- Appropriate therapy for mastitis cases during lactation
- Culling chronically infected cows.
- Good nutrition to increase the udder resistance to mastitis

Farm hygiene and udder hygiene

(a) *Farm hygiene* -It is of utmost importance in maintaining good udder hygiene. The following points are to be taken into consideration for a good farm hygiene.

- Always keep the dairy animals in clean, dry, and comfortable sheds.
- Bedding material should be clean and dry dust-free (without mould).
- Regular removal of manure and drainage of muddy areas helps to eliminate the breeding of flies.
- The floors in barns should be properly sloped to maintain a dry surface area.

- Proper ventilation helps reduce humidity and bacterial load in the barn.
- Provide adequate space to reduce stress and prevent overcrowding.

(b) *Udder hygiene-* A good udder hygiene involves pre and post milking hygiene.

i. Pre-milking hygiene

- Trimming of long udder hair
- Before handling teats, hands should be thoroughly washed with soap and water.
- Use a pre-dip solution (usually an iodine-based disinfectant) to sanitize the teats. Dip each teat in the solution and ensure it is fully covered. Leave the solution on for about 30 seconds to one minute to allow proper disinfection.
- Use a clean, disposable towel or cloth to wipe each teat. This helps remove the pre-dip solution and any debris.
- Make sure to use a separate towel or cloth for each cow to prevent cross-contamination.
- Squirt a few streams of milk from each teat into a strip cup to check for mastitis or abnormalities in the milk.
- Ensure that the teats are completely dry before attaching the milking machine to avoid slipping and to ensure a good seal.

ii. Post-milking hygiene

- Prevention of new intramammary infections (IMI) can be done by post milking teat dipping.
- The teats of all the lactating cows and dry cows (during the first 10-14 days of the dry period) are dipped regularly after every milking (Fig.199) in the germicidal teat dips such as Iodine (0.5%) solution + Glycerine @ 15% of iodine solution or Chlorhexidine (0.5%) solution + Glycerine @ 06% of chlorhexidine solution. The iodine teat dip also treats various types of teat lesions and injuries.

Hygienic milking procedures

- Establishing a consistent milking schedule to reduce stress on cows.
- Gentle and calm handling of dairy animals to promote milk letdown.
- Steps of a proper milking process: fore stripping, cleaning, attaching, and detaching milking units.

- Ensuring proper attachment and alignment of milking units to prevent teat damage.
- Machine milking- It reduces the risk of contamination compared to hand milking and properly used machines are gentle on the cow's teats, helping to prevent injuries and infections.

  Monitoring vacuum levels and pulsation rates to ensure efficient and gentle milking.

  Regular cleaning protocols for milking equipment should be followed. Regularly inspect and maintain milking equipment to ensure it is in good working condition.

  Replace worn-out or damaged parts promptly to prevent contamination or injury.
- Hand Milking- Washing of hands with soap and water before milking should be strictly adhered to. Use clean metal containers with lids for storage of milk.

  Prefer full hand milking method and avoid stripping and knuckling method of milking.
- Milkers' hygiene- A milker should be in good health and free of communicable illnesses such as tuberculosis, cholera, typhoid fever, and scarlet fever.

  Periodic medical examinations of milkers should be performed for communicable diseases such as tuberculosis, brucellosis etc.

  They should also maintain proper personal hygiene, such as frequent nail cutting, beard trimming, and haircuts.

  Milkers should always wear neat and clean clothes and caps.

  It is strictly forbidden to smoke, eat, drink, or spit when milking. Coughing and sneezing should also be avoided.

  Milking orders have a major impact on the spread of mastitis. First lactation cows should be milked first followed by cows with low somatic cell counts, then cows with high

Fig. 199: Post milking teat dipping in cow. Figure with the courtesy of Dr. D.K. Gupta, Principal Scientist, Department of Veterinary Medicine, GADVASU, Ludhiana.

somatic cell counts and in the last cows with clinical mastitis. This order of milking cows will reduce the spread of mastitis organisms from cow to cow.

Dry cow therapy

- Dry cow therapy (Fig.200) helps in the elimination of existing udder infections. It is done at the end of lactation (after the last milking) with a long-acting antibiotic intramammary preparation that maintains effective drug concentration for 6-8 weeks i.e., throughout the dry period.
- It not only eliminates the subclinical infections of previous lactation but also prevents new intramammary infections and increases milk production by about 8-10%.
- It also improves the milk quality at calving and prevents the occurrence of clinical mastitis cases during dry periods and around calving.

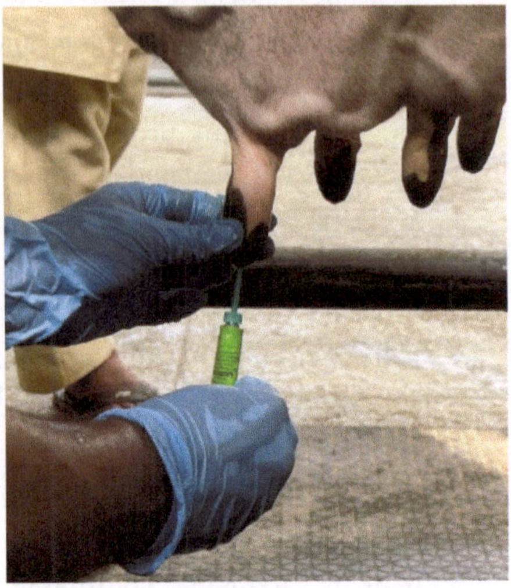

Fig. 200: Dry cow therapy. Long-acting antibiotic intramammary preparation is being infused through teat at the end of lactation. Figure with the courtesy of Dr. D.K. Gupta, Principal Scientist, Department of Veterinary Medicine, GADVASU, Ludhiana.

- Periodic checking of milk for detection of mastitis. Mastitis can be identified in its early stages (subclinical) before symptoms manifest, which helps dairy farmers avoid suffering significant losses as the disease advances into clinical stages.

  There are numerous mastitis diagnostic tests available, and they differ markedly in sensitivity, specificity, simplicity, rapidity, and cost. Among these, the Bromothymol blue (BTB) card (Fig.201), California mastitis test (Fig.202), and Electrical conductivity tests are simple and economical tests that can be performed as cow-side tests at the field level.

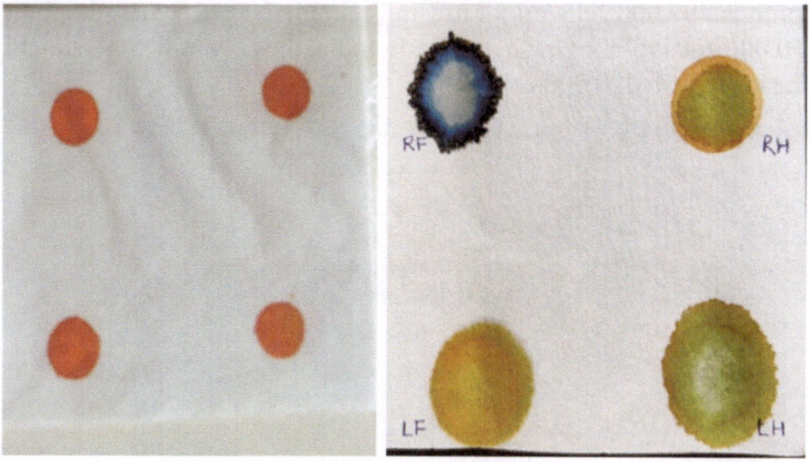

A **B**

Fig. 201: BTB card test for mastitis. Figures with the courtesy of Dr. D.K. Gupta, Principal Scientist, Department of Veterinary Medicine, GADVASU, Ludhiana.

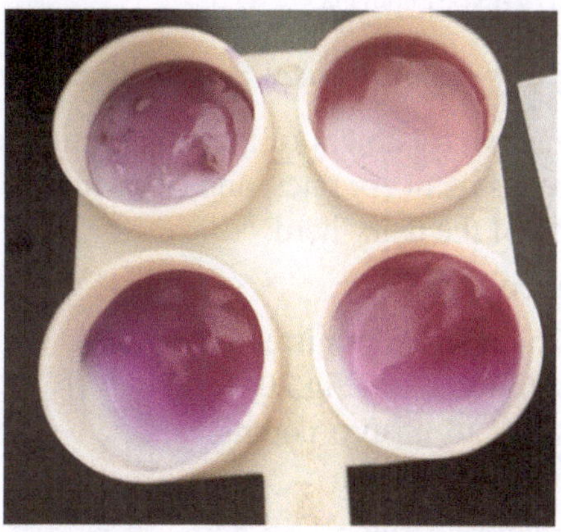

Fig. 202: CMT test for mastitis. Figure with the courtesy of Dr. D.K. Gupta, Principal Scientist, Department of Veterinary Medicine, GADVASU, Ludhiana.

- Clinical mastitis should be promptly treated with antibiotics and supportive therapies.
- Somatic Cell Count (SCC) is an important indicator of milk quality and udder health in dairy cows. It measures the number of somatic cells, primarily white blood cells (Fig.203), present in a milliliter of milk. Systematic records of SCC, monthly screening tests on individual cows, and tank milk SCC provide useful management information on the clinical status of the herd and the milk quality. Different countries have set legal limits for SCC in milk. For example, in the European Union,

the limit is 400,000 cells/ml for raw milk. In the United States, the limit is 750,000 cells/ml for Grade A milk. In India, the cow milk SCC should not be more than 400,000 cells/ml.

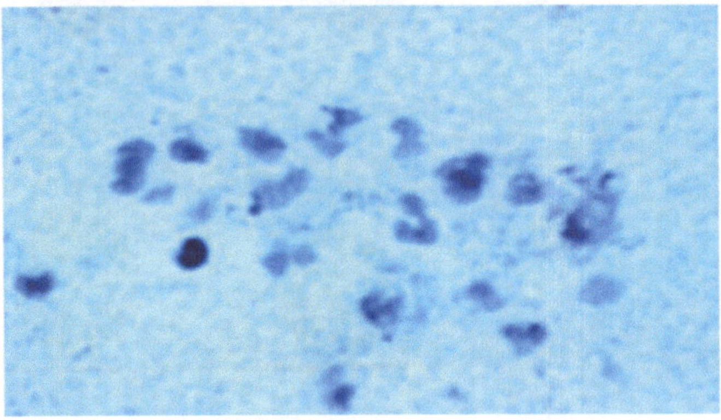

Fig. 203: Milk smear stained with Newman's stain showing inflammatory and secretory cells (100 X). Figure with the courtesy of Dr. D.K. Gupta, Principal Scientist, Department of Veterinary Medicine, GADVASU, Ludhiana.

Appropriate therapy for mastitis cases during lactation

- Prompt and effective treatment of mastitis minimizes the impact of the infection, improves recovery rates, and helps maintain the overall health and productivity of the dairy herd.
- Treatment Protocol:
  (a) Antibiotic Therapy: Administer appropriate antibiotics based on a culture sensitivity test (Fig.204) at least for 5 days to ensure complete eradication of the infection.
  (b) Anti-inflammatory Drugs: Use anti-inflammatory medications to reduce pain and swelling.
  (c) Frequent Milking: Milk the infected udder more frequently to remove infected milk and reduce bacterial load. Discard the milk from treated cows as it may contain antibiotics.
  (d) Intramammary Infusions: Use intramammary antibiotic infusions directly into the teat canal for targeted treatment.

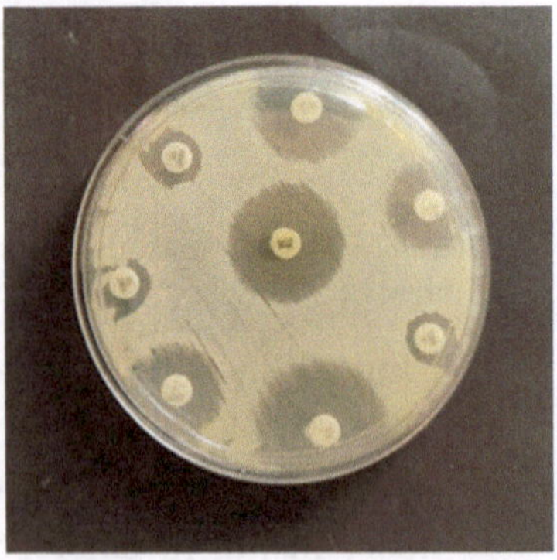

Fig. 204: Culture Sensitivity Testing of milk (Figure with the courtesy of Dr. D.K. Gupta, Principal Scientist, Department of Veterinary Medicine, GADVASU, Ludhiana)

(e) Fluid therapy and systemic glucocorticoids can be used in per-acute mastitis.

Culling chronically infected cows

- Chronic mastitis: A case is considered chronic when (i) there is the formation of fibrotic cord inside the teat canal (ii) there is a thick pus discharge, not responding to treatment (iii) there is frequent reoccurrence of mastitis in the same quarter. The treatment/surgery of chronic mastitis is not rewarding. Thus, it is best to cull these cases from the milking herd.

Good nutrition to increase the udder resistance

- Certain vitamins (Vit E, C, A and ß-carotene) and micro-nutrients (Cu, Se, Zn, Co) are reported to have a beneficial impact on the efficient functioning of the udder immune system. Supplementation of cows during dry periods and around calving (first 8-10 weeks) with the following nutrients per head per day proved beneficial in preventing mastitis/ lowering milk SCC:-

  Vitamin 53000 IU + Beta- carotene 300 mg

  Zinc-methionine (180-360 mg Zn, 360-720 mg methionine)

  Copper @ 20 ppm i.e. about 200 mg

  Vitamin E 1000 IU during dry period and 500 IU for lactating cows

  Selenium @ 3 mg during dry period and 6 mg during lactation

Foot Health

Foot is an extremely important structure of animal body. Though, animal may function with affected foot, optimum production is adversely affected. Therefore,foot health management in dairy animals is crucial for maintaining an overall herd health, productivity, and welfare. Uncared foot (Fig. 205) may lead to lameness and production losses. Lameness in dairy cattle can lead to reduced milk production, poor reproductive performance, and increased culling rates. Some key aspects of foot health management in dairy animals are given below.

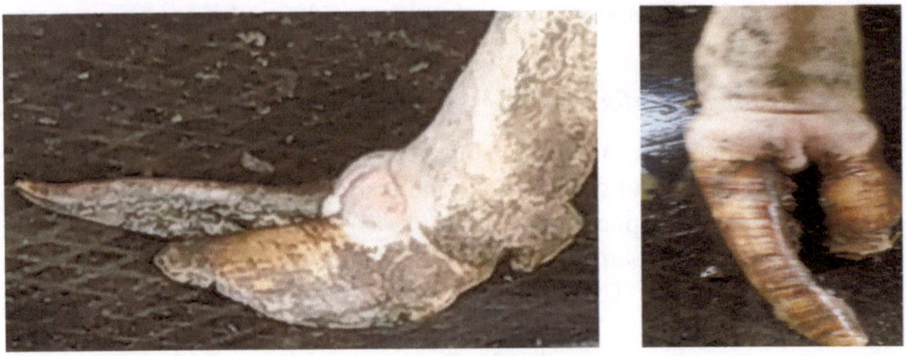

Fig. 205: Uncared foot, a common cause of lameness, showing enlarged hoof.

1. ***Regular hoof trimming***- Bovine hooves grow approximately at the rate of two inches per year. The horn of the wall grows@ 7 mm per month and horn of the sole grows @ 3 mm per month. Weight bearing in the normal hoof occurs on the outer horny portion. When weight bearing shifted from this area, lameness becomes a common complaint. Hoof trimming should be done at least twice a year, but more frequently if some issues are detected or there is some infections. The aim of hoof trimming is to provide a flat surface for weight bearing. Hoof trimming is less common in beef cattle but is more common in dairy animals. The sole of the medial (inner) claw should be left 3 mm thicker than the lateral (outer) claw.

Several hoof-trimming methods are used worldwide. Two types of hoof-trimming methods are commonly in vogue.These are based on the way they treat the angle of the sole relative to the metatarsals/metacarpals.(i) In the first method, the sole is trimmed flat, creating a perpendicular angle with metatarsals. (ii) While in the second method, the sole is trimmed to create a slope, with the whole axial sole being lower than the abaxial side. The most common methods are the Functional and White Line methods (both flat methods), and the Kansas method. Variations of these methods differ in the extent of horn removed underneath the flexor tuberosity to decrease pressure on the location where sole ulcers occur. Decreasing the pressure in this fashion is a technique known as modeling. The most common hoof-trimming method worldwide is based on the functional method developed by Dr. Touissant Raven. This method is a three-step procedure of functional (preventive) trimming, followed by two steps of corrective (therapeutic) trimming to address the treatment of horn lesions.

2. ***Routine footbath*** - Foot lesions, particularly infectious skin lesions, are very effectively controlled by formalin footbath (4%). Footbath dimensions should be 1 m wide, 3 m long, and 15 cm tall. The best and least expensive are

concrete footbaths. For foot bathing, formalin (39–40%) should be used as a 4 percent solution in the footbath (120 liters of water plus 5 liters of formalin). Every two weeks, on the first, second, and third days, animals should have foot baths. Animal's feet need to be well cleaned before being placed in a footbath. Animals should be kept on a clean pucca (concrete) floor for at least thirty minutes after receiving a foot bath.

3. *Nutrition* - A well-balanced feed with the right proportion of protein to energy should be fed to the animal. It is important to ensure that dairy cattle are provided with high-quality roughages for proper ruminating to minimize the occurrence of laminitis. Provide the animal with one-third of its dry matter intake in the form of fodders and roughages. Rumination and increased salivation are caused by roughages, and both actions are necessary for balancing off the excess acids in the rumen.

Changes to the composition of feed rations must be made gradually. During the dry period, dairy cows should be fed a small amount of concentrates, or none. Two to three weeks before calving, give a transition ration, and over the first six weeks of lactation, progressively increase concentrate. Don't ever administer more concentrate than 4 kg at once. At least 21% of NDF in rations should come from pasture. In early lactation, add extra nutritional buffers to the diet, such as sodium bicarbonate at 1% of the total DM ration. A high proportion of concentrates in feed rations, which contain more than 28% starch, has undesirable effects on the usefulness and health status of animals, including hoof diseases.

Regularly add a mineral mixture to the ration since it has been found that inadequate synthesis of hoof horn is caused by trace mineral deficiencies, including those of zinc, copper, manganese, and cobalt. It has been shown that biotin strengthens the intercellular cementum that holds keratinocytes together, thereby improving the quality of the horn. Where lameness is a significant issue on farms with high-yielding animals, biotin 2 percent at 1 gm/animal/day and/or zinc sulphate at 4 gm/animal/day should be added to the ration.

4. *Stall management*: Make sure the stall's width is at least 1.2 meters and its length, depending on stature, is between 2.5 and 2.8 meters of the cow. Make sure there is adequate drainage and great underfoot conditions. Maintain clean and dry living conditions to prevent bacterial infections like digital dermatitis and foot rot. Provide non-slip, comfortable flooring in barns to reduce trauma to the hooves.

5. *Early detection of hoof problems and their treatment*: Inspect animals regularly for early signs of lameness or hoof problems. Address hoof issues promptly to prevent them from worsening. Employ appropriate treatments like topical antibiotics or bandages when necessary.

6. ***Genetic selection***: Select for genetic traits that contribute to better hoof health. Avoid breeding from animals with a history of lameness or hoof problems.

7. ***Training and education of the staff***: Farm staff should be aware of the importance of hoof health and should be trained in hoof management techniques.

Heat Stress in Dairy Herd and its Management

Heat stress refers to any stress associated with high temperatures, which causes dairy animals to alter their thermoregulatory responses. Heat stress develops when an animal's capacity to regulate their body temperature through sweating and panting is disrupted, particularly in hot, humid, or dry weather. Dairy animals exhibit physiological and behavioural changes, such as decreased feed intake, altered metabolism, and increased energy expenditure to disperse heat, during high temperature and humidity times. Due to reduced milk yield and impaired reproductive function as a result of heat stress, farmers suffer significant financial losses. Heat stress can also lead to malfunctioning of immune system and gastrointestinal tract impacting negatively the health and productivity of dairy cattle.

Susceptible animals

Indigenous breeds of cattle are more thermo-resistant than crossbred, and exotic breeds of cattle. Buffaloes are more susceptible to heat stress due to their black skin that absorbs more solar radiation and fewer sweat glands (only 1/6th that of cattle), compromising heat dissipation through evaporative heat loss.

In India, the April to September months are more stressful for dairy animals due to the intense heat waves.

Impact of heat stress on dairy animals

- *Effects on Milk Production and Composition*- Heat stress significantly impacts both the quantity and quality of milk produced. Elevated temperatures lead to lower milk production due to decreased feed intake and altered metabolic processes. Heat stress inhibits the synthesis of milk fat and protein, resulting in lower energy-corrected milk yields. The fatty acid profile of milk also changes, with reductions in beneficial fatty acids such as polyunsaturated fatty acids (PUFAs). Heat stress exacerbates somatic cell count.

- *Poor reproductive performance-* Heat stress can disrupt oestrus cycles, reduce conception rates, and increase embryonic mortality, leading to decreased reproductive efficiency and longer calving intervals.
- *Negative effect on feed intake and digestion-*Dairy cattle experiencing heat stress may reduce feed intake, leading to nutritional deficiencies and impaired digestion, which can further impact milk production and overall health.
- *Increased morbidity and mortality-* Heat stress can exacerbate existing health issues and increase the risk of diseases such as mastitis, metabolic disorders, and respiratory problems, ultimately leading to higher mortality and morbidity rates.
- *Heat-related stress and discomfort-* Heat stress can cause discomfort, stress, and behavioural changes in dairy cattle, affecting their overall well-being and behaviour.

Management strategies for heat stress

By implementing following management strategies, dairy producers can help minimize the impacts of heat stress on their herds, maintain productivity, and ensure the health and welfare of their cows during periods of high temperatures.

1. Environmental modifications

a. *Building design considerations:* The building's orientation needs to be taken into account when designing the animal shelter. The buildings should be oriented east-west to maximize shade and prevent sun radiation. Higher sidewall heights can allow more afternoon sun exposure, which might increase the heat load inside the building. Large eave extensions may mitigate the direct radiation on the side walls.

Fig. 206: Feeding in shed to reduce solar radiation during summer.

b. *Shade provision:* Providing shade reduces the solar radiation absorbed by cattle. Shade structures (Fig. 206) can be natural (trees) or artificial (shade cloths, shelters). There are many different kinds of roofing materials, but an aluminium or galvanized reflective roof with insulation underneath is the most efficient. Compared to asbestos, thatch and agro-net can effectively block the detrimental effects of environmental factors and combat hot, humid conditions.

c. *Bedding:* The thermal properties of bedding materials can play a crucial role in alleviating heat stress during summer. For instance, at a bedding thickness of 2.5 cm, limestone and sand maintain lower temperatures (25.9°C and 26.9°C, respectively) compared to wood shavings, which register higher temperatures (28.6°C). Furthermore, sand bedding has demonstrated a higher summer heat flux compared to straw and mattresses filled with rubber granules, indicating its superior performance in dissipating heat and providing a cooler environment for the cows. This makes sand an effective choice for bedding during hot weather conditions.

d. *Cooling systems:* Fans (Fig. 207 A), misters, coller, and sprinklers (Fig. 207 B) can help lower the ambient temperature and provide evaporative cooling. Tunnel ventilation and cross-ventilation in barns are also effective.

A B

Fig. 207: Cooling arrangements during intense heat. Fan **(A)** and Sprinklers **(B)** in the cow shed.

2. Nutritional strategies

a. *Increase energy density-* Use high-quality forages and add fats to the diet to increase energy density without increasing feed intake volume. Fats are a good source of concentrated energy and generate less heat during digestion compared to carbohydrates.

b. *Avoid high fibre feeds*- High fibre diets generate more heat during digestion. Therefore, incorporating easily digestible, high-energy feeds can help reduce the metabolic heat load.

c. *Electrolyte supplementations*- Heat stress can lead to increased sweating and loss of electrolytes such as sodium, potassium, and bicarbonate. Providing supplements that restore these electrolytes can help maintain proper hydration and physiological function.

d. *Adding salt* – Addition of salt to the diet can encourage water intake and help replenish sodium levels.

e. *Water additives*- Consider adding buffers (such as bicarbonates) and other additives to water to improve hydration and buffering capacity.

f. *Provision of potable water*- Provision of ample, clean, cool, and freshwater sources for cows to drink can keep the animals hydrated and help them to regulate their body temperature through evaporative cooling. Cows can consume up to 50% more water under heat conditions.

g. *Changing feeding time*- Provide the bulk of the daily ration during cooler parts of the day (early morning and late evening) to encourage intake when the temperature is lower.

h. *Vitamins and mineral supplementation*- Supplementing diets with vitamins (A, C, and E) and minerals (like zinc and selenium) will help in reducing the oxidative stress induced by heat stress.

i. *Yeast and probiotics supplementation*- Supplementation can help improve gut health and nutrient absorption, which can be beneficial during periods of heat stress.

j. *Buffer supplementation*- Adding buffers to the diet can help maintain rumen pH and reduce the risk of acidosis, which can be a concern during heat stress when feed intake patterns change.

3. Genetic and breeding strategies

a. *Selective Breeding*- Selecting animals with better thermotolerance traits, such as shorter hair coats or higher sweating rates, can improve herd resilience to heat stress.

b. *Genomic Selection*- Utilizing genomic tools to identify and select heat tolerance traits can enhance breeding programs. Traits associated with heat stress resistance, such as those identified through genome-wide association studies (GWAS), can be incorporated into breeding strategies.

c. *Cross breeding* - Cross breeding with heat-tolerant breeds, such as crossing Holstein cows with more heat-tolerant breeds like Gir, can produce offspring with improved resilience to heat stress while maintaining productivity.

4. *Management of heat load*

Schedule activities such as feeding, handling, and milking should be done during cooler times of the day (early morning or late evening) to minimize heat load on cows. Implementing shade structures and sprinkler systems in holding areas can also help in reducing heat stress during milking.

5. *Health monitoring and disease management*

Regularly monitor cows for signs of heat stress and health issues, and implement preventive measures and treatment protocols as needed to mitigate the risk of diseases and health complications associated with heat stress.

6. *Management of reproduction*

Implement reproductive management practices to optimize fertility and minimize the negative effects of heat stress on reproductive performance, such as timed artificial insemination protocols and heat detection aids.

7. *Iodized oil*

The use of iodized oil in alleviating hyperthermia in cross-bred cows has given new dimensions. Subcutaneous administration of 2 ml iodized oil (containing 375 mg/ml of elemental iodine) for three consecutive days is a highly effective treatment of hyperthermia.

Further Readings

Costa, A., de Marchi, M., Neglia, G., Campanile, G. and Penasa, M. (2021). Milk somatic cell count-Derived traits as new indicators to monitor udder health in dairybuffaloes. Ital. J. Anim. Sci. 20:548–558.

Halasa, T. and Kirkeby, C. (2020). Differential somatic cell count: Value for udder health management. Front. Vet. Sci. 7: 609055. https://doi.org/10.3390/bioengineering9110608.

Lam, T.J., van Veersen, J.C., Sampimon, O.C. and Riekerink, R.G.O. (2011).On farm udder health monitoring. Tierarztl Prax. Ausg. G Grosstiere. Nutztiere. 39: 95–100.

Neculai-valeanu, A. and Ariton, A. (2022). Udder health monitoring for prevention of bovine mastitis and improvement of milk quality. Bioengineerin G 9:608.

Sharun, K.,Dhama, K., Tiwari, R., Gugjoo, M.B., Mohd, I.Y., Patel, S.K., Pathak, M., Karthik, K., Khurana, S.K. and Singh, R. (2021). Advances in therapeutic and managemental approaches of bovine mastitis: A comprehensive review. Vet. Q. 41: 107–136.

34

Metabolic Disease Management in Dairy Herds

Management of metabolic diseases depends on an early diagnosis. For an early diagnosis, metabolic profile test (MPT) is a diagnostic tool being commonly used in veterinary practice, particularly for dairy herds, to assess the metabolic health of the animals. It helps in an early detection by measuring various blood parameters that indicate the nutritional and health status of the cows for their effective timely management. MPT is based on the principle that the imbalance between input (feed) and output (production) is reflected in the abnormalities of key metabolites in the blood.

Key components of the metabolic profile test

1. Energy balance indicators

- *Glucose*- Reflects energy status and carbohydrate metabolism. Low levels can indicate negative energy balance or inadequate carbohydrate intake.
- *Beta-Hydroxybutyrate* (BHB)- Indicates the level of ketone bodies, high levels suggest negative energy balance.
- *Non-Esterified fatty acids (NEFA)*- High levels indicate mobilization of body fat reserves, often due to energy deficit.

2. Protein status indicators

- *Blood urea nitrogen (BUN)*- Levels of blood urea nitrogen indicates protein metabolism and balance.
- *Total protein and albumin*-Their values reflect an overall protein status and liver function.

3. Mineral balance indicators

- *Calcium* -It is essential for muscle function and milk production.Its imbalance can lead to milk fever.
- *Phosphorus*- It is important for energy metabolism and bone health.

- *Magnesium*- It is crucial for nerve and muscle function. Its imbalance can lead to grass tetany.

4. Liver function indicators

- *Aspartate Aminotransferase (AST) and Gamma-Glutamyl Transferase (GGT)*- Elevated levels of these enzymes indicate liver stress or damage.

5. Urine indicators

- Urine evaluation is done to monitor the acid-base, ketone bodies, and calcium status in dairy cattle.

6. Hemogram indicators

- Packed cell volume and hemoglobin are the important indicators of blood deficiency (anemia).

Uses and benefits of metabolic profile test

- It helps in an early detection of subclinical metabolic issues before they become clinically apparent.
- An early detection of metabolic imbalances can help in adjusting diet and feeding strategies to prevent metabolic disorders.
- Regular testing can monitor the effectiveness of dietary and management changes.
- Regular monitoring of metabolic profile can ensure optimal health of the cows leading to better milk production and overall herd performance.

Test procedure

- This test is conducted on a herd basis by selecting a representative sample of cows. This helps to obtain a comprehensive view of the herd's metabolic status.
- Blood samples are collected from 3 groups (7 animals in each group) of dairy cows/buffaloes i.e. dry animals, medium yield, and high yield animals.
- Samples are collected at least thrice in a year (summer, autumn, and winter) or when nutritional imbalance is suspected.
- The samples should be collected at the same time of the day at each collection.
- Ensure that animals are calm and handled gently to reduce stress, which can affect blood parameters.

- Collect blood samples from the jugular vein or tail vein using appropriate sterile techniques.
- Five ml of blood is collected in EDTA vial for estimation of PCV and hemoglobin and oxalate-fluoride vials for glucose and serum inorganic phosphorus estimation. Twenty ml of blood is collected in heparinized vials for the estimation of other biochemical parameters.
- Samples should be submitted to the laboratory within a few hours.
- Urine samples can also be collected and tested for urinary ketone bodies and results are correlated with the blood glucose level of each animal.
- The following parameters are monitored in Compton metabolic profile (CMP) test.
  - Packed cell volume (PCV)
  - Hemoglobin
  - Blood glucose
  - Blood urea nitrogen
  - Serum inorganic phosphorus, calcium, magnesium, sodium and potassium
  - Total serum protein, albumin and globulin
  - Serum copper
  - Serum iron
  - Plasma non-esterified free fatty acid
  - Beta hydroxy butyrate
  - Liver function test- aspartate aminotransferase (AST), Sorbitol dehydrogenase (SDH), gamma-glutamyl-transferase (GGT), total bilirubin, and cholesterol.
- A mini-profile test has been recommended in cows between 4 and 10 weeks after calving to assess the adequacy of energy and protein intake by measuring levels of blood glucose, serum urea nitrogen, and albumin.
- Interpretation of results- The mean values of each lactational group of cows for each parameter are calculated and compared with predetermined population mean values. Animals whose values are 2 SD (standard deviation) or more different from the population means are considered abnormal.

Energy balance

- Plasma non-esterified fatty acids (NEFA) concentration provides the most sensitive indicator of energy balance, especially in the last 2 weeks of gestation. High plasma concentrations of NEFA indicate a negative energy balance which could be a risk factor for development of ketosis and abomasal displacement in dairy animals.
- Plasma Beta-hydroxybutyrate is the gold standard test for diagnosis of subclinical ketosis in dairy animals. Increased concentration of plasma beta-hydroxybutyrate (above 1.0 mmol/L) in the first two weeks postpartum is associated with reduced milk production, increased incidence of clinical ketosis, and abomasal displacement.
- Decreased plasma glucose concentration is a specific indicator of negative energy balance. Plasma glucose concentrations are usually lower in early lactation due to the heavy demand for glucose and during winter months the energy intake is likely to be lowered than the requirement. Blood glucose levels are influenced by various factors viz. fluctuation in daily feed intake, physical form of feed, roughage content infeed, excitement, environmental temperature.

Protein evaluation

- Several indices such as milk urea nitrogen, plasma/serum urea nitrogen, creatinine, albumin, and total protein are used to evaluate the protein status of dairy animals.
- A low concentration of blood urea indicates minimal protein intake and an early warning that low protein status may develop later if protein intake is not increased.
- Low levels of albumin and hemoglobin are suggestive of long-standing protein deficiency.
- Milk urea nitrogen (MUN) concentration is also a useful indicator of protein status particularly when the diet contains adequate energy. Elevated MUN is indicative of excess protein feeding to the dairy animal for a given level of milk production. The target MUN levels are 8.5-11.5mg/dl.

Liver function

- The enzymes such as aspartate aminotransferase (AST), Sorbitol dehydrogenase (SDH), gamma-glutamyl-transferase (GGT), and alkaline phosphatase are used to evaluate the presence of liver injury.

- Plasma AST greater than 162U/L is suggestive of hepatic lipidosis.
- Liver function is also assessed by plasma/serum total bilirubin, cholesterol, and albumin concentrations.
- The ratio of NEFA to cholesterol on a molar basis is a clinically useful index to assess the degree of hepatic lipidosis and predict the incidence of metabolic disease in postpartum dairy animals.

Macromineral evaluation

- Abnormal serum concentration of four macrominerals, calcium, phosphorus, magnesium, and potassium during the transition period of dairy cattle is related to clinical milk fever (clinical hypocalcemia), postpartum hemoglobinuria, hypomagnesemia, and hypokalemia respectively.
- Estimation of serum concentration of calcium during the first 24 hours after calving can provide useful insight for the prevention of milk fever.
- Decreased concentrations of serum inorganic phosphorus, and magnesium are associated with long-term insufficient dietary intake.
- Serum magnesium levels are usually low during winter months and subclinical hypomagnesemia exists in many dairy animal herds. This can be converted into clinical hypomagnesemia with the sudden drop in environmental temperature or deprivation of feed.
- Low serum sodium concentration (less than 135 mmol/l) is associated with depraved appetite, polyuria, and polydipsia. Supplementation of the diet with phosphorus, magnesium, and sodium is protective.

Hemogram

- In most dairy herds a low PCV may be an indicator of suboptimal energy and protein nutrition. The mean values of PCV, hemoglobin, and serum iron are generally higher in nonlactating cows than in lactating cows. The cow should be grouped by lactation stage when evaluating the hematologic values. Blood loss due to parasites could result in low Hb and PCV.

Urine analysis

- Urine appears to be the optimal fluid to monitor acid-base, ketone bodies, and calcium status in dairy animals.
- Urinary acetoacetate is a sensitive indicator for early diagnosis of ketosis.

- Feeding rations with low dietary cation-anion difference (DCAD) to dairy cows for at least 2 weeks before calving decreases the incidence of milk fever which is reflected by low urinary PH and hypercalciuria.

Other recent approaches

- Recently, a more holistic approach has gained attention which includes the evaluation of body condition score (BCS), ultrasound-guided back fat thickness measurement (BFT) along metabolic profiling for the prediction of metabolic diseases in transition cows.
- BCS provides an indirect estimate of energy balance in cows. Proper BCS management can maximize milk production, improve reproductive efficiency, and reduce the risk of metabolic and peripartum diseases.
- BCS is the assessment of the body fat reserves of dairy animals by visual and manual inspection of the thickness of fat cover and prominence of the bone at the tail head and loin region.
- It uses a 5-point scale with quarter-point increments to assess the relative amount of subcutaneous body fat. The scores range from 1 (very thin) to 5 (excessively fat), with 3 representing an average body condition.
- As a thumb rule, cows should calve with a body score between 3.0 and 3.25. Cows having very low (<2.5) or very high BCS (>3.5) are prone to various production diseases.
- An increased risk of subclinical ketosis is also noticed when back fat thickness (BFT) is greater than 12 mm on real-time ultrasonography. Practically, if the average BFT in a herd is 12 to 14 mm, 2 to 3 weeks before calving, cows should be evaluated for feed intake and diet should be evaluated for energy density.

Management of Metabolic Diseases

After ascertaining the deficit, the metabolic or production diseases can be effectively managed through correcting the deficits and adopting scientific feeding practices.

- Ketosis can be prevented with balanced ration (particularly energy intake) maximizing dry matter intake before, at and after calving; adapting dry cows to milking-cow diet by lead feeding; avoiding overfat cows at calving (BCS>6 out of 8); lowering potassium and calcium levels in the dry cow; providing niacin and feeding monensin in the last month before calving. Clinical cases of ketosis can be managed with intravenous administration of about 500 ml of a 50 percent glucose with insulin. Use of corticosteroids for a few days is also followed to boost blood glucose levels.

- Milk fever can be prevented by stimulating cow's calcium mobilization before calving; feeding a transition diet three weeks to calving; including anionic salts in the diet; avoiding feeds high in potassium, sodium and calcium in the diet; increasing dietary calcium immediately after calving; avoiding fat cows at calving; administering vitamin D injections 2–8 days before calving for cows with a history of milk fever.
- If there is an indication of Mg deficit, an addition of 10-15 g of magnesium into the diet of pregnant cow and 30 g into the diet of lactating cow per day can help in preventing hypomagnesemic tetany. Mg being unpalatable and least soluble in water, it can be mixed with concentrate (60 g of magnesium oxide in 0.5-1 kg of grain).
- Cows on negative energy balance and low levels of selenium, vitamin E and beta-carotene prepartum are likely to suffer from retained placenta and poor immunological response. Risk of retained placenta in cows can be drastically reduced by increasing energy consumption and adding beta carotene, selenium or vitamin E in the diet during the last weeks of the dry period and transition period management. Cows with subclinical or clinical parturient peresis do have increased chances of retained placenta. Therefore cows with impending hypocalcemia should be managed to reduce the occurrence of retained placenta.
- Rumen acidosis can be prevented by giving a balanced diet in respect of starch and fibre. Sudden dietary changes and slug feeding of grain and/or molasses are to be avoided. Roughages should be provided with or immediately after feeding grain/molasses. Feeding buffers such as sodium bicarbonate or ionophores are also recommended to prevent acidosis.
- Increased incidence and longer duration of lameness has been observed in cows fed on higher level of crude protein possibly due to increased amount of degradation products of the protein in the rumen. In such cases, supplements with biotin and zinc have shown beneficial results.
- Laminitis due to acidosis on high grain ration can be prevented by feeding of good quality roughage, reducing grains and adding buffers. Supplementation of biotin and zinc (Zn) assists in reducing lameness through improving claw horn quality. Vitamin A and E is also important in maintaining claw integrity. Iodine, selenium, copper and manganese also play important role in improving integrity of claws.

- Feeding of low concentrate ration in dry period followed by too rapid increase of concentrate after calving increases the risk of abomasal displacement. Estimation of beta hydroxyl butyric acid (BHBA) and aspartate aminotransferase (ASAT) activity seems to have predictive value for the development of LDA. Development of LDA can be prevented by maintaining the forage to concentrate ratio of the diet fed in late gestation and early lactation.

35

Prevention and Control of Infectious Diseases

Infectious diseases are illnesses caused by pathogens such as viruses, bacteria, fungi and parasites. Infectious diseases usually spread in animals through contact, contaminated food or water and through bug bites. Some infectious diseases are minor while others are very serious. Infectious diseases in the production animals are of great economic concern owing to losses though morbidity, mortality, reduced production, increased food insecurity, public health concern, increased cost of veterinary care etc. Therefore, prevention and control of infectious diseases of animals is very important for food security and safety, public health, animal welfare, access to market and alleviation of rural poverty.

Key Factors for Effective Pevention and Control of Infectious Diseases

The key factors for effectiveness of prevention and control strategies of infectious animal diseases are, (i) effective surveillance (active i.e. planned; or passive i.e. event based), (ii) an early detection of infection (making an early diagnosis with laboratory support), (iii) transparency and rapid response mechanism; and(iv) good governance.

- ***Effective surveillance-*** It is upstream of animal disease prevention and control activities. There should be both active (planned) and passive (event based) surveillance. The components of surveillance system are data and sample collection, analysis and communication, information management and disease reporting, epidemiological and laboratory investigations. The strategy needs optimum communication and collaboration between all stakeholders of animal production chain (animal owner, veterinarian, laboratory personnel and state and national veterinary authorities). When there is a suspected or confirmed outbreak of a scheduled disease, veterinarians and animal health officials are required to report it to the relevant state veterinary authorities. State governments are responsible for notifying the Central Government through the Department of Animal Husbandry and Dairying (DAHD) about such outbreaks.

- ***Early detection of infection***- An early detection system is necessary for timely detection and identification of an incursion or emergence/re-emergence of a disease/infection in a locality, compartment, zone, state or country. There is a need to have an effective response mechanism for sample collection and laboratory analysis for a rapid and effective diagnosis.
- ***Transparency and rapid response mechanism***- Once a disease diagnosis and pathogen is confirmed, national and international community should be well alerted and disease should be notified to WOAH (World Organization for Animal Health formerly known as the Office International des Epizooties-OIE). Bio-security policies and measures should be taken to protect human and animal health from biological risks.
- ***Good governance***- Good governance of veterinary services is an important requisite to undertake any control programme to improve animal health worldwide. There should be a training programme for veterinarians, veterinary para-professionals and other stakeholders for detecting and reporting animal health incidents. Private veterinarians should be under legal obligation to report the occurrence of infectious disease outbreak to national veterinary authority.

Measures for Preventing Infection

Measures for preventing infection in animals can broadly be categorized in three heads viz. animal husbandry, biosecurity and vaccinations.

Good husbandry practices

Good animal husbandry practices play significant role in preventing infections. Components of good husbandry include nutrition (balanced, adequate and hygienic); clean potable drinking water; clean, comfortable, ventilated and safe housing with adequate light and enough floor space; grooming (Fig. 208), daily dung disposal and manure management; quarantine and isolation facility for sick animals (Fig. 209), stress reducing measures; improving internal and external biosecurity; and regular veterinary advise and ability to undertake effective epidemiological investigations and reporting.

Fig. 208: Grooming of the cows with automatic grooming machine.

Fig. 209: Isolation of sick animal in a separate ward/shed

The infected animal should be isolated and kept in a distant isolation ward/ place far away from regular animal houses. The isolation ward should not be at a higher level than the healthy sheds. If separate isolation building is not feasible, a buffer zone of at least 6 feet should be kept between healthy and isolated animals. Separate feeding, watering, and other equipment should be used for sick animals. Attendants working on sick animals and equipment used for them should not be used for a healthy herd. If this is not practicable, the sick animals should be attended after the healthy stock. Strict biosecurity measures such as handwashing, and disinfection of boots, clothing, and equipment should be followed. The isolated animals should be brought back into the herd only when the outbreak ends and they are fully recovered.

All new or returning animals should be quarantined separately from herds to minimize the risk of entering infection in the existing herd. The quarantine period depends on the incubation period of a disease. In practice, a quarantine period of 30 days covers almost all diseases except rabies for which quarantine period should be about six months. During the quarantine period, animals should be thoroughly screened for parasitic infestation and infectious diseases. If needed, de-worming is carried out on the $23^{rd}/24^{th}$ day of quarantine. The animals should also be subjected to dipping or spraying on the $25^{th}/26^{th}$ day to remove ectoparasites if any.

Biosecurity

Biosecurity is intended to protect animals and humans from disease causing materials of biological origin (bacteria, virus, ecto and endo parasites, fungi etc). Animal farm biosecurity is adoption of all management practices that reduces the chances of entry and spread of the infectious diseases on to the farm either by animals or humans. Animal biosecurity planning focusses on preventing infection (by various pathogens) of animals by preventing transmission of infectious agents between animals and the persons in contact with animals. Management practices adopted on the farm to reduce the entry of infections in the farm fall under the category of external biosecurity. And all management practices intended to prevent the animal contacts within the farm fall under the category of internal biosecurity. Continuous improvement in biosecurity measures is essential for healthy animal production.

Ensuring good hygienic practices (such as hand washing, implementation of cleaning and disinfection routines); testing and /or isolation of new animals or sick animals to reduce the spread of infectious agents in to the herd; cleaning and disinfection of all materials, related to animals, entering into the farm/ premises; controlling the visitors contact with animals; controlling contact of wild life, neighboring livestock and pets; separating groups of animals based on risk; rodent control practice; vector control practice (control of ticks, flies and mosquitoes); deworming practices: use of all-in all out concept with cleaning and disinfection between batches of animals; disease monitoring and record keeping; and effective communication, training and assessment are of great value in preventing infection. For parasite control, use the correct product in correct dose, in correct species and at correct time with periodic assessment of efficacy.

Farm biosecurity starts from the entrance. Therefore footbaths containing disinfectant should be provided at the entry so that persons frequently visiting and vehicles coming to farm may go through these foot baths for reducing the entry of the infection. There should be proper fencing of farm boundaries.

A proper record of visitors entering into the farm should be maintained. Common disinfectants recommended for use on farm are phenol (0.5 to 5.0%), potassium permanganate (1-2 mg/lit.), sodium carbonate (2.5- 4.0%), sodium hydroxide (1.0-5.0%), bleaching powder, boric acid (4.0-6.0%), formaldehyde solution (5.0 -10.0%), copper sulphate (5mg/l), iodophors, aldehydes, alcohol, chlorhexidine, hydrogen peroxide or quaternary ammonia compounds.

Vaccination

Vaccinations are valuable in herd health programme. They stimulate body's immune system. Though, vaccination of animals is a great asset in the prevention and control of many infectious diseases, on its own it will not usually achieve the desired results unless it is a part of an integrated control strategy utilising a combination of control measures. Before launching a specific vaccination policy, preconditions (vaccine quality and exit strategy- conditions under which vaccination be stopped) for a successful outcome should be ensured. For the effectiveness of the most of the vaccines, maintaining a cold chain is essential. Important vaccines for cattle and buffaloes are foot and mouth disease (FMD), haemorrhagic septicaemia (HS), black quarter (BQ), brucellosis, theileriosis, anthrax, infectious bovine rhinotracheitis (IBR), buffalo pox and rabies (post exposure). Vaccines commonly used in sheep and goats are sheep pox, goat pox, PPR, enterotoxaemia, blue tongue and orf. In pigs, swine fever vaccine is commonly used. Equine influenza and equine herpes are the vaccine used in horses. Camel pox vaccine is recommended in camel. Ranikhet disease/ Newcastle disease, fowl pox, and infectious bursal disease vaccines are routinely used in poultry. Details of vaccination schedule is given in chapter 37.

36

Approaches to Disease Outbreak Investigations and its Control

Sometimes veterinarians have to deal with disease outbreaks and unexplained changes in herd production. In this situation, broad objectives of the clinician are:-

i) To halt the progress of disease
ii) To determine the reasons of outbreak
iii) To recommend the procedures to reduce the risk of future outbreaks

Terminology used in Disease Outbreaks

The following terms are commonly used

Outbreak - It defines as a sudden increase in the occurrences of a disease when cases are in excess of normal expectancy in the population for the location, season or species. It may affect a small and localized group or may affect a large number of animals.

Epidemic- It has been described as an unexpected increase in the number of cases of a particular disease in a specific geographical area. In another words an epidemic is an rapid spread of a disease to a large number of animals or hosts in a given population within a short period of time.

Endemic- It defines the baseline incidence of disease in a population. Only emerging diseases have zero incidence, all others are present in small proportions of the population and can propagate at any point in time, given the appropriate environmental conditions. A disease outbreak is called endemic when it is consistently present but limited to a particular region.

Pandemic- It is an epidemic that affects multiple regions.

Case - It is an animal affected with the disease or condition under study.

Control - It is an animal that is not affected with the disease or condition under study.

Index case- It is the first known affected animal.

Steps Involved in Investigation of an Outbreak

There are six steps in an outbreak investigation that should be followed in order to ensure appropriate conclusions and can provide with further insight into a specific disease process.

Step 1. Case Definition

- The first step in outbreak investigation is to clearly define the case, so that animals can be classified as cases (affected) and control (unaffected). For example, if investigating an outbreak of possible Pasteurellosis in cattle, the case definition could be simply based on clinical signs (an animal with dysnea, an animal with hypersalivation, or an animal with nasal discharge and a temperature of $\geq 104^0$F as cases, and animals with absence of these signs as controls). Case definition can also be based on other diagnostic tests (culture, serology, clinical signs, or a combination of two or more of these.

Step 2. Magnitude of the Problem

- The second step is to determine the magnitude of outbreak that will require an accurate data regarding the baseline incidence of disease.
- If investigating a new disease, the baseline incidence should be zero. To calculate the magnitude of the problem, the affected proportion (AP) of animals is calculated.It is commonly also called as attack risk. It is calculated as number of cases divided by population at risk.

Step 3. Spatial and Temporal Pattern

Spatial pattern

- It is necessary to determine the place in farm where affected animals (cases) were housed and pastured in relation to unaffected animals (controls).
- A sketch of housing facilities, pastures, facilities and watering sources should be prepared.
- Any potential areas of concern including chemical storage,garbage dumps or industrial facilities should be noted.
- It is important to note a place where the index case was found, record its movements and subsequent cases should be mapped, preferably with dates.

Temporal pattern

- To determine the temporal pattern of the disease, a histogram of cases per day (or cases per hour if it is something more sudden) is built to reveal the epidemic curve. This will help determine if we are dealing with a potentially contagious situation or not.
- There are two distinctive types of epidemic curves: the point-source curve and the propagated curve.
- Point source curve: This is the typical epidemic curve of food borne and waterborne toxicity or poisoning, where all animals are exposed at one point in time.

  It is typically an epidemic of short duration as animals will show clinical signs within 1- 2 hours of exposure. It is difficult to point to one single index case.
- Propagated epidemic curve: In this, cases appear slowly but constantly throughout time period, dragging the epidemic curve over a certain period of time that commonly lasts days, and sometimes weeks. The index case is usually easy to determine.

Step 4. Analyze Potential Risk Factors

- All the data collected in previous steps is analysed for the key determinants of the disease using following methods.

  I. Method of difference: If the frequency of disease is different in two different circumstances and a factor is present in one circumstance but is absent from the other, then that factor may be suspected of being causing disease.

  II. Method of agreement: If a common factor is present in a number of different circumstances in which the disease is present, then that factor may be the cause of the disease. This method is frequently used to identify possible causal factors in outbreak investigations.

  III. Method of concomitant variation: The method involves a search for a factor, the strength of which varies continuously with the frequency of disease in different situations e.g. occurrence of bovine hypomagnesaemia and pasture levels of magnesium or association between severity of toxicity signs with amount of fodder/feed ingested.

Step 5. Provide a Written Report to the Owner

- Preliminary reports summarizing initial findings should be issued within a few weeks of visit to outbreak site. A final report can be issued later when all laboratory results are available. The report should contain the background to the investigation and initial complaint; objectives ; location of the site ; history ; methodology; description of facility and herd management; results of clinical and pathological examination; discussion of results and diagnosis; assessment of financial impact of problem and final recommendation.

Step 6. Follow up

- It is important to include new cases in the analyses as information that will appear during the course of investigation.

  If a conclusive diagnosis is reached, it is important to divulge the information so that everyone can learn from it, especially with emerging diseases.

  Document all the work, write up a report, and publish it!

Strategies for Control and Eradication

Control- It describes efforts directed towards reducing the frequency of existing disease to a level biologically and/or economically justifiable. The control means reducing the morbidity and mortality from disease and can be achieved by treating diseased animals or by preventing disease.

Eradication- The term eradication describes the efforts to eliminate selected causative agents from a defined area. The eradication can be at local, national or global level e.g. rabies, glanders and pleuropneumonia has been eradicated from European countries.

Prevention- The term prevention is generally applied to measures designed to exclude the disease from an unaffected population e.g. by quarantine, vaccination etc. The prevention can be applied at individual or population level.

The following Steps are taken for control and eradication of diseases.

- ***Doing nothing***- The incidence of disease might reduce due to natural changes in host-parasite relationship without the intervention of man e.g. blue tongue does not occur in winter because vector of causal virus - culicoides species cannot survive.

- ***Quarantine***- It is enforced physical separation of infected, potentially infected or suspected animals, their products or items they might have contaminated from the healthy population.
- ***Slaughter***- It is deliberate killing of infected, potentially infected or in-contact animals in an attempt to remove source of infection. The slaughter is usually accompanied by other procedures to reduce the risk of disease transmission like disinfection of premises, destruction of carcasses by burning or burial; this is termed as 'stamping out'. The slaughter can be *selective slaughter* (on individual animal basis; identification of positive animals using immunological or other screening tests and slaughtering the test positive animals, hence termed 'test and slaughter' policy. The method is followed in early stages of eradication programme or if agent spreads very slowly) or by *complete depopulation* (on a herd/ flock/ area basis; whole population including infected as well as uninfected and in-contact animals are sacrificed. Depopulation is preferred over selective slaughter if an agent is exotic to the area or spreads very quickly). The slaughter policy in cattle cannot be followed in India due to religious taboos.
- ***Vaccination***- It is administration of vaccine (active immunization) to an animal so that specific immune response is mounted and animal becomes resistant to infection. Mass immunization programmes have been used in successful control of various farm animal diseases such as rinderpest etc.
- ***Prophylactic chemotherapy***- Antibiotics, anthelmintics or other drugs can be administered at the times of high risk to prevent the disease e.g. routine deworming to prevent liver flukes in sheep.
- ***Movement of hosts***- Animals can be moved from high-risk areas where infections are endemic e.g. The Fulani tribe in West Africa migrate from south to north in the wet season to avoid tsetse fly.
- ***Mixed, alternate and sequential grazing***- The level of pasture contamination with some nematodes can be reduced by mixed, alternate and sequential grazing. The mixed grazing means grazing of susceptible animals with genetically or immunologically resistant animals e.g. calves (susceptible) can be grazed with adult cattle (immune); sheep (susceptible) can be grazed with cattle (resistant to *Ostertagia*). The alternate grazing means grazing of same pasture by different species of livestock e.g. annual alteration of cattle and sheep. The *sequential grazing* means grazing of resistant and susceptible animals of same

species at different times e.g. grazing of cows and calves at different times reduce Trichostrongyle infection in calves.

- ***Control of biological vectors***- The infectious diseases transmitted by biological vectors can be controlled by elimination of vectors e.g. removal of snails to prevent fasciolosis.
- ***Control of mechanical vectors***- The living organisms that mechanically transmit infectious agent can be controlled by destruction and disinfection.
- ***Disinfection of fomites***- Fomites such as farm equipment, surgical instruments should be disinfected to prevent the transmission of infectious agents.
- ***Niche filling***- The presence of organism within a niche prevents its occupation by another organism e.g. feeding of endogenous intestinal microbes to one-day old chicks prevented colonization of their digestive tract by virulent *Salmonella* spp. and *E. coli.*
- ***Genetic improvement***- by genetic screening and selective breeding.
- ***Improvement in environment, husbandry and feeding***- Details given in other chapters.

Further Reading

Thrusfield, M. (2007).The control and eradication of disease. In:Veterinary Epidemiology, 3rd edn. Blackwell Publishing. pp. 383-398.

Villarroel, A.(2015). Outbreak investigations. In: Practical Clinical Epidemiology For The Veterinarian. Wiley Blackwell. pp. 109-116.

37

Vaccines and Vaccination in Animals

Diseases of livestock cause massive economic losses in terms of reduced productivity and production and some of them have zoonotic potential posing a serious risk to public health. Vaccination is the most efficient and cost-effective method of controlling infectious diseases both in humans and animals. The eradication of smallpox and rinderpest from the globe; the elimination of hog cholera and brucellosis from many countries; and the control of diseases such as foot-and-mouth disease, canine distemper, rabies, influenza, and pseudorabies would not have been possible without the use of effective vaccines. Vaccination not only protects the vaccinated animals but also reduces disease among non-vaccinated animals through herd immunity. They also reduce antimicrobial usage and minimize the risk of antimicrobial resistance.

Vaccines

Vaccines are prepared antigens to be used to prevent diseases caused by bacteria, viruses, fungi, mycoplasma, protozoa, and parasites.

Types of vaccines

a. *Live vaccines*- Modified live vaccines infect host cells and undergo viral replication in the body. This may be hazardous because the vaccine viruses may themselves cause disease or persistent infection (called *residual virulence*). Live vaccines are of three types.

(i) Fully virulent vaccines: e.g. contagious ecthyma vaccine.

(ii) Heterologous vaccines: Closely related antigens share some isotopes. So, vaccine produced against one disease may protect the animals from other diseases. Such vaccines are called heterologous vaccines. For example, measles vaccine prevents canine distemper in dogs; feline panleukopenia vaccine prevents parvoviral gastroenteritis in dogs; herpes virus of turkey prevents Marek's disease in poultry.

(iii) Attenuated vaccines: Attenuation is process of reduction of virulence of organisms without causing its death. Examples of attenuated vaccines are BCG strain of *M. tuberculosis* that is rendered avirulent by growing it for 13 years on bile-saturated medium. The strain of anthrax is rendered avirulent by growing it under an atmosphere rich in CO_2 so that organism loses its ability to form spores. Rinderpest virus is attenuated by growing it in goats but this caprinised virus retains virulence for exotic and cross-bred cattle. To overcome this limitation, the virus is attenuated in rabbits and this laprinised vaccine has less residual virulence.

b. *Killed vaccines*-Killed organisms, in contrast, act as exogenous antigens. This may not be the most appropriate response for some organisms, but it is safer. They are produced by inactivating the infectious agent so that it cannot replicate in the host but its immunogenicity is maintained. Chemicals like formaldehyde, betapropiolactone, alcohol, ethylene oxide, ethylene amine and acetyl ethylene amine are most commonly used for inactivation.

c. Other types of vaccines- Although both killed and modified live vaccines have been successful in controlling many infectious diseases, modern molecular techniques have produced new and improved vaccines.

(i) Antigens generated by gene cloning (Category I):Gene cloning can be used to produce large quantities of purified antigen in culture. In this process, DNA coding for an antigen of interest is first isolated from the pathogen. The DNA is then inserted into a bacterium or yeast in such a way that it is functional and the recombinant antigen is expressed in large amounts. The first successful use of gene cloning to prepare an antigen in this way involved foot-and-mouth disease virus. The first commercially available category I recombinant veterinary vaccine was made against feline leukemia virus.

(ii) Genetically Attenuated Vaccines (Category II):This involves attenuation of organism by prolonged tissue culture, and is considered a primitive form of genetic engineering. The procedure results in the development of a strain of organism that cannot cause disease. The risk with these vaccines is reversion of organisms to virulent form. These vaccines are available against the Pseudorabies in swine.

(iii) Live Recombinant Organisms (Category III): In these vaccines, genes coding for protein antigens are cloned directly into a variety of organisms. The recombinant organism may itself be used as vaccine. The most widely employed vaccine for this purpose are poxviruses such as vaccinia, fowl pox, and canarypox. These viruses are easy to administer by dermal scratching or by ingestion. Example of this vaccine includes vaccinia-vectored rabies.

(iv) Polynucleotide Vaccines (Category IV): Another method of vaccination involves injection of DNA that encodes foreign antigens.

Vaccination

Vaccination is the process of administering vaccine (s) for preventing and controlling the occurrence of a disease (s) and reducing the transmission of the pathogenic agent. Therefore vaccination contributes to improved animal and human health, animal welfare, agricultural sustainability and to reduction in the use of antimicrobial agents in animals.

Canine Vaccination

The vaccination guidelines groups have adopted the concept of classifying vaccination of companion animals into three categories.

i. Core vaccine - The "core" vaccine is one that every dog or cat should receive because it provides immunity against infectious diseases that may be lethal or cause high morbidity e.g. rabies, canine parvovirus, canine distemper, etc.
ii. Noncore vaccine- A "noncore" vaccine is one that every dog or cat does not need to receive e.g. canine influenza, Bordetella, Lyme, etc.
iii. Not recommended vaccine - These vaccines do not have sufficient scientific evidence to justify their use in the field e.g. Giardia

Vaccination Schedule

There is no single universally acceptable vaccination schedule. The passive immunization by maternal antibodies should always be kept in mind while vaccinating the young ones. It is not practically feasible to vaccinate the pups repeatedly at short intervals from birth to 12th weeks of age, a procedure that would ensure complete protection. Moreover, certain parameters of cell-mediated immunity are not fully mature until the puppy reaches six weeks of age. So, two situations can be there:

(i) *Pups from vaccinated bitch*: One possible protocol is to vaccinate the pups at 8-10 weeks of age; followed by second vaccination at 12-13 weeks of age, which is a booster for early vaccinated pups and will vaccinate those not earlier vaccinated. A booster may be given at 5-6 months of age (optional) and then given yearly.

(ii) *Pups from unvaccinated bitch or If pups fail to take colostrum*: In pups complete immunological maturity is not achieved till 3-4 weeks of age because of immunosuppressive effects of endogenous corticoids, suppressor T-cells, immaturity of B-lymphocytes and initial hypothermia.

Modified live virus (MLV) vaccines may cause encephalitis in puppies before 3 weeks of age (CD-MLV); and cardiomyopathy before 4 weeks age (CPV-MLV). If bitch is not vaccinated, then vaccination against canine distemper (CD) and canine parvo (CPV) should be done at 4 weeks of age followed by a booster at 6 weeks, 8-10 weeks, 12-13 weeks, and at 5-6 months of age, and then yearly vaccination is done.

MLV vaccines should not be used in pregnant bitches.The first dose of any vaccine can be given at any age over and above manufacturers guidelines but booster and repetition doses are to be given exactly at a recommended time interval.

There should be a gap of 14-21 days between two different vaccines

Vaccination Schedule in Dogs

Vaccination schedule for dogs is given in the table 27.

Table 27: Vaccination Schedule in Dogs

Age	Schedule
3 weeks	Deworming for roundworms and hookworms
4 weeks	Vaccination against CD and CPV (if bitch was not vaccinated)
6 weeks	Vaccination against CD and CPV (if bitch vaccinated otherwise act as a booster)
8-10 weeks	Vaccination against CD, CPV, CAV-1, 2, Leptospirosis, Parainfluenza and Bordetella.
12-13 weeks	-Booster of above-
3 months or above	Rabies and repeat annually
5-6 months (optional)	Vaccination against CD, CPV, CAV-1, 2, Leptospirosis, Parainfluenza and Bordetella and repeat annually.

Causes of vaccination failure

Vaccination failure may be due to host related and vaccine relataed factors.

- Host-related factors: Maternal antibodies, intestinal worm load *(Toxocara canis, Ancylostoma caninum)*, immunosuppression, immunosuppressive drugs (Chloramphenicol, Cyclophosphamide, methotrexate), concurrent infections (CD, CPV, Demodectic mange), congenital, malnutrition, stress (anaesthesia, surgery, thermal), aging, debility, or vaccination given in incubation period of the disease.
- Vaccine-related factors: Improper handling, wrong administration, cold chain disruption, different virus strain, excessive attenuation, or improper concentration.

Adverse Reactions to Vaccination

Though, the vaccines are safe, nevertheless following adverse reactions may occur as vaccines are foreign proteins

- Immediate hypersensitivity or anaphylactic reaction (Fig. 210) may occur. It may be manifested by vomiting, diarrhoea, facial swelling, hives, dyspnoea or collapse. It is a serous reaction and is to be treated immediately.

Fig. 210: Dog showing swelling on the face and around eyes after anti rabies vaccine.

- Transient thrombocytopenia may occur following MLV-CDV vaccine.
- Canine adenovirus can cause renal and ocular lesions.
- Post-vaccinal canine distemper encephalitis has been reported under different circumstances.
- Serious adverse reactions to vaccine should be immediately attended with corticosteroids, antihistamines, epinephrine. Fluid therapy needs to be instituted in cases associated with shock.
- Care should be exercised in revaccinating the patients who have suffered from serious adverse reactions during primary vaccination as potentially fatal reaction to vaccine may occur again.

Mild Reactions to Vaccination

When the immune system is stimulated by a vaccine, mild side effects are common. If these side effects last longer than a day, veterinarian should be consulted. When there is mild temperature, there is no need to give antibiotics. If need be, give only antipyretics.

- Tenderness, bump or pain at the site of injection.
- Abscessation.
- Low grade temperature.
- Fatigue.
- Reduced appetite.
- Coughing or sneezing in case of intranasal vaccine.

Feline Vaccination

The core vaccines for cats are universally considered to be those that protect against Feline Panleukopenia virus (FPV), Feline Herpes Virus1 (FHV-1), and Feline calicivirus (FCV). In countries where rabies is an endemic, cats should also be vaccinated against rabies even if not mandated by law. Vaccination schedule for cats is given in the table 28.

Table 28: Vaccination schedule in cats

CoreVaccine	Kitten Protocol	Adult Cat Protocol
FPV, FHV1, FCV	Starting at 6-8 weeks of age with booster every 3-4 weeks till 14-16 weeks	Annual vaccination
Rabies	Above 12 weeks and a 12-month booster	Revaccination every year

Non-core Vaccine	Kitten protocol	Adult Cat Protocol
Bordetella bronchiseptica (infectious intranasal vaccine)	1 dose from 8 weeks of age	Annual booster
Feline Leukemia Virus	2 doses 3-4 weeks apart starting at 8 weeks; booster at 12 months	Not more often than every 3 years if sustained risk of infection
Chlamydia felis	2 doses 3-4 weeks apart starting at 9 weeks; booster at 12 months	Annual booster, if sustained risk of infection
Feline Immunodeficiency Virus	3 doses 2-3 weeks apart starting at 8 weeks; booster at 12 months	Annual booster, if sustained risk of infection

Adverse Reactions to Feline Vaccination

1. Mild and transient pyrexia, lethargy and anorexia for 2-3 days
2. Anaphylaxis
3. Induction of the feline injection-site sarcoma

Bovine Vaccination

Vaccination schedule for bovine is given in the table 29.

Table 29: Vaccination schedule for bovine

Disease	Age	Revaccination
Foot and Mouth Disease (FMD)	Primary at 4 months of age and booster after a month	Twice a year
Haemorrhagic Septicemia (HS)	Primary at 6 months of age	Twice a year in endemic area
Black Quarter(B.Q)	Primary at 6 months of age	Annually in endemic area
Brucellosis	4-8 months age group (only for female animals)	Once in a lifetime
Anthrax	4 months and above	Annually in endemic area
Theileriosis	3 months of age and above	Annually. Only required for crossbred and exotic cattle.
Lumpy skin disease	6 months and above	Annually in endemic area

Small Ruminant Vaccination

Vaccination schedule for small ruminants is given in the table 30.

Table 30: Vaccination schedule for small ruminants

Disease	Age	Revaccination
Peste Des Petits Ruminants (PPR)	Primary at 3-months of age	Every three years (Annual in endemic areas)
Sheep pox (Sheep only)	Primary at 3-months of age	Annually
Foot and Mouth Disease (FMD)	Primary at 4-months of age	Twice a year
Enterotoxemia	Primary at 4 months of age (if the dam is vaccinated), otherwise at 1-week of age (if the dam is unvaccinated). Booster 2-3 weeks after primary vaccination	Twice a year
Bluetongue	Primary vaccination at 4 months and booster after 28 days	Annually in endemic area

Swine Vaccination

Vaccination schedule for small ruminants is given in the table 31.

Table 31: Vaccination schedule for swine

Disease	Age	Revaccination
Classical Swine fever	Primary at 3-months age	Twice a year. Avoid vaccination in late pregnant sows
Foot and Mouth Disease (FMD)	Primary at 2-months age	Twice a year
Haemorrhagic Septicemia (HS)	Primary at 6-months age	Every six month

Equine Vaccination

Vaccination schedule for equine is given in the table 32.

Table 32: Vaccination schedule for equine

Vaccine	Foals/yearlings	Brood mares
Tetanus Toxoid	At 3,4, 12-months, if mare is vaccinated followed by annual vaccination At day 1 (Tetanus antitoxin + Tetanus toxoid), 4 weeks (Tetanus toxoid), 4, 12-month (Tetanus toxoid)	Annually (preferably in the last month of gestation)
Equine rhino pneumonitis	At 4 months, booster 4 weeks and 6 months later	5th, 7th and 9th month of pregnancy